Organic Farming: Science and Practice for Profitable Livestock and Cropping

C[illegible] iety, 2004

Proceedings of the BGS/AAB/COR Conference held at the Harper Adams University College, Newport, Shropshire, UK 20 - 22 April 2004.

Edited by A. Hopkins
Institute of Grassland and Environmental Research
North Wyke, Okehampton, Devon, EX20 2SB, UK

Published by the British Grassland Society, PO Box 237, Reading RG6 6AR

COR

Colloquium of Organic Researchers

OCCASIONAL SYMPOSIUM NO. 37

British Grassland Society

Acknowledgements and disclaimers
This Conference was organized by a committee representing the British Grassland Society, the Association of Applied Biologists and the Colloquium of Organic Researchers. This comprised Steve Peel (DEFRA, Chairman), David Younie (SAC) John Downes (farmer, Shropshire), Elizabeth Stockdale (Rothamsted) Debbie Sparkes (University of Nottingham), Lois Phillips (Elm Farm) and Jan Crichton (BGS Chief Executive Officer). Committee members also assisted in the editorial process, in addition to their contributions in the conference planning and organization. Sponsorship for the Conference was provided by the Department for Environment, Food and Rural Affairs (DEFRA) and the Biotechnology and Biological Sciences Research Council (BBSRC).

British Library Cataloguing-in-Publication Data available

2004

ISBN 0905944 844

ISSN 0572 7022

Cover design by Jo Chisolm; incorporates photographic images which are the copyright of David Younie or Alan Hopkins

Printed by Bartlett Printing, Exeter EX4 1HU (www.bartlett-printing.co.uk)

CONTENTS

SESSION 1: PROFITABLE ORGANIC SYSTEMS - CURRENT AND FUTURE ISSUES

Theatre papers

Poster papers

SESSION 2: PRACTICAL FORAGE AND LIVESTOCK PRODUCTION

Theatre papers

Poster papers

SESSION 3: CROPPING SYSTEMS

Theatre papers

Poster papers

FOREWORD

This volume includes the theatre and poster papers presented at what was an innovative conference in a number of ways. For the BGS and the AAB it was the first meeting they had held specifically devoted to organic farming, the first meeting they had held together, and the first time either had worked with COR. For COR, only formed in 2000, this was only their second meeting. Co-operation between societies in this way is a welcome trend; it may be born from the increasing pressures on time and resources that learned societies face, but it should result in better understanding and integration across sectors in land use, and I hope along the food chain.

The aim of the conference was to focus on practical issues and applied research for registered and aspiring organic farmers. This volume was printed in advance of the conference and given to participants on arrival. Some additional invited contributions were given which do not feature here. Session 1 began with an assessment of the market realities, and the conference was concluded on with a paper by Lawrence Woodward OBE. In addition there were two farm case studies. Day 2 of the conference was particularly aimed at farmers; session 2 filled the morning and in the afternoon there were visits to three organic farmers: Ed Cavenagh-Mainwaring, Ed Goff and JCB Wootton Farms.

A further innovation was that on the evening of day 1 there was a public meeting with the theme 'Organic food – is it worth it?' The aim was to offer the general public, and the conference participants, an opportunity to put questions to a panel consisting of Helen Browning, organic farmer, Robert Duxbury, previously with J Sainsbury, Catherine Reynolds, Institute of Food Research, and Oliver Walston, farmer and journalist. The meeting was chaired by Kay Alexander of the BBC.

It has been a privilege to chair the organizing committee, who have worked with imagination and commitment to make this event a success. On behalf of the three societies I express grateful thanks to them, the editor, the speakers and host farmers, the panellists and the sponsors. I hope that those present at the conference, and the public meeting, found it useful and enjoyable, and that this volume will be a valuable reference for some years to come.

Stephen Peel
Chairman of Organizing Committee
BGS President 2002 – 2003.

SESSION 1

PROFITABLE ORGANIC SYSTEMS: CURRENT AND FUTURE ISSUES

The Development and Prospects for the Organic Vegetable Market in the UK

C. FIRTH and U. SCHMUTZ
HDRA-IOR, Ryton Organic Gardens, Coventry, CV8 3LG, UK

ABSTRACT

The vegetable market is one of the largest sectors within the UK organic food market. This market grew by 30% per annum in the late 1990s and early 2000s, although it is now slowing down to a rate of 10-15% per annum. This growth has been caused by a greater consumer awareness of food safety, health and environmental issues and has been further stimulated by the increased presence of the multiple retailers in the market. In contrast, the production of UK organic vegetables has been slower and constrained due to the time it takes growers to convert land to organic production, by lack of knowledge of organic systems and higher costs of conversion faced by many vegetable growers. However, by the year 2000, growers had responded to improved economic and policy incentives, and now the UK is 58% self sufficient in organic vegetables. In the future, market growth and the numbers of farmers converting are likely to be slower and any market growth will depend on broadening the customer base, expanding different market channels and increasing home production especially at the beginning and the end of the season, thus enabling a substitution of imports.

INTRODUCTION

The vegetable market is one of the largest sectors within the UK organic food market. It has been reported that much of this growing market was made up by imports; however, contact with organic growers in the 2001/02 season indicated that there was a more even balance of UK supply and demand since they were having difficulty in selling their produce. This lack of clarity of the market was causing uncertainty amongst growers who were considering conversion or expanding their production. This paper examines the growth in this market and the corresponding response of UK farmers and growers to this growing market.

This paper is based on results obtained from a DEFRA-funded project on the organic vegetable market, conducted from 2001-2003. It has also drawn from another DEFRA funded project 'Conversion to Organic Field Vegetable Production'. That project monitored the agronomic and economic performance of ten farms, which converted to organic production during1996-2002.

MATERIALS AND METHODS

The overall aim of the DEFRA-funded organic vegetable market study was to provide detailed market information on the demand and supply of individual UK vegetable crops throughout the UK growing season. This information is used to clarify the supply and demand for individual vegetables and thus identify opportunities for UK growers to expand production and to facilitate the balanced growth of the market. The project was led and conducted by HDRA, in collaboration with the Soil Association, Elm Farm Research Centre and the Institute of Rural Studies, Aberystwyth. Data have been collected from UK packers and wholesalers of organic vegetables on the amounts, value and source of organic vegetables traded during the 2001/02 and the 2002/03 seasons. This was supplemented with data from the organic certification bodies on the area of organic vegetable crops grown in each season. All data has been cross-referenced with other published sources of information for the same seasons.

RESULTS AND DISCUSSION

In the 2002/03 season the total market for organic vegetables in the UK was estimated at 122,000 tonnes, with a retail value of £165M. Of this, 72,000 tonnes (58%) were grown in the UK. When measured by value (£) the level of UK self-sufficiency is 46%, with a retail value of £76M. The organic vegetable market represents 3% of the total market for all vegetables. By value, 73% of the vegetables were sold through multiple retailers, 14% through wholesalers, 11% sold directly and 4% sold to processing outlets. For a range of twenty-five organic vegetables, which can be grown commercially in the UK levels of self sufficiency, or market share, have risen to previously reported levels of 40% (Hamm *et al.*, 2000) to an average of 58% for all vegetables. This did result in the oversupply of some vegetables at certain times of the year in the 2001/02 marketing sêason. When considered on a crop by crop basis, however, there are large variations of market share, ranging from 96% for swedes to 36% for onions. For staple crops such as potatoes, carrots and cabbage the UK share is 68%.

During the late 1990s and early 2000s the growth of the organic vegetable market was very rapid. In the past six years the growth rate has averaged 30% per annum, although is now beginning to slow down to growth rates of 10-15% per annum.

Figure 1. Growth of the organic vegetable retail market in the UK, 1997-2003.

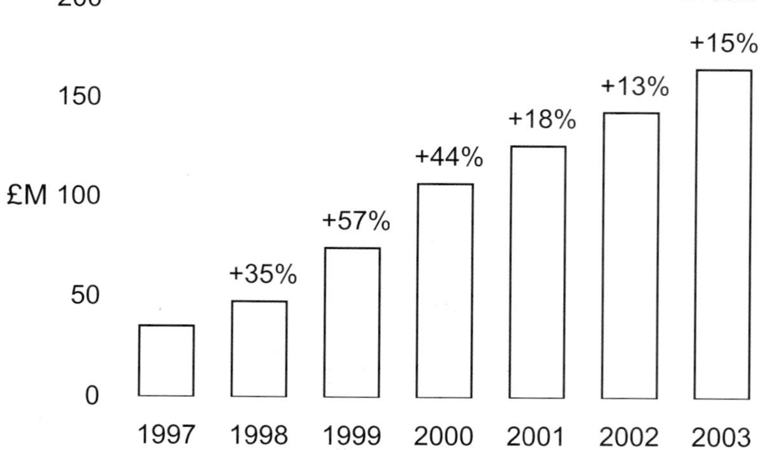

(%-increase relating to previous year, Source: HDRA and Organic Monitor, 2001)

The rapid growth in the organic vegetable market has been in parallel with the growth of the whole organic food market in the UK. This has been due to a number of factors, which have resulted in greater consumer awareness of food safety, health and environmental issues, and concerns about the use of Genetically Modified Organisms (GMOs) in conventional production. These have all caused more consumers to buy organic vegetables.

The stronger supermarket presence in the market towards the end of the 1990s also assisted market development. They have stimulated consumer demand and awareness by high profile advertising and by removing some of the supply side impediments to growth of the market, which included inconsistent supplies and quality (Organic Monitor, 2001). The supermarkets and their suppliers, the pre-packers, now dominate the organic vegetable market.

Despite rapid growth in the demand for organic vegetables, there has been a relatively slower and delayed increase in UK supply (Figure 2). A large proportion of the increased demand was initially met by imports, reported as meeting 60% of the UK market for vegetables in 2000 (Hamm *et al.*, 2002) or as high as 85% when combined with fruit (Soil Association, 2002). The initial slow response in the increase in UK supply during 1998/99 can partly be explained by the two-year conversion period, which is required before organic production can begin, thus delaying growers' ability to respond rapidly to meet changes in demand. Other constraints to growers converting to organic vegetable production were the unfamiliarity with, and lack of, published knowledge on organic production systems. These included uncertainty of how they would control weeds, pests and diseases. There are also higher 'costs of conversion' that vegetable growers have to incur, in comparison with conversions to other organic farming systems. These costs are mainly related to the need to take land out of production during the conversion phase and the subsequent loss in income during this period. The more intensive the production system is, the higher the conversion costs are likely to be (HDRA, 2000).

Figure 2. UK area of organic vegetables and flowers 1997-2002.

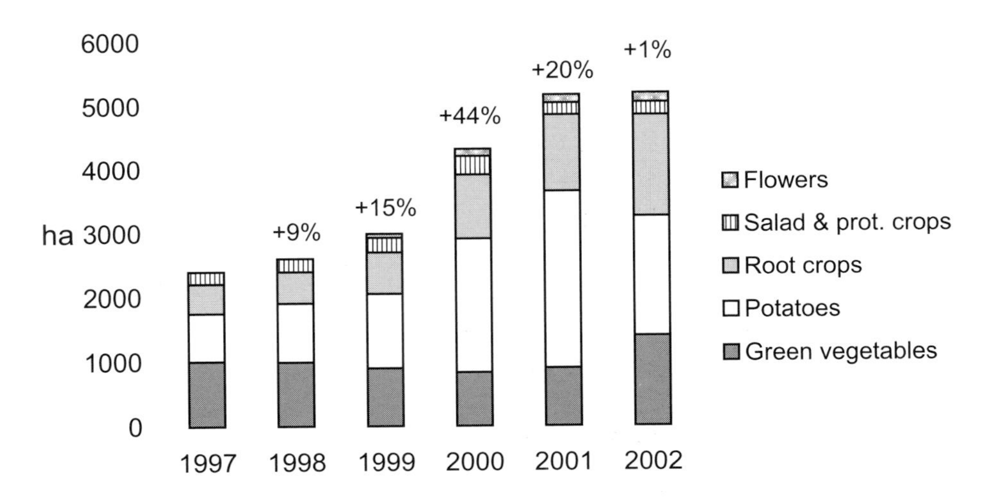

Source: Soil Association (2003).

Until the mid 1990s much of the UK organic produce had been supplied from smaller farms, often not located in conventional vegetable producing regions. The rapid increase in the area of organic vegetables experienced from 1999 to 2001 (Figure 2), was closely linked to the fact that in April 1999 the UK government increased the payments available for organic conversion. The increase from £250/ha to £450/ha acted as an incentive to conversions. In addition, the supermarkets were pressing the growers, who supplied them conventionally, to convert and supply them with organic produce. In the late 1990s there were high 'price premiums', over those paid for conventional produce, available for organic produce, enabling good financial returns to be obtained by organic vegetable growers (HDRA, 2000). This coupled with low prices and returns in the conventional sector further stimulated growers to convert to organic production. In the late 1990s this resulted in a wave of new conversions from large-scale conventional vegetable producers.

FUTURE DEVELOPMENT

In future, the market is expected to grow at a slower rate than the past five years, predicted by marketers (personal communication) at between 10-15% per annum. A wide range of factors, including changing consumer trends, level of disposable incomes, occurrence of food scares and government policy will determine the future growth of the market. Presently, the organic market is dominated by a core 8% group of 'committed' consumers who account for 60% of the market, the remaining 92% of organic consumers, 'the dabblers,' buy on a less frequent basis and account for 40% of the market (Soil Association, 2002). For the market to grow the challenge is, firstly to encourage the 'committed' to spend more, and secondly to convert the 'dabblers' to become more frequent buyers of organic food. Additionally, there are new opportunities for growers to supply catering and institutional outlets, to expand into the chilled and prepared sector and to increase focus on regional and local foods, all which have potential to grow (Firth *et al.*, 2003). The Food and Farming Report of the Policy Commission on the Future of Food and Farming in the UK is favourable to the development of organic farming, with its recommendation for a strategic plan for organic farming (Crown, 2002). The Government's Organic Action Plan (DEFRA, 2002) has begun to take some of these issues forward; notably in seeking to develop the domestic supply to match levels, which are achieved in the conventional sector. Increased UK production is likely to come from existing farmers expanding production, especially at the beginning and end of the season.

REFERENCES

CROWN (2002) Food and farming -a sustainable future. Report of the Policy Commission on the future of farming and Food, January 2002.
DEFRA (2002) Action plan to develop organic food and farming in England, July 2002, Department of the Environment and Rural Affairs, London.
FIRTH C., GEEN N. and HITCHINGS R. (2003) The UK Organic Vegetable Market. HDRA
HAMM U., GRONFELD F. and HALPIN D. (2002) Analysis of the European Food Market. The University of Wales, Aberystwyth.
HDRA (2000). Conversion to Organic Field Vegetable Production. Final report to MAFF.
ORGANIC MONITOR (2001) The UK market for fresh organic vegetables. Organic Monitor. London.
SOIL ASSOCIATION (2002 and 2003) Food and Farming Report, Soil Association, Bristol.

Profitable Organic Systems: A Farmer's Perspective of Current and Future Issues

J. BURDETT
Cockhaise Farm, Lindfield Heath, West Sussex, RH16 2QP, UK

ABSTRACT

Experience of changes in farm management made in the period of conversion from conventional to organic production on a dairy farm in South-East England are described. Organic forage production is based on leys of hybrid ryegrass plus red or white clover. Autumn grazing of grass-clover swards has reduced conserved feed requirements. Herd size is 180 cows, milk yield 5800 l, cross-bred away from Holsteins. Heifer calves are reared outdoors until December in groups of thirty. There are higher direct costs compared with conventional production, but these need not exceed 20%. Challenges for organic dairying nationally are to produce enough milk to reduce processing costs, and to increase efficiency of production at the farm level.

INTRODUCTION

Every one I know approaching 50 has some kind of change in their lives; some are the proverbial mid-life crises, some are just a change of priorities. I took the decisions to go organic when I was in the pre-crisis stage and am reaping the benefits in the post-crisis era. When I started conversion in 1999, my goals were to pay off mortgages and to see the children through university. Now I require some quality of life, whether at work or rest, and to catch up on some of the years spent underneath the back end of cows.

The novelty of a new system of farming has undoubtedly switched me on in the last five years. Since 1986 we have farmed the same 600 acres (240 ha) of Sussex Weald, firstly as dairy contractors and, since 1994, as owners. Our dairying system has always been simple and profitable. A flagging interest and sagging profits needed a fresh challenge, which has been well provided by the triumphs and disappointments of an organic conversion.

In 1992 we had our first advisory visit from ADAS about organic matters, which showed that our simple system would convert easily to organic standards as they then were, but we knew nothing about farming without artificial fertilizers and spray chemicals. By 1998 we had grown enough clover leys and red clover to realize that it was possible to do without most of the above and in 1999 we started conversion.

Before we started conversion, I made some highly optimistic forecasts leading me to think we would only attract an extra 2p/litre of costs. Taking into account the extensification required, our current costs since conversion are up by about 1p for labour and over 2p for bought-in concentrates (nearly double the price for lower quality material).Our milk protein was initially down by 0.3%, worth nearly 1p (no maize and poorer quality concentrates), but this is being improved by extended grazing. The farm milk sales are currently down by 15% (79 tonnes of fat and protein, compared with 92 tonnes 10 years ago).

All other farm costs are down, but they are not actually down on a per litre basis, due to lower farm output. The price currently being paid to us by the Milklink organic contract gives a premium over conventional milk of about 5p per litre.

IMPROVEMENTS IN FARM MANAGEMENT

To survive organically we have made changes on the farm which are more to do with modern management techniques. Although the grass leys we now have came about by the need to farm without artificial fertilizers, they are not specific to organic farming.

We have consolidated our forage system around good arable leys, consisting of hybrid ryegrasses with red and white clover. These can either form a three-year fertility building break in our arable rotation, or stay down for up to six years in our dairy grazing area as a cut and graze ley. Last autumn, when we did not need a third cut, the milking herd grazed our red clover for seven weeks in September and October, producing some of the cheapest milk in the drought-stricken South East of England.

In June 2004 it will be 10 years since we bought the farm and repayment pressures will ease. We have been paying in excess of 5p/l per year to the bank, which is, in effect, the cost of our Milk Quota. We now have the opportunity to put money back into the infrastructure of the farm's aged buildings and equipment. We currently milk 180 autumn-calving cows giving 5800 litres each. We have plans to update our milking parlour and buildings to milk up to 240, which would probably be the maximum the farm would take under organic low-input management.

CHANGES THROUGH ORGANIC CONVERSION

Grassland and Arable

The last 10 years have seen a lot of advances in UK grazing management and we have done our best to keep up with these. We have put down some excellent new cow tracks and established rotational grazing. Our average turn-out date between 1994 and 1998 was 5 April. The average of the last five years was 10 March. Grazing good clover into the autumn has also cut down our conserved feed requirement. All in all, this has saved us over 300 tons of silage-making each year and has given us an 8-month grazing season.

The leys already mentioned have become the cornerstone of our forage system. They provide quality silage, seemingly little affected by different growing seasons, just as maize has done for conventional systems. Grazing leys are more of a problem, not so much for their output, but for the way they encourage docks, particularly, to thrive. In this area we have a lot to learn, if we are not to spend the next decade with a Lazy Dog dock digger in our hands.

The 300 acres (121 ha) of IACS ground has provided the elasticity we needed to maintain our milk output, while learning to grow organic cereals and pulses. In 1995 we sold 600 tons of wheat and oats off the farm, but that was an exceptional year. By 1999, that was down to about 400 tons, at a price of less than £70/t, turning in an operating loss. We now harvest about 160 acres (65 ha) with an equivalent yield of about 200 tons (at a slightly smaller loss), the majority being fed as whole-crop or crushed for a young stock supplement. Our total bought-in feed is in the region of 250 tons, and by 2005 when all our diets will be 100% organic, we should be self-sufficient, allowing for a proportion to go off the farm to be processed and returned as concentrate feed.

Cattle
To see us going about our daily business now would not appear very different from our pre-organic existence. In the parlour we have rid ourselves of a dependence on dry cow therapy. About 10% of the cows receive dry cow antibiotics and they are almost exclusively the cows which pre-date organic management. We are very fortunate in living on a farm with a low risk of summer mastitis, but the success or failure of the dry period depends very strongly on the husbandry skills of the herdsman. Three years ago we found that only homeopathic remedies would lower the cell counts of some antibiotic-resistant cows, but the pressure of high cell counts has eased a bit as we have progressed further in organic management (again a factor of better husbandry).

We have deliberately cross bred our cows away from Holstein, using Ayrshire, Swedish Red and New Zealand Friesian semen, producing a more compact lower yielding cow to suit the system.

Perhaps the biggest difference in our animal husbandry has been to rear the August to October born heifer calves outside, in groups of thirty. They have adapted extremely well to deteriorating weather in well-sheltered fields and are housed in early December when they have grazed all the available clean forage. By this time they resemble Highland cattle with skirts on their bellies up to a foot long. The result has been more robust but not necessarily beefier heifers, which calve down well and settle quickly into the herd. Some would say that this change has nothing to do with being organic; the fact is that a more extensive regime with more ungrazed silage leys away from the herd grazing area, has proved ideal for the experiment.

In 2003 the yearling heifers benefited from grazing our unneeded third cut silage ground until mid-October when they were housed ready for November service. All heifers are trained to self feed silage from a few months old, and spend most of their time respecting anything that looks like a horizontal string suspended about 18 to 30 inches above the ground, depending on their growth stage. This has enabled us to contain them on previously unfenced arable ground at reasonable cost. Last autumn they were most likely to break out in the early hours around the time of the full moon, and we had many theories from the sublime to the spiritual why this should be. Whatever the aggro, we made savings in silage and straw consumption and the heifers were in prime condition at housing.

FITTING THE FUTURE
Organic farming has become like a good book to me. If every day is a page, it has been well worth turning the leaf to find out how the story unfolds. The future then is intrinsically fascinating. Is our cropping sustainable? Will we suddenly hit brick walls that set us back severely? The conventional view is that our performance is bound to deteriorate, while those who have been farming organically for a good while assure me that the farm will maintain a good level of fertility once it has reached its organic potential.

To continue in organic dairying we have to accept what our current figures show: that we have higher direct costs, but that these need not be more than 20% above conventional production, and can be lower. The challenge is for there to be enough organic milk produced nationally for it to be processed at similar cost to conventional milk, and this requires a concerted act of faith by all concerned. The wisdom now is that in 2005 many will finish with organic milk production because of the cost of 100% organic rations. This may bring the current surplus down to manageable proportions, but it will do nothing to take organic milk away from its expensive niche market.

The organic movement has sought the high ground of environment, welfare and nutrition, while campaigning to convince the consumer of the benefits of its wares. I agree that we should occupy that ground, but as in politics, the rest of the food industry is being very quick to encroach through assurance schemes, making the advantages of organic food appear smaller than they were. We cannot fight on the low ground of cheap mass production, and to take even higher ground merely leaves more room for higher class conventional foods to insert themselves into *our* market.

At farm level, we can only respond to this situation by greater efficiency of production so that the wholesalers and retailers have a lot of good quality goods, year round, to sell on. Over the last five years it has been consistently apparent that if a quarter of the R and D that has gone into conventional farming had gone into organic production, we would be happily feeding the UK on all organic, home grown foodstuffs, and I am looking to this conference to confirm that. Although a long way behind in some departments, we believe we can reach that target.

The Contribution of Organic Producer Initiatives to Rural Development: a Case-Study Approach

PETER MIDMORE[1], MARY BECKIE[1], CAROLYN FOSTER[1]
MARKUS SCHERMER[2]
[1]University of Wales, Aberystwyth, Ceredigion, SY23 3AL, UK
[2]University of Innsbruck, Austria

ABSTRACT
Agri-environment policy support in Europe has led to significant increases in crop and livestock activities managed under organic systems of production, and although there have been parallel increases in consumer demand, marketing problems have occasionally prevented a match between the different sides of the market. Rapid growth has outpaced traditional selling strategies, based on direct contact with consumers and niche-scale products. In order to retain the farm-level benefits of organic agriculture for development of the wider rural economy, organic marketing initiatives involving producers are emerging in a key position. This paper uses a comparative case-study approach, carried out as part of a larger EU-funded project on organic marketing initiatives and rural development, to investigate the rural development impact of four organic marketing initiatives, in Vorarlberg in Austria, Burgundy in France, Marche in Italy and Lancashire in the United Kingdom, drawing on documentary sources, semi-structured interviews and stakeholder involvement to establish the most important influences contributing to their contextual performance, and drawing on Actor-Network Theory, placing organic agriculture within an overall set of cultural, social and economic relations. The paper concludes by developing best practice recommendations to maximize the impact of the whole organic food supply chain on rural regeneration.

INTRODUCTION
In the past two decades the organic sector in Western Europe has experienced transformation through rapid growth, as a result of support schemes to compensate for the costs of conversion, and also buoyancy of consumer demand. Organic farming is attractive from a policy perspective, particularly in Less Favoured Regions, because premium prices improve farm incomes and employment; *a priori*, the principles on which it is based also interlock closely with sustainable rural development objectives, such as resource conservation, and self-reliance. Organic producers who subject themselves voluntarily to standards and regulation have stronger common bonds, especially as pioneers in a new area of activity; increasingly, these attributes are recognized as significant for "softer" aspects of rural development, where resilient identity may lead to improved self-confidence and better entrepreneurial performance, particularly if they can be encouraged to spill over into the rest of the food chain.

Yet significant problems are emerging which may disrupt the potential contribution of organic farming to rural development. Markets in European countries are developing differently and have diverse national characteristics (Hamm *et al.*, 2002), and occasionally local supply has outrun demand with resulting reduced or absent premiums. Also, as markets have grown, larger agri-food businesses and multiple retailers have begun to apply their expertise and scale economies in conventional food marketing to the organic sector. Finally, and partly in response to the greater profile of organic foods in conventional retailing, consumers' attributes and expectations are changing. These all present challenges to the development of appropriate marketing strategies and collaborative structures.

Research reported here is part of a wider undertaking[1] which has also investigated European organic market development, consumer motivation and preference, and general issues of organizational performance in a restricted number of regions. Here, we report on the most in-depth element of the project, in which five organic marketing initiatives involving different products were studied in detail. We describe our method, key results and policy conclusions, and finally discuss the value of the comparative case-study approach in investigating such complex social issues.

RESEARCH METHOD

Selection of case studies involved a multi-stage procedure, drawing on earlier studies of regions used to review broader organizational performance. Criteria applied included past economic success, type of organization, size of turnover and membership, objectives and strategies, marketing and distribution channels, major products, location and regional tradition in organic farming, socio-cultural contributions, formal and informal network functioning, and the overall institutional environment. To represent maximum diversity, we chose a cheese-making dairy in Vorarlberg, Austria; a meat processing and marketing cooperative in Burgundy, France; two contrasting cooperatives with highly interdependent characteristics producing pasta in Marche, Italy; and a vegetable box scheme in Lancashire, England.

We drew on a wide range of sources to prepare for fieldwork, describing the background on the economy, social structure and regional policy frameworks, and relations between the initiatives and their broader community. This description included a preliminary SWOT analysis for each case study. We also set up advisory committees of key stakeholders to identify key issues, locate additional stakeholders, and highlight important socio-economic and political networks.

Initial desk study summaries formed the basis for development of common interview guidelines used in each of the case studies, applying a semi-structured format and list of relevant themes. Respondents were encouraged to speak openly on these themes, but also other relevant topics that were prompted. The ultimate aim was to encourage them to categorize and explore internal and external actors and entities that have shaped the initiative, and conversely, its role as an agent of change in the network it is situated in and a part of.

Interviews were taped and transcribed verbatim, and combined with other material accumulated as fieldwork proceeded (including notes of regular discussions among research teams). Qualitative data were organized through coding, and their validity explored through the application of source criticism (Alvesson and Sköldberg, 2000).

[1] Organic Marketing Initiatives and Rural Development, a project co-funded under the European Commission's Fifth Framework Research and Technological Development Programme. The authors acknowledge financial support but views expressed are their own and do not anticipate the Commission's future policy in this area.

Draft reports were discussed with the advisory committees, and the final versions (Midmore *et al.*, 2004) form the basis from which this summary is drawn.

CASE STUDY SUMMARIES

Sulzberger Biobauern (SBB) is a small cooperative of 14 milk producers, manufacturing organic mountain cheese (Bergkäse) in a village dairy. Traditionally Vorarlberg milk is processed into cheese in such small village dairies, but in recent decades numbers have declined to achieve scale economies and cope with increasing levels of part-time farming. The case study initiative came into being after a merger between the original dairy in Sulzberg and another local dairy; before this merger, organic farmers were in the majority, but an influx of new conventional members and a change of management made it appropriate to develop the organic business separately. The new cooperative rented vacant dairy premises in the next village. One member is also a specialist cheese marketer, and the business has developed up to capacity with relatively small capital overheads. There is an associated business delivering organic foods, which exploits synergies with the dairy, sharing premises and locally directly marketing some output.

BioBourgogne Viande (BBV) is a larger cooperative consisting of livestock producers, marketing organic meat from Burgundy (mainly Charolais cull cows). It sells mostly to supermarkets through an agro-industrial group with an established presence in the organic beef sector, but also to local organic butchers, a consumer cooperative, mail-order sales, and a wholesaler specializing in frozen organic meat. Its development was disrupted due to BSE crises in France, causing wild fluctuations in demand from which it has been difficult to recover; it has since undertaken capital investment in a cutting and boning plant to extend its influence to develop a separate organic meat supply chain. A currently high level of costs locks it into supermarket sales (capacity is under-utilized), and though it has had problems with its own direct sales outlets it is attempting to diversify its sales base. It faces a decision about whether to collaborate more widely in order to gain more security, at the expense of reduced autonomy in decision-making.

In Italy, two closely related although rather different initiatives were studies in tandem, *Alce Nero* (AN) and *la Terra e il Cielo* (TC). Both produce pasta from their members' production of organic cereals. Producers in the Marche region were among the earliest to adopt organic farming in Italy, although after recent rapid development of the sector in Italy as a whole, faster growth is occurring elsewhere, driven less by export demand than development of the domestic market. AN was established in 1977 in a former monastery at Isola del Piano, a remote village between Pesaro and Urbino; it mills cereals and produces pasta, but also extends its brand to bought-in organic products such as tomato paste, and sells mostly to Rapunzel, a German organic wholesale chain. TC provides both similarities and contrasts; it is also a cooperative established in 1980, but although its registered office is in Senigallia (a relatively large coastal town) in 1999 its operations moved to an industrial estate in Arcevia, a rural town further inland; its premises include warehousing, a torrefaction plant for coffee and barley, and packaging equipment; it outsources milling and pasta production to independent processors.

Growing with Nature (GwN) is an enterprise aiming to link local Lancashire consumers and growers through a vegetable box marketing scheme. Unlike the other case studies it is not a cooperative, though there is significant collaboration between producers. Before its establishment most of the core holding's output was sold to an organic pack house supplying supermarkets, but increasing competition led to the shift of emphasis to direct marketing in 1992, and with growth came the involvement of several more growers.

ANALYTIC RESULTS

While the comparative case study of success cannot be exhaustive, our definition and means of selection to reflect diversity allows us to explore key contributions to rural development. Perhaps one of the most striking results of all four case studies is that the initiatives' direct contributions to rural development, in terms of income and employment generation, are relatively modest, although their indirect or "softer" contributions are considerable, supporting and embedding confidence and raising regional profile. They also provide a model for improved impacts in the future, particularly if marketing management is improved.

In general terms, most (apart from GwN) are in peripheral LFAs, valued for their culture and landscape; most also (apart from AN and TC) have good transport links to thriving markets; and most (apart from GwN) produce traditional, typical regional products. Institutional conditions provide a key dimension to rural development success. For GwN, location outside the upland area means that inexperience limits access to support from EU Structural Funds; organic producers have tended not to take advantage of schemes promoting rural economic development. Political and agricultural institutions in Lancashire are not openly hostile, but at best reticent or uninformed of the potential of organic farming. Only at local level was any interest being shown. After foot and mouth disease in 2001, rural issues are more prominent and a need for a more integrated approach is recognized. In contrast, strong regional identity in the Bregenzerwald area is reflected in institutions, especially in the REGIO, a vigorous rural development agency. Its *Käsestrasse* (cheese route) project established a broad marketing platform for a core regional product. However, this and other initiatives capitalizing on the region's image have bypassed organic producers who have tended to concentrate on more specialized opportunities. Local agricultural institutions in Burgundy played an important role in the development of organic farming in the region. Good relationships between them and BBV were decisive in securing financial support, but tensions exist as to whether organic agriculture should be promoted as a mainstream opportunity or market niche. Equally, in the Marche region, the case study initiatives have benefited from longstanding strong political involvement in the promotion of organic food and farming in the region.

To explore the interdependence between the initiatives and regional institutional actors in rural development processes, we have used Actor Network Theory (ANT). Despite indications that there is potential for OMIs to contribute to sustainable regional development, little is known of the mechanisms encouraging or hampering the convergence of the two concepts. The main process analysed by the ANT is the growth and extension of spheres of influence and power, through "processes of translation" or "enrolment" (Clegg, 1989): recently ANT has been increasingly used for analysis of rural change processes (Murdoch, 2000), showing how rural networks function and exploring their ability to involve various stakeholders of organic and rural initiatives into a common set of interests.

Actors are defined as all entities that are able to connect texts, humans, money etc. to build more or less effectively a world that is filled with other entities having their own history, identity and relations (Callon, 1986; 1991). Translation follows four stages: problematisation, an actor analysing a situation, defining the problem and proposing a solution; interessement, other actors becoming interested in the solution proposed and changing their affiliation to a group in favour of the new actor; enrolment, the solution accepted as a new concept and a new network of interests generated; and mobilization, the new network operating to implement the proposed solution. Stable relations and target orientation are achieved through an "obligatory passage point", channelling all group interests, where translation processes run automatically and are not renegotiated. We use this framework to analyse who acts as a powerful (macro) actor, how the situation is problematised, how others are enrolled in the suggested solution and their final mobilization. The dynamic perspective of ANT needs starting point, usually the "critical event" leading to the foundation of the OMI.

For GwN, the initial problematisation was how to market their produce successfully and efficiently in a way that was compatible with their value system. Growing competition from supermarkets was the point of contention, and direct marketing offered an opportunity to regain power by improving links between consumers and producers, and raise awareness about the politics of food in general. For SBB, the initial problematisation emerged from restructuring and liberalization of the dairy sector prior to accession to the EU from 1992 onwards. Opening protected agricultural markets would inevitably result in falling prices; scale economies were not seen as a viable solution, or off-farm employment, because full time farming is seen as a cornerstone for sustainable maintenance of cultural landscapes, quality of life and tourism. Actions proposed were converting to organic agriculture, shifting from Emmental to Bergkäse production, and supplying market partners directly rather than through the conventional cheese marketing board. For BBV, the problematisation emerges as economically pragmatic, even though the founders are very committed to the principles of organic farming. Prior to establishment, their meat was badly marketed and premiums were poor, so the initiative was seen as a way to take control of the marketing in order to compete successfully and eventually become viable players in the region's meat sector. Development of a separate organic meat chain would bypass the constraints of being a minor player on a conventional playing field. In contrast, establishment of both AN and TC was initially driven by ideology; both founders saw, through the establishment of the initiatives, an opportunity to translate their vision into action. Influenced by their rural background, they wanted to revitalize marginal areas by providing opportunities for continued farming, and upholding and reviving traditions, skills and culture, although for TC it is also an alternative way of life.

GWN's direct marketing channel opened new possibilities for a small number of neighbouring organic growers, and by creating a new, committed customer base allowed the founders to communicate their ideas of sustainability. SBB's proposed solution involved splitting from the original dairy; when management changed to reflect the new conventional farming majority, it became incompatible with the nucleus of organic farmers. Once SBB was established, members fully embraced the project and strong commitment emerged. The proposed solution for BBV also generated high levels of interest in and commitment to the project, providing an opportunity to market outside conventional structures and offering good prices through its own processing facility at Avallon. Engaging interest from local producers was relatively easy for AN and TC, as vigorous demand for organic pasta opened up opportunities for organic producers to remain in farming in the immediate area, and also some more geographically dispersed members of TC. Both founders were influential in the emerging regional organic network and ideas developed in a close group of likeminded people. AN drew support from extended family, but this later caused tensions and the departure of some members.

Enrolment, in terms of extension and adoption of proposed solutions within emerging networks, varies between the case studies. For GwN, influence extends mostly over organic growers and consumers with little engagement with regional development institutions. Recently the box scheme has been marketed as an "organic vegetable club", intended primarily to strengthen consumer loyalty as competition from supermarkets increased, although influencing power seems too weak to effect further mobilization. SBB became a closed group of likeminded farmers with strong internal cohesion; to refurbish the dairy members had to pay a capital contribution, further increasing their affiliation, and they also share duties and take up part time employment in dairy processing and the delivery service. This inward-looking nature has prevented growth; only two new members joined since 1996. In contrast, BBV has recruited 70 new producers since foundation. Strong membership cohesion has enabled it to withstand several crises, and high fixed capital costs of its processing facility stimulated extension of its networks. Investment in processing facilities helped to focus attention on short as well as long supply chains, enabling it to develop a local customer base and expand into new areas such as catering and mail order, and to cooperate with organic producer groups outside the region. Links with conventional producer groups, however, remain weak.

For both AN and TC, styles of management (hierarchical with strong family ties for AN; democratic and participatory for TC) generated strong commitment and internal cohesion. For AN, though, there has been tension between management and non-family employees, and new members and employees do not always share its ideals and goals. In contrast, TC seeks out potential new members who share in its vision. Beyond internal cohesion, enrolment and mobilization of interested actors within the Marche region has not been straightforward. AN started as a local community development project, but relations with the immediate community were strained by collapse of a related financial cooperative; also, on foundation its rural development ideas were in advance of its time, and it has since remained slightly out of step with mainstream agricultural institutions and networks. Nevertheless, it has a high profile in the region due to its numerous activities, and has been an influential actor in commercial, technical and cultural terms. In contrast, TC's rural relocation in 1999 allowed development of favourable cooperation with local institutions to set up various initiatives, yet influence on conversion in the immediate area of Arcevia has been limited.

None of the initiatives has been particularly active in introducing their ideas and solutions to institutional structures; but the prevailing institutional climate plays a key role in facilitating and directing the nature and extent of the cooperation. In Lancashire it is difficult to identify a single agency for rural development activities; although the urban has been prioritized over the rural, after FMD rural regeneration became a priority. Surprisingly, however, the organic sector has not explicitly become part of this process; according to an interviewee from local government, the organic agenda remains "a kind of a blank area". In Vorarlberg, a well established development agency leads in the region and after accession to the EU it took up various programmes for the benefit of the whole area, particularly the Käsestrasse which provided a common marketing platform. However, as conventional agriculture was perceived to have low environmental impact (relying on traditional extensive systems of alpine agriculture and silage-free feeding practices) again organic agriculture was not promoted as a mainstream opportunity.

The Green movement has had a major influence on Burgundy and rural development issues have had a high profile, linked to broader issues such as food quality and health. There is a strong tradition of organic farming in the region, and contact with public authorities has always been good, partly as a result of strong personal relationships. The development organization for organic agriculture (SEDARB) is supported by public funds, and has been an important actor in the rural development process. While Chambers of Agriculture have a generally positive attitude towards organic farming, there is competition between SEDARB and the traditional public institutional structures. The Marche region began, in the 1980s, to reject the industrial agricultural model for marginal rural areas, relying instead on cultural assets and, as in Lancashire, public authorities favour a more integrated regional agro-rural system approach; the importance of establishing horizontal and vertical networks features strongly in the Rural Development Programme. This favourable climate and the multifunctional nature of the region have opened opportunities for organic sector development; the regional government recognizes it as an integral part of its agro-rural development plans. This is reflected on a more local level, in the communes where AN and TC are located; both mayors attribute an important role for organic farming in their strategies for local development.

CONCLUSIONS

Using the ANT framework, contribution of the case study initiatives can be assessed on three levels: effectiveness, efficiency and impact. Effectiveness is closely connected with the achievement of objectives, which in terms of ANT can be judged by the form of problematisation, the solution proposed and its realization. Internal objectives mainly concern economic success, achieved all cases; external objectives are based on wider contributions to regional development. For BBS the main goal was to help conserve the cultural landscape, with organic farming as a means for farmers to remain full-time. On a very local level, among the members, the initiative has succeeded, but failed on a wider regional level. GwN appears similar, in that its external objectives to effect changes in regional food chain management and local food security remain restricted to a very local level. The difference, though, is that GwN by its very nature never intended to extend much beyond the very local level. BBV has successfully developed a regional organic meat supply chain, allowing organic farmers to be independent from conventional marketing structures. However, operations remain small scale and no notable horizontal networks in the region have developed (outside the organic sphere) which that enable it to contribute to the broader rural development goals of the BioBourgogne association. AN's objectives of enhancement of farming's status and the prevention of abandonment of the countryside have to a certain extent been achieved, although its cultural and educational activities have done more to enhance the image of the area. TC established more extensive goals, linking economic objectives more strictly with its ethical and cultural ideals. This wider mission has been adopted by local institutional actors, although so far not at a regional level; in the latter's view, the regional organic sector is held back by its weak and disorganized capacity for project development and management.

Efficiency in ANT refers to the translation process, how far the network's relations have become embedded into a powerful new macro-actor. For SBB this sphere of translation is restricted mostly to members, albeit powerfully, whereas GwN has a more diverse range of involvement, with attachment by its growers remaining essentially on the level of business relations. BBV has worked outside its producer membership, broadening its network to include a range of vertical stakeholders, although no horizontal networks have been established. AN and TC have fostered strong internal cohesion among their producers and employees, and have also successfully shared their vision with a wide range of outside interests through their cultural and educational activities. Identifying also the direction of translation, efficiency in the case of GwN is more directed to customers, whereas the other initiatives are more producer orientated.

Evaluating impact requires an assessment of how far the case studies represent a functioning model for sustainable rural development. This depends on a combination of the acceptance and interest of the surrounding institutional environment and also on the active role played by the initiatives themselves to engage external interest and support. While each case study offers a quite successful local model, the signals sent out were so far not effective, especially with GwN and SBB. In terms of ANT, the collaborative organic marketing solution has not been fully accepted on a regional scale and the problematisation does not match all potential partners. In the case of BBV, AN and TC, although their solution is more effectively communicated to a wide range of interests, levels of convergence and integration with rural development processes in the region remain fairly low.

This brief paper shows how ANT can be used to assess the influence of initiatives on rural development in a qualitative way. On all three levels of evaluation, it provides insights into processes, purposes and motivations, and the dynamic aspect allows clearer recognition of driving forces and barriers for extension of concepts. It illustrates how more precisely targeted strategies can be devised, if the exemplar of the organic marketing initiative is considered useful for further extension.

REFERENCES

ALVESSON M. and SKÖLDBERG K. (2000) Reflexive methodology: new vistas for qualitative research. London; Thousand Oaks, Ca.; New Delhi::Sage Publications.

CALLON M. (1986) Some elements of a sociology of translation: domestication of the scallops and the fishermen of St. Brieuc Bay. In: Law, J. (Ed.) *Power, Action and Belief: A New Sociology of Knowledge?* Sociological Review Monograph 32, London: Routledge Kegan Paul, 196-223.

CALLON M. (1991) Techno-economic networks and irreversibility. In: Law, J. (Ed.) *A Sociology of Monsters*. London: Routledge, 132-161.

CLEGG S. (1989) *Frameworks of Power*. London; Thousand Oaks, Ca. Sage Publications.

HAMM U. GRONEFELD F. and HALPIN D (2002). *Analysis of the European market for organic food*. Organic Marketing Initiatives and Rural Development: Volume One, Aberystwyth: School of Management and Business (University of Wales).

MIDMORE P., FOSTER C.J. and SCHERMER M. (2004) *Organic farming and rural development: four case studies*. Organic Marketing Initiatives and Rural Development: Volume Three, Aberystwyth: School of Management and Business (University of Wales).

MURDOCH J. (2000) Networks: a new paradigm of rural development? *Journal of Rural Studies*, 16, 407-419.

The Economics of Conversion to Organic Field Vegetable Production

C. FIRTH, U. SCHMUTZ, R. HAMILTON and P. SUMPTION
HDRA-IOR, Ryton Organic Gardens, Coventry, CV8 3LG

ABSTRACT

Lack of data and knowledge of the transition to and the economics of organic vegetable production is often cited as a major reason why farmers have been reluctant to convert to organic systems with vegetables. This paper outlines the initial results of a DEFRA funded project investigating the economic implications of conversion to organic vegetable production. Whole farm financial data has been collected and analysed for a group of 5 farms from 1996-2001. The findings show that net farm income declined by an average of 66% during conversion, although it recovered to within 36% of pre-conversion levels once organic vegetable production began. This was a result of falls in output and smaller overall reductions in costs. In contrast costs of casual labour rose sharply following conversion. The costs of conversion, for this group of farms, is estimated at a total of £556/ha in comparison with organic aid payments available at £450/ha over 5 yrs. In conclusion the economics of conversion are very much dependent on the starting financial position of the farm prior to conversion, the rate at which the farms converts and the price of organic vegetables received once conversion is completed.

INTRODUCTION

Despite the growing organic vegetable market, which grew by an average 30% per annum in the late 1990s (Firth, 2003), UK farmers, especially field vegetable growers, have until recently, been reluctant to convert their land to organic production. One of the major reasons often cited is the lack of data and knowledge on the economic performance of organic vegetable production.

What are the costs of converting to organic vegetable systems, and how economic is organic vegetable production once you are converted? The conversion (or transition) from conventional to organic farming systems is subject to several physical and financial influences. The process is complex, involving a significant number of innovations and restructuring of the farm system as well as changes in the production system. Previous studies indicate that costs of conversion may include, output reductions, new investments, information and experience gathering, variable costs reductions, and fixed costs increases (Lampkin et al. 2002), although up until now specific studies of farm systems including vegetables have not been completed.

This paper is based on some initial results obtained from a DEFRA funded 'Conversion to Organic Field Vegetable Production Project'. This project has monitored the agronomic and economic performance of 10 farms, which have converted to organic production from 1996-2002.

MATERIAL AND METHODS

Presently data from 5 of the farms have been analysed. The farms varied in size from 20-1900 ha and comprise 3 general cropping, one dairy and one intensive vegetable farm (Table 1).

Table 1. Farm details

Farm	Farm size (ha)	Farm classification	Conversion started (year)	Rate of conversion[1] (% per year)
A	1557	General cropping	1997	20
B	1900	General cropping	1997	6
C	38	Lowland dairy	1997	38
D	998	General cropping	1998	7
E	20	Horticultural	1998	10

[1] Average area of land converted per year as a percentage of the total farm area

For each farm, accounts have been collected and analysed according to Farm Business Survey (FBS) procedures (Crown, 2002), for the year prior to conversion and then for between 3-5 years following. This has enabled us to monitor the effects of changes, during the conversion period and initial organic cropping, on farm financial output, costs and net farm income. Information presented here is based on averages for all the farms. Data on time spent information gathering and details of conversion specific investments has also been recorded. Each farm's data has been compared with best match published conventional data available from regional farm business centres within which each of the farms is situated, for example taking into consideration farm size and farm type. The costs of conversion on each farm has been estimated by subtracting each years net farm income from the pre-conversion year. In order to take account of the fact that part of the farm was still conventional during conversion, corresponding changes in the conventional sample's net farm income have been subtracted from this cost. Conversion specific investments have been added to this total to arrive at an estimated cost of conversion per hectare. This has been compared with the Organic Farming Scheme aid payments, which were available on each farm.

RESULTS AND DISCUSSION

Prior to conversion (Figure 1) all the farms were profitable with net farm income of £397/ha (range £108 –682/ha). In the year conversion began output, costs and net farm income per hectare fell on all the farms. All of the farms, except the livestock farm, took land out of crop production and put it in to a two-year grass clover, fertility- building ley. Therefore there was no crop to sell from this land for the first two years in conversion phase. This was the major factor, which led to a decline in financial output of 32% during the first two years following conversion. Output was still 24% below the pre conversion year in the third year when organic production began. The rate of decline of output on most of the farms was closely related; firstly to the amount of land which each farm put into conversion per year, this ranged from 6-38% (Table 1). This is referred to here as the rate of conversion. The faster the rate of conversion the more rapid the decline in output. It is estimated that half of the output decline can be attributed to the fact that conventional prices also fell during this period, and this effected the part of the farm which was still conventional. The rate of decline in output was also related to the intensity of production, or level of output prior to conversion. The intensive vegetable farm had the highest output prior to conversion and this fell by 50% in the two years following conversion. This farm also received a lower rate of organic aid (£350/ha) and was unable to claim set aside on its fertility building leys as it was not registered for arable area payments.

Figure 1. Average output, total costs and net farm income from converting farms

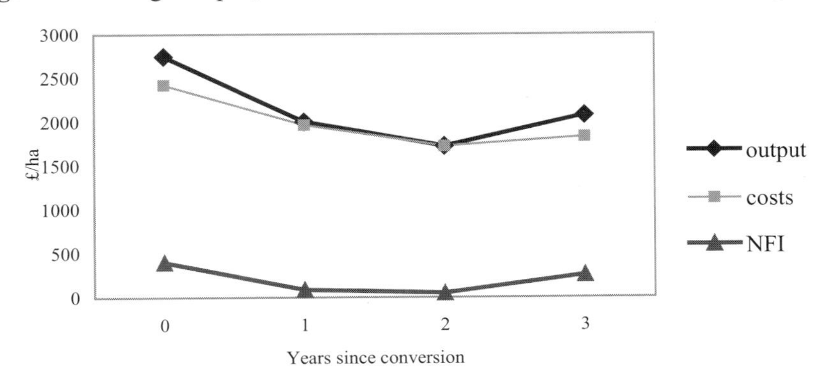

* Year 0; conventional, year 1&2; in-conversion, year 3; organic

Farm costs also fell during the conversion period by an average of 24% in year over the first two years following conversion. Since a part of the farm was not in cash cropping naturally less variable costs were incurred and these fell by an average of 33% during the in conversion period. Fixed costs also fell by 20% the most notable cost to fall was contracting which would be related to the area of crops grown. Once organic production began variable costs were still 12% below pre conversion levels and fixed costs 17% less. Some of the highest costs on this group of farms were labour accounting for 37% and machinery costs accounting for 20% of total costs. During the conversion process overall labour costs stayed at the same as pre-conversion levels on most of the farms, but rose to 35% higher following conversion. Regular labour costs did not increase but casual labour costs rose significantly, this was especially so for the lager arable farms where the average wage bill for casual labour rose by a factor of 6 in comparison with pre conversion levels. The additional casual labour was mainly employed for hand weeding and to harvest the wider range of crops grown. The increase in casual labour, which occurred mainly on the large arable farms, was mainly gang labour, this caused many management problems in its sourcing and organization. Overall machinery costs fell during conversion and into the first year of organic vegetable production by 10%. Investigation of the different farms reveals some variations; where the farms had grown vegetables prior to conversion the total costs fell or stayed at a similar level, but on farms where vegetable production had been limited there was large increases in machinery costs (+30% on one farm). This was due to the need to make investments in new machinery.

Net farm income on average fell during conversion by 66% in the first two years following conversion (Figure 1), with two of the farms registering negatives net farm income during these years. Following conversion and the growing of the first crops, net farm income rose, although it was still 36% below pre conversion levels. A comparison with the conventional sample's net farm income (Figure 2) shows the importance that the sale of organic vegetables has made to income on the converting farms. This represented an average of 16 % of total output in that year. It should be noted that this upturn in the farms performance occurred in 1999 (3 farms) and 2000 (2 farms) when organic vegetable prices were high. In the following year organic vegetable prices fell by an average of 30% and this would therefore have made a difference to the economics of conversion if the conversion process had begun later.

Figure 2. Net farm income; comparison of converting group with conventional sample

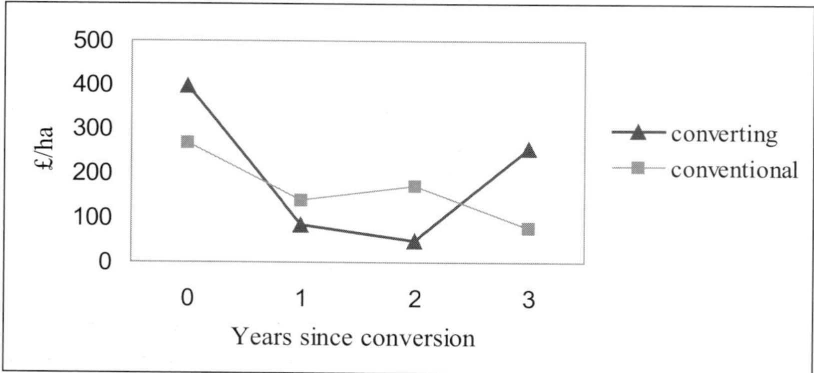

The average costs of conversion for all the farms over the first 5 years, in terms of decline in net farm income and the costs of new investments, is £556/ha (Table 2).

Table 2. Estimated 'costs of conversion'

Farm type	Costs of conversion (ha)	Organic aid (ha)
Arable	550	450
Mixed	499	450
Intensive	631	350

CONCLUSIONS

Conversion to organic vegetable production has led to a decline in both output and net farm income during the conversion phase. The decline in output is related to the rate of conversion and to the level of intensity of the system prior to conversion. Generally costs have also fallen, although at a lower rate; however, casual labour has increased rapidly especially on the larger farms. The costs of conversion taking into account decline in income and costs of new investments is estimated to be £556 for the group of farms. This compares with the organic aid available of £450/ha over 5 yrs, at this rate it goes to paying for the majority of the costs. There is considerable variation within the group of farms, with farms converting the fastest and those converting from intensive vegetable systems having the greater costs. The economics of conversion is very dependent on the starting financial position of the farm, the rate of conversion and the price of organic vegetables received once converted. Conversion to organic production with vegetables can put the farming finances under pressure. The costs can be minimized with careful planning, and it is important for farms to examine their costs prior to conversion.

REFERENCES

CROWN (2002) *Farm Business Survey*: Instructions for collecting the data and completing the farm return, Farm and Animal Health Economics Division, DEFRA, London

FIRTH C., GEEN N. and HITCHINGS R. (2003) The UK Organic Vegetable market: HDRA.

LAMPKIN N., MEASURES M. and PADEL S. (2002) 2002/03 Organic Farm Management Handbook: Institute of Rural Studies, University of Wales, Aberystwyth.

Monitoring the Effects of the Organic Farming Scheme in Wales: Preliminary Findings

D. FROST and D. ARDESHIR
ADAS Pwllpeiran, Cwmystwyth, Aberystwyth, Ceredigion, SY23 4AB, UK

ABSTRACT

There is evidence that organic farming benefits biodiversity. Positive impacts ascribed to organic farming derive particularly from the withdrawal of synthetic pesticides and the increased uptake of rotational land use. These features are primarily associated with mixed lowland farming. Less attention has been given to the effects of organic systems on the biodiversity of all-grassland farms, despite this being the most important land use of organic and in-conversion land in the EU. A review of the research literature found that organic farming could assist in the conservation of bird and mammal species in grassland where mixed farming could bring back the benefits of arable land. It also found that a number of Biodiversity Action Plan species and habitats would benefit from wider uptake of organic farming in Wales. A baseline survey of farms entering the Organic Farming Scheme (OFS) undertaken in 2002, indicated that the OFS delivers greater environmental benefits when combined with a whole-farm agri-environment scheme.

INTRODUCTION

In Wales, the National Assembly has encouraged the development of organic farming by funding three policy measures, the Organic Conversion Information Service (OCIS), a centre of excellence for organic farming - Organic Centre Wales, and the Organic Farming Scheme Wales. These measures are all predicated on the assumption that organic farming delivers environmental and socio-economic benefits. With increasing uptake of the Organic Farming Scheme and other agri-environment schemes, a series of projects has been initiated to monitor the associated changes. A three-year project undertaken by ADAS and funded by the Agriculture and Rural Affairs Department of the Welsh Assembly Government started in 2002 to investigate the environmental and socio-economic effects of organic farming. This paper presents some preliminary findings.

METHODS

Initial work has comprised a review of the implications of organic farming, with particular reference to UK Biodiversity Action Plan (BAP) habitats and species, and a baseline environmental survey of a sample of farms entering the Organic Farming Scheme (OFS). These farms will be resurveyed in 2004 to monitor changes associated with organic conversion.

Desk study: A Review of the Environmental Effects of Organic Farming

In order to answer the particular question set by the Welsh Assembly Government, namely "to what extent do systems of organic farming benefit key species and habitats that have been identified in the Wales Biodiversity Action Plan (BAP)?" an exercise was undertaken to review the regulatory framework of organic farming and to relate this to the actions proposed for BAP species and habitats in Wales. Four distinct levels of regulation at which environmental benefits may accrue were identified for Wales. These were:

- Prescriptions arising from the implementation of EU / UKROFS (Advisory Committee on Organic Standards, ACOS, as from 2003) organic farming standards, e.g. prohibition on the use of chemical pesticides and herbicides;
- Prescriptions arising from the implementation of sector body standards that include environmental prescriptions not required by UK baseline standards, e.g. Soil Association Certification standards and Organic Farmers and Growers Ltd standards for creative habitat management, maintenance of traditional field boundaries, etc;
- Prescriptions arising from the environmental management prescriptions in the Organic Farming Scheme Wales, e.g. prohibition on the cultivation or application of manures within one metre of boundary features;
- Practices arising from general organic farming recommendations but which are not mandated by organic standards, e.g. changes in the direction of mixed farming.

In order to identify species and habitats most likely to benefit from organic farming, a comparison was made between the actions proposed for BAPs in Wales and the effects of the wider adoption of organic farming at each of these four levels.

The Baseline Environmental Survey

Thirty farms participating in the Organic Farming Scheme were surveyed during September and October 2002; these farms will be surveyed again in 2004. The farms surveyed are located in south, mid and east Wales. At the time of the survey, farms had been in the OFS scheme for between 18 months and two years. Mixtures of farm types were surveyed, with the highest proportion, 50%, being sheep and beef enterprises. At the time of survey, half the farms surveyed were also in another agri-environment agreement (such Environmental Sensitive Area (ESA) schemes and Tir Gofal) and half were not.

The core of the survey was based around a one-day field visit, comprising a habitat survey, botanical quadrat survey and boundary assessment. The habitat survey was based on Phase 1 Habitat Survey methodology; it used existing information where available and noted the presence of BAP habitats. Quadrat surveys were carried out on between two and four fields on each farm. The fields chosen were those most likely to experience change. The boundary assessment concentrated on hedges.

RESULTS

The literature review revealed considerable evidence for the general biodiversity benefit from organic farming, but many studies have been on a small scale and primarily in the lowlands. There are very few studies of the effects of organic systems on the biodiversity of all-grass and upland farms. This is a significant limitation since, according to the Eurostat Farm Structure Survey, grassland is by far the most important land use of organic and in-conversion land in the EU, covering 55% of the total organic land compared with only 40% of total land in the survey (Hau and Joaris, 2003). Furthermore, the EC evaluation of agri-environment programmes noted that the environmental benefits of organic farming, compared with intensive conventional farming, derived particularly from the absence of synthetic pesticides and from sustainable rotation of land use (EC VI/7655/98). These are features of mixed farming, arable farming and horticultural organic systems, but are less significant for the kind of all-grassland systems that prevail in Wales.

The comparison of the actions proposed for BAPs in Wales and the effects of the wider adoption of organic farming found that, although many BAP actions were concerned with the promotion of advice and effective habitat management for specific species, there were a number of examples where BAP actions and organic farming methods converged. In some cases BAP actions directly paralleled organic standards, e.g. "use alternative less harmful methods of stock parasite treatment to ivermectins" - Ivermectins are prohibited by organic standards. In other cases BAP actions are a weaker version of organic standards, e.g. "more cautious and targeted use of herbicides and fertilizers on farmland" - herbicides and most artificial fertilizers are prohibited in organic farming, though restricted use of permitted sources of P and K is allowed. In a third category, BAP actions reflect wider organic farming philosophy, e.g. "sympathetic management of hedgerows and farmland scrub for the benefit of farmland birds"- the higher densities of birds reported for organic systems are associated with better quality non-crop habitats, especially hedgerows (see, for example, Azeez, 2000; Stockdale *et al.*, 2001). Overall, the review found that organic farming methods are broadly compatible with actions for BAP species and habitats in Wales. A number of bird and plant BAP species, and certain BAP habitats, would benefit from the wider uptake of organic farming in Wales. Some of these would benefit directly from organic standards; others would benefit from the general effects of organic farming. Where benefits arise from the general principles associated with organic farming, they rely on the voluntary adoption of good organic farming practice rather than regulation. This emphasizes the importance of the Organic Farming Scheme environmental prescriptions for many BAP species and habitats.

All the farms surveyed in the baseline environmental survey contained at least one BAP habitat. Farms with another agri-environment agreement tended to have more BAP habitats than those without, and farms in the Tir Gofal scheme had on average the greatest number. The most commonly recorded BAP habitats were wet woodlands, upland oakwood and lowland meadows; in total fourteen BAP habitat types were recorded on thirty farms.

On a number of farms without other agri-environment agreements, several lowland meadows were recorded on field parcels classified as 'Land to be Converted' (LTC); these lowland meadows are a declining BAP habitat. On LTCs, intensive farming practices are permitted; this means that lowland meadows which are classified as such could potentially be damaged or destroyed. Lowland meadows can be a particularly difficult habitat to recognize; therefore it is important that competent botanists carry out surveys before land is designated LTC.

Boundaries and, in particular, hedges are important environmental features on organic farms (Unwin *et al.,* 1995). Agri-environment schemes such as ESA and Tir Gofal provide funding for hedge maintenance and re-instatement work. The baseline survey found that while some fencing work was carried out on farms not in such schemes, on average over three times as much fencing work was carried out on farms in these schemes. Considerable hedge laying, hedge coppicing and some planting work was carried out on organic farms that were also in other agri-environment schemes. With regard to hedge maintenance and re-instatement, it is clear that the OFS and whole-farm agri-environment schemes are delivering more environmental benefit than the OFS on its own.

There is relatively little arable farming in Wales, and farming systems tend to be predominantly pastoral. Organic farming systems place an emphasis on crop rotation, often including unsprayed cereal crops. The introduction of unsprayed, spring-sown cereal crops, which are not under-sown, could therefore create a more diverse environment. There is little evidence from the baseline survey that arable crops are being introduced on farms recently entered into the OFS. On one dairy farm, oat and barley crops had been introduced, and this was supported by Tir Gofal payments. In these fields, there was wide plant diversity and a large number of arable weeds, which demonstrated the benefits of such an approach. On the other hand, another farm which previously had an arable enterprise had ceased to grow crops once in the OFS, and on a further farm the switch from arable to pasture was funded by Tir Gofal under a reversion to semi-improved pasture option. Interestingly, one farm was found to no longer grow cereals, the farmer stating the reason as the difficulty in growing cereals organically, due to difficulties in weed control. This raises a fundamental issue, as it is the weed content that provides the crop with much of its environmental value.

DISCUSSION

Results from the review and from the survey indicate that the OFS and whole-farm agri-environment schemes can combine well together. The baseline survey found examples of how the OFS delivers considerably greater environmental benefits when combined with a whole farm agri-environment scheme. This is particularly true in the areas of hedge maintenance and restoration and the protection of BAP habitats.

The resurvey of farms in 2004 will enable further assessment to be made in the areas investigated so far. Other factors for investigation will be the change in grassland habitat type. In particular, it will be important to monitor any change in the ratio of improved to unimproved grasslands, and to investigate whether improved and semi-improved grasslands are becoming more diverse and whether BAP habitats are being sufficiently protected. Resurvey will also allow a comparison of stocking rates to be made and to assess the extent of changes in farming practice associated with conversion to organic systems.

ACKNOWLEDGEMENTS

The authors would like to acknowledge the National Assembly for Wales Agriculture and Rural Affairs Department for funding this work.

REFERENCES

AZEEZ G. (2000) The Biodiversity Benefits of Organic Farming. Bristol: Soil Association.

EC (VI/7655/98) DGVI COMMISSION WORKING DOCUMENT STATE OF APPLICATION OF REGULATION (EEC) NO. 2078/92: *Evaluation of Agri-Environment Programmes*

HAU P. and JOARIS A. (2003) *Organic Farming* [www document] http://europa.eu.int/comm/agriculture/envir/report/en/organ_en/report_en.htm

STOCKDALE E. A., LAMPKIN N. H., HOVI M., KEATINGE R., LENNARTSSON E. K. M., MACDONALD D. W., PADEL S., TATTERSALL F. H., WOLFE M. S., and WATSON C. A. (2001) Agronomic and environmental implications of organic farming systems. *Advances in Agronomy*, 70, 261–327.

UNWIN R., BELL B., SHEPHERD M., WEBB J., KEATINGE R., and BAILEY S. (1995) *The effects of organic farming systems on aspects of the environment.* A review prepared for Agricultural Resources Policy Division of MAFF, ADAS.

Farm Auditing for Sustainability

MARK MEASURES
Elm Farm Research Centre, Hamstead Marshall, Newbury, Berkshire, RG20 OHR, UK

ABSTRACT
Policy makers have now established sustainability as the new aim for UK farming. The development of the Farm Audit for Sustainability involved identifying the objectives of sustainable farming, based on the principles of organic farming as set out by the International Federation of Organic Farming Movements (IFOAM) and establishment of indicators to assess the effectiveness of individual farms in meeting these objectives. On-farm use of the Farm Audit demonstrated that the tool was able to provide a comprehensive assessment of sustainability of the farming system and that it is an information and advisory tool which is potentially useful in benchmarking and development of the farming operation.

INTRODUCTION

Government Policy in the UK has commenced a programme of change for British farming towards what is loosely described as sustainable farming, one which not only ensures that the production of food is a commercially viable business, but also one which delivers across a broad range of public goods and services. This policy is being driven by changes in EU policy and support and is being vigorously encouraged through the Report of the Policy Commission on the Future of Farming and Food which has been largely adopted by DEFRA. The imminent application of new support measures following the Mid-Term Review will more or less facilitate aspects of this process of change on the farm. A clear understanding of the real, practical meaning of sustainable farming on the ground is, however, lacking, although there have been efforts to identify the desirable outcomes on a national scale, e.g. *Towards Sustainable Agriculture - A Pilot Set of Indictors* (MAFF, 2000).

Organic farming is the only system of agriculture which has a track record of setting a clear aim of sustainable farming, (IFOAM Standards 2000); one which meets societies' wider objectives for farming including: human health and welfare, environmental care, resource conservation and animal welfare in what is self evidently a finite world. It achieves this through the operation of farming practices that are characterized by an emphasis on biological systems and management techniques, rather than the use of inputs which characterize conventional farming.

The Organic Advisory Service (Elm Farm Research Centre) has set up a new initiative, the Organic Systems Development Programme (OSDP), which is seeking to help farmers develop their farm management in order to better meet the overall objectives of organic farming. The OSDP, headed by Mark Measures, is working with a group of nine mixed, well established organic farms which are committed to going beyond the absolute minimum set by organic standards to better address the broader needs of society in the way in which they produce food and to progressively develop more sustainable systems.

METHODS

The literature on the use of sustainability indicators was reviewed (Bell and Morse, 1990) and existing procedures for monitoring assessed (Wackernagel and Rees 1996; Haas *et al.* 2000; LEAF, 2001; Rigby *et al.* 2001; Leach and Roberts 2002). In the light of this, a new auditing system was formulated in order to meet the needs of the farmers involved, one which assessed their achievements, through measurement as far possible and which could be applied quickly and with the involvement of a farm adviser to provide independent assessment.

Development of the Farm Audit involved a meeting with Elm Farm Research Centre staff in order to consider how the work related to their research programme, which had already identified key issues relating to sustainability and developed techniques to address these issues. This was followed by a meeting with the farmer members of the OSDP to assess the relevance of the approach and to engage their input. The Farm Audit was conducted on five farms during routine advisory visits. During the following year the Farm Audit was used as part of on-farm group meetings to highlight the performance of the host farm and to refine the procedures.

Creation of the Farm Audit required the development of an audit procedure and a spreadsheet to calculate farm-gate energy and nutrient balances, preparation of a farm record sheet and collation of standard data for comparative purposes. The latter is still in the process of compilation as more farms are audited.

The Farm Audit identifies all the key objectives of sustainable farming, it does this by focusing on the key criteria or objectives set out by the IFOAM Standards and then aims to select indicators for each criterion which can be measured, or some meaningful assessment made, of the degree to which the farm is sustainable. It does not therefore endeavour to asses every component of every criterion, such an approach risks being excessively time consuming, neither does it focus on monitoring activities (much of this is already being undertaken by the organic certification procedures) but instead attempts to monitor the outcome of the farming system and practices.

AUDIT PROCEDURE

The auditing procedure summarized in Table 1 was used on all farms

AUDITING IN PRACTICE

The application of the Farm Audit was relatively straightforward, requiring between one and two hours to conduct, plus a variable amount of time by the farmer to access the information which was generally readily available. The use of benchmarking for factors other than those directly related to financial performance is unfamiliar to most farmers. However, the Farm Audit was effective in highlighting those areas in which a farm was particularly effective. This was very encouraging for the farmer concerned, for example one farmer achieved a veterinary cost of 20% of the national average, which was rewarding and indicated that there were farm practices from which others could learn. It also highlighted some shortcomings which was of real help in focusing the attention of the farm owners, manager and adviser. It might be argued that these shortcomings would have been identified anyway, but the Farm Audit does help prioritize areas for development.

Table 1. Auditing procedure

Objective	Monitoring to Identify Performance
Maintain a closed system	Nutrients imported on farm as a % of total exported
	Feedstuffs imported as % of total consumed Subjective assessment of manure plan and management
Maintain soil fertility	Annual soil analysis of the same 2-3 fields
Avoidance of pollution	Nitrate leaching scored on the basis of key risks Ammonia and methane risks Identify excessively high phosphate soils Identify and quantify pesticide use Identify and quantify farm waste Identify other risk areas
Food production	Energy output per hectare Quality monitored through sales data
Non-renewable resources	Energy is expressed as % of energy out Identify use of renewable energy Water sourcing, recycling and efficiency of use
Livestock management	Average age of herd Health: Mastitis incidence per 100 cows Medicine bill per head Other, e.g. distance to slaughter
Social function	Profitability of the farm Number of labour units per hectare Staff training Community engagement
Use of appropriate technology	Observations on both appropriate technology such as minimal tillage, reed bed, composting, and also inappropriate actions
Decentralization	Information where available on input miles and product miles
Biodiversity and landscape	Conservation plan in place and acted on. Presence of 'red or amber list' bird species Diversity: number of crop and stock types and % permanent pasture

The use of the energy and nutrient budgeting tool is in its infancy as the facility was not available at the start of the programme; however, for the first time it is providing farmers with some indication of how efficient they are. Understanding their energy efficiency and improving it is something which this group of farmers is keenly interested in. Early indications are that they are already relatively efficient due to their non-use of nitrogen fertilizers but there is clearly great scope for improvement and this information will begin to provide them with data by which to measure their progress.

An important outcome of the work has been to focus farmers' attention on the impact of their day-to-day practices on sustainability, it has also identified major shortcomings in information available to farmers and provides a useful means of identifying research needs.

ACKNOWLEDGEMENTS

Input to the development of the Farm Audit by Abbey Home Farm, Bagthorpe Farm, Commonwork Organic Farm, Duchy Home Farm, Lower Pertwood Farm, Luddesdown Organic Farm, Manor Farm Godmanstone, Sheepdrove Organic Farm and Woodlands Farm, staff from Elm Farm Research Centre and students of the Sustainable Development Advocacy Programme (Holme Lacy College PGC) is gratefully acknowledged.

REFERENCES

BELL S. and MORSE S. (1990) *Sustainability Indicators - Measuring the Immeasurable?*

HAAS G., WETTERICH F. and KOPKE U. (2000) Life cycle assessment of intensive, extensified and organic grassland farms in southern Germany. *Proceedings 13th IFOAM Conference p157.* Earthscan Publications Ltd.

IFOAM (2000) Basic Standards for Organic Production and Processing.

LEACH K.A. and ROBERTS D.J. (2002) Assessment and improvement of the efficiency of nitrogen use in clover based and fertilizer based dairy systems. 1. Benchmarking using farm gate balances. *Biological Agriculture and Horticulture,* 20, 143-155.

LEAF (2001) *The LEAF Audit.* Stoneleigh, UK: LEAF.

MAFF (MINISTRY OF AGRICULTURE FISHERIES AND FOOD) (2000) *Towards Sustainable Agriculture – A Pilot Set of Indicators.*

RIGBY D., WOODHOUSE P., YOUNG T. and BURTON M. (2001) Constructing a farm level indicator of sustainable agricultural practice. *Ecological Economics,* 39, 463-478.

WACKERNAGEL M. and REES W. (1996) *Our Ecological Footprint – reducing Human Impact on the Earth.* New Society Publishers.

The Use of Indicators to Assess the Sustainability of Farms Converting to Organic Production

[1]I.N. MILLA, [2]P. HARRIS and [3]C. FIRTH
[1]Escuela Técnica Superior de Ingenieros Agrónomos, Universidad Politécnica de Valencia, Camino de Vera S/n. 46022, Valencia, Spain.
[2] School of Science and the Environment, Coventry University, Coventry CV1 5FB, UK
[3] HDRA, Ryton Organic Gardens, Coventry, CV8 3LG, UK

ABSTRACT

A key feature of farm sustainability is the need to protect and make optimum use of limited natural resources within an economically efficient and socially acceptable agricultural system. Organic farming is often presented as a sustainable solution for agriculture. It is a challenge to measure sustainability in a practical way. This paper is based on a project which aimed to discover and test a suitable method for assessing changes in sustainability at the farm level. It studied the changes in sustainability on three conventional farms that had converted to organic production systems in the UK. A set of indicators for evaluating farm sustainability was devised, based on the French 'Method IDEA' (Indicators of Sustainability in Agriculture), which works with the three dimensions of sustainability: environmental, social and economic. The main conclusions from this project were that conversion to organic production increased the social sustainability of the all the farms studied. The environmental dimension has tended to either increase, or stay at a high level. In contrast to this, the economic dimension has decreased or increased slightly from a low level. The technique has proved a useful way of obtaining a broad ranging snapshot of a farm's sustainability. The project was limited by the small sample of three farms, and applying the technique to a larger sample of farms would certainly strengthen the validity of the conclusions.

INTRODUCTION

Sustainability in relation to agriculture is becoming increasingly important as the linkages between the economy, society and the environment are more widely recognized (Van der Werf and Petit, 2002). Sustainability is a term with diverse interpretation; however, it is commonly applied to ecological or environmental, social and economic aspects of farming systems. Sustainable agriculture is essentially concerned with the ability of agro-ecosystems to remain productive in the longer term.

In order to judge whether an agriculture system or farm is sustainable or not, an easy to use tool or method is required that can provide information understandable to practitioners such as farmers and to other stakeholders such as policy makers. The aims of this MSc project were to discover and to test a suitable method of assessing changes in sustainability. The method was applied to a group of farms that had recently converted to organic production in the UK during 1996-2001, in order to look for changes in sustainability that had occurred due to conversion.

MATERIALS AND METHODS

After examining a number of techniques used to measure sustainability, it was decided to adapt and test the Method IDEA (Indicators of Sustainability of Agriculture) (Vilain,

2000). This is a quantitative evaluation system devised for assessing sustainability at the farm level in France. In the system, the three main dimensions of sustainability are; agro-ecological, socio-territorial and economic. Each of these is subdivided into three or four main components (Table 1) and these are further sub divided into indicators, which are based on measurable data or values. During this project 37 different measures were made to score each farm on a wide range of farming practices, land organization, product quality, social interactions on and off the farm, economic performance and efficiency.

Table 1. Sustainability dimensions and components

Dimensions	Components
Agro-ecological	Diversity Land organization Agricultural practice
Socio-territorial	Quality of the products and territory Employment and services Ethic and human development
Economic	Viability Independence Ownership Efficiency

Three different farms were chosen from a DEFRA-funded 'Conversion to organic vegetable production' project. This project has studied the agronomic and economic performance of a group of ten farms, which have converted to organic production. The farms chosen represent different farm types from the group; mixed, arable and intensive vegetable farms. For each of the three farms, data were assembled for the year prior to conversion, and then again 3 to 5 years later, to determine the value of the indicators for each of these years. The data were obtained from farm records and accounts, and were supplemented by a questionnaire and interview with each farmer and by observations undertaken when visiting the farm. This enabled a snapshot of the sustainability of the farming system, pre- and post-conversion to organic production.

Table 2. Descriptions of the farms

	FARM I	FARM II	FARM III
Location	Warwickshire	Lincolnshire	Bedfordshire
Type of business	Family	Company	Family
Farm size (ha)	36	1956	24
Farm type	Mixed	General cropping	Horticulture
Type of conversion	Single step (1999)	Phased over 7 yr	Phased over 6 yr
Data based from	1996-2001	1996-2000	1998-2001

RESULTS

Through the process of conversion to organic production and adopting less intensive agricultural systems, two of the farms (Figure 1) improved their environmental sustainability. This was achieved through stopping the use of artificial fertilizers and pesticides and by growing a wider diversity of crops; all of this was necessitated by the adoption of practices to comply with organic standards (UKROFS, 2001). One of the farms (farm 2), was already using less-intensive farming methods, prior to its conversion, and contained a lot land in set-aside and environmental schemes. This farm was also converting more slowly that the others; therefore its environmental changes were less marked.

All of the farms increased the social dimension of their sustainability through the conversion process. This was related to adopting new marketing practices such as direct marketing, with benefits to the local economy and interactions with the consumers. It was also due to them employing more labour and contributing more employment to the local economy.

Only one of the farms had increased its economic sustainability. This was the small mixed farm where the premiums to be obtained from sale of organic products enabled this farm to become more economically viable. All of the farms incurred costs of converting to organic production, through reduced financial output during the conversion period as they incurred lower yields and were unable to sell their products as organic. This is reflected in their reduced economic sustainability. Output levels rose again when the farms started to sell organic products. However, after only 3-5 years, it may be too early to assess the full economic changes.

Figure 1. Changes in the dimensions of sustainability

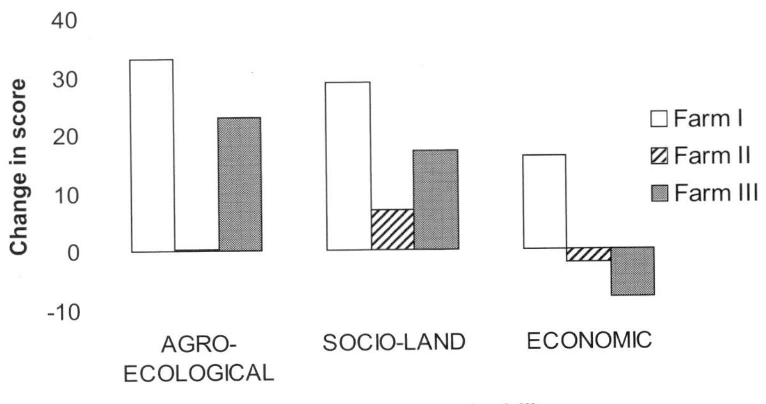

CONCLUSIONS AND DISCUSSION

The main conclusions from this project are that conversion to organic production has increased the social sustainability of the all the farms studied. The environmental dimension has tended to either increase or stay at a high level. In contrast to this, the economic dimension has decreased or increased only slightly from a low level. It is thought that the costs of restructuring the businesses and their decline in income during the conversion period have contributed to this decrease. Greater environmental and social sustainability has been achieved by growing a greater diversity of crops, stopping the use of artificial inputs, marketing more produce locally and employing more labour. It is often the case that these benefits to the environment and society do not necessarily have great rewards to the farmer in terms of higher on–farm economic returns.

The method has provided a useful illustration of changes in sustainability by considering a wide range of factors. This assessment has been achieved in a relatively short period of approximately two weeks per farm. It was necessary to make a number of changes to the French system of evaluation to adapt it to UK conditions. The project was limited by the small sample of three farms, and applying the technique to a larger sample of farms would certainly strengthen the conclusions. This project considered two separate years' data for each farm. The choice of the years chosen could have affected the economic results, as economic performance does vary from year to year according to variations in the weather and prices etc. In future it would probably be more advisable to choose average results from a farm in order to make the economic evaluations. Finally, it became apparent during the project that a period of 3-5 years from conversion is too short a time to assess the full impact of conversion to organic production on the sustainability of a farm.

REFERENCES

VAN DER WERF H.M.G. and PETIT J. (2002) Evaluation of the environmental impact of agriculture at the farm level: a comparison and analysis of 12 indicator-based methods. *Agriculture, Ecosystems and Environment*, 93, 131-145.

VILAIN L. (2000) La Methode IDEA. *Educagri éditions.*

UKROFS (2001) UKROFS Standards for organic food production. DEFRA UK.

Consumer Attitudes in North-West England to Organic and Regional Food

SUSANNE PADEL
Institute of Rural Sciences, University of Wales, Aberystwyth, SY23 3AL, UK

INTRODUCTION

The market for organic foods in the UK is considered to be one of the leading ones in Europe, and world-wide is estimated to have a value of £1 billion (SA, 2003). The main reason why consumers buy organic food is considered to be health benefits, followed by taste. Of growing importance also are concerns for the environment and animal welfare, but these more altruistic concerns remain less important then personal ones. The rapid increase of demand for organic food over the last few years is seen to be the result of various food-scares (Farodoye, 1999; MINTEL, 1999; SA, 1999; Datamonitor, 2002). The development of the market appears, therefore, driven primarily by consumers, but retailers, especially the multiples, have had a key role in furthering growth, promoting products, increasing range and aiding farmers to convert. The majority of what are considered to be "heavy" organic buyers are in Greater London and the South East.

Organic farming is also frequently associated with the promotion of food production and short supply chains, i.e. consumption close to the place of production. This implies that organic food must be purchased by consumers in rural areas near to where most of it is grown, and last year the sales of organic food through farmers' markets, box schemes and farm shops reached a dynamic growth rate of nearly 30 per cent, higher than the organic sector as whole. However, most organic food in the UK remains purchased through supermarkets (SA, 2003). Consumer studies do not differentiate between urban and rural consumers, so that the available knowledge reflects general, rather than specific, trends.

This paper present results of a series of focus groups carried out in North-west England in the city of Lancaster, in order to supplement existing survey knowledge of consumer attitudes to organic food in relation to more rural areas and attitudes to organic marketing initiatives. The work was carried out as part of a larger EU-funded project on Organic Marketing Initiatives and Rural Development (OMIaRD QLK5-2000-01124), in support of a case study of one Organic Marketing Initiative in the region.

BACKGROUND TO THE REGION AND THE WIDER CASE STUDY

Agriculture is a cornerstone of, and the major land use in, the rural economy of Lancashire, North-west England, ranging from mixed, arable and horticultural production in the costal plains, to extensive beef and sheep grazing systems in the uplands. There is no strong tradition of organic farming, and in terms of land area farmed organically and the sales of organic products, the North West of England lags behind other areas of UK, notably the South (Lees and Rogerson, 2001 cited after Foster and Beckie, 2003). Organic food in Lancaster is also sold in a number of other outlets: several multiple retailers belonging to national companies as well as one regional chain, a whole food shop, organic stalls on the weekly markets and one vegetable box scheme. Production and consumption of organic food in the north of England lags behind the rest of the country (see also Midmore *et al*,. (2004, this volume) for further details and results of the wider case study).

METHODS

The focus groups were conducted between the 12 and 14 November 2002 in Lancaster, a small university and market city in the county of Lancashire. The qualitative method of focus groups was aimed at gaining an understanding of factors that might influence buying behaviour in rural areas, such as consumer knowledge of, expectations and attitudes to, organic food, attitude to and preference for different distribution channels, the origin of food and organic food. The research also provided the customer perspective of one particular Organic Marketing Initiative (OMI) for other work in the project (a vegetable box scheme located nearby that delivers bags of organic vegetables for a fixed price each week, (Midmore *et al.*, 2004: this volume) .

Fifty-one participants in six groups were recruited from existing and potential customers of a vegetable box scheme, and with variation in terms of some social characteristics, but being a qualitative research method, and not random sampling, this cannot be taken as in any way representative of the UK population. One regular customer of this Organic Marketing Initiative acted as regional contact, organized the venue and carried out the recruitment.

The author moderated the groups with assistance from the University of Applied Sciences in Hamburg, the German project partner in the project. Moderation followed a common guide covering unprobed "top of mind" associations with organic food and origin of product, and consumer expectations of the specific OMI including the use of projective techniques. All discussion groups were taped and transcribed and, with the help of a VHS recording, statements could be attributed to the type of participant (gender, previous contact with the initiative). The results highlight general discussion as well as conclusions in relation to organic marketing initiatives in rural areas.

RESULTS

The results of the focus groups in Lancashire give some important insights into attitudes to organic food in North-west England, although they can not be seen as representative of the population as a whole. However, some participants had a higher than expected level of environmental awareness, which may be a reflection of the recruitment carried out in association with one particular Organic Marketing Initiative, although not only customers were specifically invited to participate.

In the first round of discussion, all six focus groups strongly associated 'organic' with fresh fruit and vegetables and tended not to mention other products (meat, dairy etc.). The association of organic vegetables with soil was largely seen as positive, the proof of a genuine, wholesome and healthy product that has an earthy feel to it (*"dirty but pure")*. Many participants associated organic with *"no chemicals/pesticides"* and healthy, but also with attributes that were not related to the farming method, such as *"friendly service"* and *"unpackaged"*. Some initial associations with organic products were negative, frequently mentioned were *"expensive"* but also *"elitist"* and *"disfigured and discoloured vegetables"*.

Furthermore, despite generally high levels of education among the focus group participants, there was a low level of knowledge and understanding of inspection and certification systems and the legal protection of the term organic on food products. *"Is organic an actual definable term or is it just another bendable thing" (M)?*

This was reinforced by widespread confusion about labelling issues.

The results confirm that consumer motivations for buying organic products vary. 'Health' was a main reason, but also more altruistic reasons entered the discussion (e.g. '*better for the environment*'), especially among the more regular consumers. Interestingly, some consumers made a connection between local and organic food as they gave '*supporting local producers*' as a motivating factor for buying 'organic' food. However, the expressed preference for local over organic suggests that they may be willing to switch to other (non-organic) suppliers, if they better correspond with their personal values, and hence they may not be as fully committed to organic food as consumers who buy for health reasons.

There was widespread support for buying local food among most of the participants, followed by support for buying British, but participants did not express a preference for other regional products from outside their own area. Although no firm conclusions can be drawn from this, it does suggest that careful consideration and market research should accompany any regional food branding initiatives that seek to target a consumer base outside the home region.

The most significant barriers to buying organic food were price and price perception, access and availability, visual product quality and presentation, and mistrust of organic food in supermarkets. Eating habits and lack of cooking skills were also identified. Many participants expressed a preference for buying organic produce from markets and specialist shops. Some went so far as stating that they would never buy organic produce in multiples *("If my option is to buy organic from the supermarket, I have no faith in it, so I just go in and buy the regular stuff" (female)*. However, others preferred shopping in supermarkets because of the convenience of being able to do all shopping in one place.

The participants were also asked to identify strengths and weaknesses of the local Organic Marketing Initiative, a vegetable box scheme. As main strengths they mentioned good quality, fresh organic vegetables; trust; personal commitment of the owner to organic principles; home delivery service; commitment to seasonal and local production. Main weaknesses and barriers to joining the scheme were lack of knowledge about the OMI; elitist and exclusivity; limited product range and lack of choice, flexibility; difficulty with home delivery during working hours and for social reasons; absence of a shopping experience for vegetables (e.g. touching, smelling; bargain hunting). Contributions of the OMI to rural development were identified in the area of employment, economic benefits to the growers and retaining wealth in the region, as well as setting an example of a locally based but successful small business with a commitment to environmental issues. However, these impacts were felt to be indirect and limited by the small size of the sample.

CONCLUSIONS

The most important, and in some ways most surprising results, of these six focus groups in a rural market town in Northwest England were:

- The strong association of organic with vegetables and with "earthy", with poor visual quality and with high price.
- The relative ignorance of what the term organic means in terms of the legal production standards, inspection and certification procedures for all operators.

- The relative ignorance of what the term organic means in terms of the legal production standards, inspection and certification procedures for all operators.
- The lack of awareness of the local Organic Marketing Initiatives and other outlets that sell organic food.
- The strongly expressed commitment to local food.

Participants made some recommendations for to improve such an Organic Marketing Initiative:

- More advertising outside the 'usual' circles.
- Trial bags.
- More information in advance about the content of a typical bag.
- Introducing an element of choice, potentially for a higher price.
- Extending the concept of home delivery to other products.

The OMI has already had experience of the latter two recommendations with varying degrees of success, so it would have to weigh these options (or variations of these options) carefully within the framework of its future development plans.

REFERENCES

DATAMONITOR (2002) *The Outlook for Organic Food and Drink.* Datamonitor. London.
FARODOYE L. (1999) Focus on organic food. *Meat Demand Trends* (3), 3-10.
FOSTER C and BECKIE M. (2003) Draft internal project report on the OMIARD UK case study in Northwest England. University of Wales, Aberystwyth.
LEES D. and ROGERSON K. (2001) Northern Organic Development Strategy: Feasibility Study. Soil Association, Bristol.
MIDMORE P., FOSTER C. and SCHERMER M. (2004) The Contribution of Organic Producer Initiatives to Rural Development: a Case-Study Approach. In: Hopkins A. (ed.) *Organic farming: science and practice for profitable livestock and cropping*. Occasional Symposium of the British Grassland Society, No. 37, pp 11-18.
MINTEL (1999) *Organic Food and Drink Retailing.* UK Economist Intelligence Unit. London.
SA (2003) *The Organic Food and Farming Report 2003.* Soil Association. Bristol.

Expert Perspectives on the Future of the Organic Food Market: Results of a Pan-European Delphi Study

SUSANNE PADEL, CAROLYN FOSTER and PETER MIDMORE
Institute of Rural Sciences and School of Management and Business,
University of Wales, Aberystwyth, Ceredigion, SY23 3AL, UK

ABSTRACT
A Delphi Inquiry was carried out to assess the prospect and conditions affecting the overall growth in the European Market for organic products in the coming decade, and to provide support for research. Countries were classified as established, growing and emerging, according to the state of development of their organic market. The survey confirmed the importance of factors influencing the development of the organic food market: the supply base, the role of supermarkets as sales channels and of government support. Organic Producer Initiatives were seen as important in securing a fair deal for organic producers but managerial capacity and professionalism are key challenges for such organizations.

INTRODUCTION
The complex and diverse nature of influences affecting organic food markets in different parts of Europe renders it suitable for a Delphi questionnaire of expert opinion. This paper reports on a study established on interaction between specialists in European Organic Food Markets from a commercial, farming sector (both organic and conventional), government and academic background in eighteen European countries through the framework of an iterative survey of opinion.

In essence, the Delphi process allows a group of experts to participate jointly in defining and analysing complex problems/issues where information is fragmentary or inaccessible by contributing to successive rounds of information gathering, receiving feedback and then refining the information gathering process in the subsequent rounds). The process is well suited to situations where perspectives might differ substantially according to background, although it does not necessarily yield a unified consensus at the end. It has the advantage that each participant can reflect on and take into account views based on the range of experience of the other panel members (Linstone and Turoff, 1975).

DETAILS OF THE SURVEY
This Delphi Inquiry was carried out as part of the OMIaRD Project (Organic Marketing Initiatives and Rural Development - EU QLK5-2000-01124). It had two main aims:

1) to assess the likely prospect for, and conditions affecting, the overall growth in the European Market for organic products in the coming decade, and

2) to provide support for the research process in the broader project.

It consisted of three rounds, the first containing six open questions regarding the current state and development of the organic markets in Europe and threats to, and opportunities for, future growth. The results formed the basis for a structured questionnaire for the second and third rounds, divided into thematic sections related to the specific country of the respondents, future development of the organic food market, the role of national government in future development and organic marketing initiatives and their impact on rural development. Results of successive rounds were fed back to respondents as a basis for further comments and revision. The results presented here refer to the first aim; a full report of all results can be found in Foster *et al.* (2001) and Padel *et al.* (2003).

In the first round, 252 experts (identified by the OMIaRD project partners throughout Europe) were contacted with response rate of 85%. For the second round, these 213 respondents were contacted again receiving both the first round report and the second questionnaire, achieving a response rate of 80%. For the third round, 170 questionnaires were sent and 127 responses (76%) were evaluated. From the first to the third round this represents an overall response rate of 51%. Response rates varied between countries with a very high return in some countries (for example 90% in Austria) and between 0 and 20% in others where only a small number of experts were contacted in the first place. As it is not known how many experts of the organic food market exist throughout Europe, it is not possible to assess what proportion of a possible total sample was covered.

The share of respondents from each type of occupational background varied between the three rounds, but remained overall relatively balanced with approximately 30% from commercial backgrounds, followed by researchers (28%), government and organic organizations (both 26%) and non-organic organizations (18%). Cross tabulation of the data showed that responses to only a small number of questions were influenced by occupational background. The average involvement of respondents in the organic sector was between 10 and 12 years, a majority of respondents were male and between 30 and 44 years old, and most bought organic food for themselves.

RESULTS

Based on the responses to the first round, the European countries were classified according to the state of development of their organic market. This 'soft' classification - based on the subjective attitudes of market experts - into three major groups (*established*, *growing* and *emerging* organic markets, see Table 1) was confirmed in the second and third round and by other project findings (Hamm *et al.*, 2002). Attitudes and observations of experts in some areas appear influenced by the market development of their country. A small proportion of experts in the UK and Belgium did not agree with the proposed classification of their countries.

Table 1. Countries clustered by stage of organic market development

Established market countries	Growing	Emerging
Austria	Finland	Belgium
France	Italy	Czech Republic
Germany	Netherlands	Greece
Switzerland	Norway	Ireland
United Kingdom	Portugal	Slovenia
		Sweden
		Spain

The first round also provided a first impression of factors influencing this, such as the role of supermarkets as sales channels for organic products and the importance that consumers attach to environmental protection and animal welfare. There was widespread agreement among respondents that the integrity and quality of organic products must be safeguarded.

Within a single country not all markets for organic food are equally developed. Experts consider the markets in urban areas and for cereals, dairy products and fruit and vegetables to be better developed than those for meat and convenience products and for those in rural areas. Food scandals and the media were considered to be important driving forces for the development of the organic market, but the majority of respondents considered that government policy had also had a positive impact.

Experts were asked to estimate future growth rates. Overall expected growth varied between countries, with lowest rates anticipated in Denmark (approximately 2%) and highest rates in Germany and the UK (7 to 8%). Rates did not appear to be directly related to the state of market development but reflect specific country conditions. Expected growth also varied for product categories, with lowest growth expected for cereals markets and highest for meat and convenience products.

Of the different marketing channels, multiple retailers were considered most important, a result confirmed by the analysis of the European food market by Hamm *et al.* (2001). It appears that the importance of alternative channels (e.g. direct marketing, specialist organic shops) diminishes as an organic market becomes more advanced, whereas that of multiple retailers increases. Experts anticipated the position of multiple retailers as the main outlets for organic products to continue in future, both in urban and rural areas, but expressed concerns in relation to the impact of cut-price policies on organic producers. In rural areas direct marketing was considered to be the next important outlet, in urban areas specialist organic shops. Catering and public procurement was not expected to overtake any of the other outlets in terms of importance in the near future.

Experts considered national and regional government support to have had an important impact on the development of the organic market. Some observed variation in the answers appears to reflect different governmental policies identified elsewhere. Respondents clearly supported the need to develop EU standards in areas not yet well regulated (e.g. horticulture and fish), and to consider the environmental impact of trade and the role of production incentives in helping to overcome problems in the supply of organic raw materials.

Participants considered the integration of organic agriculture with other rural development initiatives important both for the organic market and for rural development. However, substantial variation in answers to several questions in relation to rural development indicate that the surveyed marketing experts associate a variety of different issues with rural development and have no common understanding of the contribution that organic farming can make apart from improved soil fertility, local environment and landscape. Statements that attracted most agreement were related to the fact that the same business and marketing principles apply to organic and other marketing initiatives. Producer co-operatives were seen as having an important role to play in securing a fair price for organic producers, but experts considered a lack of 'quality of management' to be most important in a list of given barriers for organic marketing initiatives to achieve their objectives (See Figure 1).

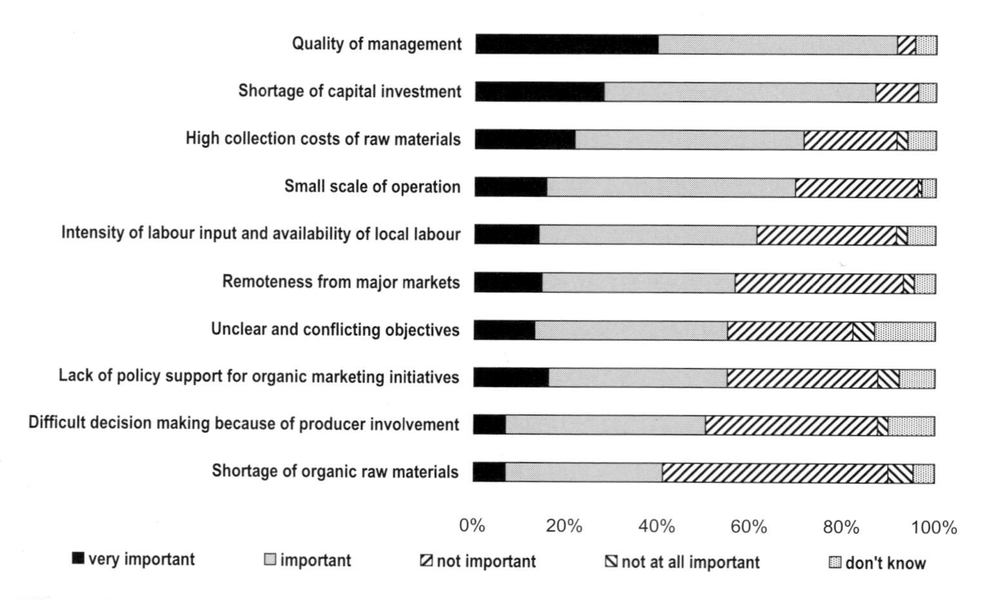

Figure 1. Importance of barriers preventing Organic Marketing Initiatives from achieving their objectives (128 Respondents in the third round)

CONCLUSIONS

Overall, the survey confirmed that the importance of factors influencing the development of the organic food market; these include the supply base, the role of supermarkets as sales channels and of government support. Organic Producer Initiatives were seen as important in securing a fair deal for organic producers but managerial capacity and professionalism are key challenges for such organizations. There was widespread agreement among respondents that the integrity and quality of organic products must be safeguarded and for common EU wide standards for the currently unregulated areas such as aquaculture and horticulture.

REFERENCES

FOSTER C., HYDE T., MIDMORE P. and VAUGHAN A. (2001) SWP 1.2: Summary report of the first round of the Delphi Inquiry on the European market for organic food. OMIARD Project Report. School of Management and Business Studies, Aberystwyth.

HAMM U., GRONEFELD F. and HALPIN D. (2002) Analysis of the European market for organic food. Organic Marketing Initiatives and Rural Development, 1. School of Business and Management, University of Wales, Aberystwyth.

LINSTONE H.A. and TUROFF M. (eds.) (1975) *The Delphi Method: Techniques and Applications*. Reading, Mass.: Addison-Wesley.

PADEL S., SEYMOUR C. and FOSTER C. (2003) SWP 5.1: Report of all three rounds of the Delphi Inquiry on the European Market for Organic Food. OMIARD Project Report. School of Management and Business Studies, Aberystwyth.

Organic Farming in Cornwall. 1. Description of the Systems and Enterprises

J. BURKE, S. RODERICK and P. LE GRICE
Organic Studies Centre, Duchy College, Rosewarne, Camborne, Cornwall, TR14 0AB

ABSTRACT
Cornwall has unique climatic conditions well suited to organic production. However, there has been scant information on the extent, range and levels of organic production required for development of a research strategy to support the sector and improve marketing. A survey of 119 Cornish producers, conducted during 2002 and 2003, provided data on organic enterprises, output, key constraints as well as social, economic and environmental issues. There were 8,778 ha of organic land in Cornwall. Most of the major farm enterprises were represented, including organic beef (66.4% of holdings); arable (45.4%); sheep (36.1%), dairy (26.1%), poultry (23.5%), field vegetables (16.0%), protected crops (10.9%) and pigs (10.9%). Permanent pasture, temporary ley and cereal production accounted for 36.3%, 33.8% and 12.3% of organic land area respectively. Forage legumes and field vegetables also accounted for significant areas. There were 3,148 dairy cows, 1,463 suckler cows, 6,206 breeding ewes and 72 breeding sows. Estimated annual poultry production was 2.5 million eggs and 22,000 table birds. There were 19.1 ha and up to 15.0 ha of organic potatoes and cauliflower, respectively, and a wide range of other vegetables. The production constraints identified included animal feeding and health, marketing, crop protection and soil fertility.

INTRODUCTION
Although organic farming has undergone a very rapid increase in scale over recent years, detail of regional and enterprise production remains scarce. The development and implementation of a research programme and technical and marketing support is constrained by this absence of detail. This is particularly relevant in Cornwall, which, within the UK, has unique climatic and market conditions for the development of organic agriculture, and yet is geographically distanced from the main channels of communication.

The Organic Studies Centre in Cornwall has a specific role to provide technical information, through research and training, to the producers in the region. Partly to address the shortage of regional production statistics and management information in Cornwall, the Centre undertook a survey of organic farmers during the latter half of 2002 and early 2003. The survey provided baseline data on the numbers of producers and a wide range of production parameters for each sector of organic production, as well as providing information about social, economic and environmental issues and broader farm level management factors, described by Roderick *et al.* (2004). Details on enterprise management and output are summarized in this paper.

MATERIALS AND METHODS
During the period August 2002 to February 2003 all registered organic producers residing in Cornwall were interviewed. Of the 130 persons registered with the Soil Association or Organic Farmers and Growers Ltd., 120 were actively producing organic or in-conversion

food and managing organic or in-conversion land at the time of the survey. One hundred and nineteen (99.1%) of these were interviewed. The one unit not included was a small-scale horticultural unit, as it was not possible to make contact with the farmer. The interviews were conducted using a pre-tested questionnaire that included closed and open-ended questions on a wide range of farming-related questions, including land-use and enterprise details, management and husbandry, economics, marketing, income, labour-use, the environment, organic standards and certification and research and training requirements. One hundred and twelve questionnaires were completed as one-to-one interviews conducted by a team of seven trained enumerators with the remainder completed by means of telephone interviews. On average, the interviews took 1 hour 24 minutes to complete.

RESULTS

Land use and enterprises

In total, 8,778 ha of land in Cornwall were either fully certified organic or in-conversion. Of these, 7,659 ha (87.3%), were fully converted. Most major farm enterprises were represented, the most prevalent being: organic beef production (66.4% of holdings); arable (45.4%); sheep (36.1%) and dairy (26.1%). Other organic enterprises included egg production (16.8% of holdings), field vegetables (16.0%), horticultural vegetables (16.0%), orchards (13.4%), protected crops (10.9%), pigs (10.9%), soft fruit (9.2%), herb production (8.4%) and table birds (6.7%). Many farms had more than one enterprise type. Permanent pasture and leys accounted for 36.3% and 33.8% of organic land, respectively.

Livestock production

Enterprise mix on livestock farms included beef and sheep (27.7% of holdings), beef only (26.9% of holdings), dairy only (13.5%), dairy and beef (8.4%), dairy, beef and sheep (4.2%) and sheep only (4.2%). Stockless holdings and farms that kept either pigs or poultry as the only livestock enterprises accounted for the remaining 15.1% of holdings.

Milk production. There were 31 organic dairy herds totalling 3,148 cows with an average herd size of 101 cows. Average milk yield was 5,937 litres per lactation representing a 3.4% average reduction post-conversion. Predominant breeds were Holstein (16 herds), Friesian (9), Jersey (4), Guernsey (3) and Ayrshire (2). Average somatic cell count was 206,000 cells/ml compared with 169,000 cells/ml pre-conversion.

Sheep Production. There were 41 organic sheep producers (6,206 breeding ewes) of which four (197 breeding ewes) were in-conversion. Average lambing percentage was 149% (range 90% to 190%) with an estimated population of 8,620 organic lambs born in 2002 plus 286 lambs born from in-conversion ewes. Most (83%) flocks of lambs were finished at home. The most prevalent breeds were Suffolk (22%) and Dorset or Dorset crossbreds (17%). There were three flocks of Devon and Cornwall Longwool sheep.

Beef Production. There were 79 organic beef producers, nine of which were very small scale. The majority of commercial producers (57) kept suckler herds. There were ten farms rearing dairy-bred calves and three buying and finishing store cattle. There were 1,463 suckler cows with an average herd size of 24 (range 1 to 79 cows). Most were spring calving. Of the estimated 1,100 finished animals, 60% came from suckler herds. Excluding in-conversion herds, total output was expected to be 750 head, with approximately 75% from suckler herds. Including crossbreds, the most popular breeds were Aberdeen Angus (15 herds), South Devon (10 herds), North Devon (8 herds),

Hereford cross (7 herds) and Limousin (6 herds). There were three herds each of Charolais, Welsh Black and Galloway.

Pig production. There were 13 organic pig herds with a total of 72 breeding sows with an average herd size of 6.5 sows (range 1 to 20). Breeds included Gloucester Old Spot (4 herds), Saddleback or Saddleback cross (3 herds) and Large white/Landrace cross (2 herds). Eleven farms were finishing pigs (approximately 750 per annum). Six herds were considered as part of a crop rotation and using home grown feeds. Marketing included direct sales (9 farms), local retailers (6 farms) and abbatoir sales (3 farms). Predators, muddy conditions, fencing difficulties, difficulties in finishing and feed price and availability were cited as the main problems.

Poultry production. There were 20 organic laying flocks ranging in size from 7 to 6,800 birds, including two flocks of 1000 birds or more. Columbian Black Tail (largest flocks) and Black Rock (most prevalent) were the main breeds. Average production, estimated from five flocks, was 261 eggs per bird per year (range 180 to 300). It is estimated that annual Cornish production was approximately 2.5 million eggs. Eleven layer flocks were in mobile housing, eight were fed entirely purchased feed and three small flocks were fed entirely home-grown feeds. Most had a range of local market outlets, including farm-gate and contract sales (the largest flocks). There were eight table-bird producers, with batch size ranging from 20 to 1000 birds and between one and twelve batches reared per annum. Approximately 22,000 birds were reared in the county during 2002 and annual farm output ranged from 35 to 12,000 birds. Five farms used mobile housing. Identified constraints included the shortage of a local feed mill and finishing slow-growing breeds on home-grown feeds. Larger flocks were produced to contract, whilst the smaller flocks tended to be sold direct to the consumer.

Animal Health. Mastitis in dairy herds, helminth infestation of lambs and predation in pig and poultry systems were the most frequently recorded health and welfare problems. 64% felt their veterinarian was supportive and 15% felt unsupported. 62% of livestock farmers used homeopathy, although only 13% received local sources of homeopathic advice.

Forage production. Seventy-nine farms made a total of 1973 ha of first cut silage, 923 ha of second cut silage (41 farms) and 454 ha of third cut silage (20 farms). At least 67 farms made at least 397 ha of hay.

Crop production

Cereal and protein crops. 48% of farms grew approximately 1083 ha of organic cereals, of which 90% was fully organic, equivalent to 12.3% of Cornish organic land area. 402 ha of the cereal area was grown as triticale, 255 ha as spring. 123 ha was winter wheat, 103 ha as spring wheat, 103 ha as winter oats, 83 ha as spring oats and 15 ha as winter barley. Details of the proportion of grain and whole crop were not available. The area of organic legumes grown in 2002 as protein feed crops was 375 ha (4.3% of Cornish organic land area). The most prevalent crops were forage peas (280 ha), lupins (86 ha), winter beans (13 ha) and spring beans (12 ha). A further three farms grew pea/cereal mixtures but did not specify the area. 22 farms made whole crop silage, of which 11 were dairy producers. These were cereal only (8 farms), cereal/peas (9 farms), forage peas (2 farms), lupins (3 farms), and maize (1 farm). Ten farmers (including six dairy farms) were crimping grain. The most popular crimped crop was triticale, followed by barley. There were comments from producers regarding the most appropriate cereal varieties for Cornish conditions.

Vegetables. There were 19 field-scale vegetable producers, of which organic potatoes (13 farms; 19.1 ha) and cauliflower (10 farms; 15.0 ha) were the most common. 56% of the potato area was grown as second earlies. Nineteen farmers grew other commercial vegetables, not always on a field scale. Given the occasional small scale and the wide range of crops, it was not possible to estimate area for each. Data describing commercial crops included: spring greens (5 farms, 8.1 ha); leeks (8 farms, 3.2 ha); swedes (4 farms, 2.8 ha); carrots (5 farms, 1.4 ha); cabbage (5 farms, 1.3 ha); onions (9 farms, 1 ha); broccoli / calabrese (4 farms, 0.8ha); parsnips (4 farms, 0.5 ha); brussels sprouts (3 farms, 0.3 ha) and garlic (6 farms, 0.2ha). A further four farms grew "a mixture of vegetables", but did not give details of area. Thirteen of the vegetable growers also had protected cropping, of which there was little production data provided.

Pests, weeds and diseases. Within the survey, 77% of organic farmers identified at least one specific weed, disease or pest problem, with many identifying more. Weeds in grass/clover (32% of all responses) and cereals (19%) were the most commonly reported, particularly docks and thistles. Pests of cereals and disease in potatoes (especially late potato blight) were also commonly cited.

Organic crop inputs. The most common organic inputs used (classed as regular or occasional inputs) were farmyard manure (76.5% of farms), sea-sand or seaweed (51.3%), lime (44.5%), slurry (23.5%), phosphate (22.7%), potassium (17.6%), compost (14.3%) and organic compound fertilizers (9.2%). The most commonly cited comment regarding inputs were concerns regarding negative nutrient balance. A summary of manure management on the surveyed farms is provided by Roderick *et al.* (2004).

DISCUSSION

This survey has provided data that not only describe current organic production in Cornwall, but which also provide a platform for enterprise and product development, in respect of technical improvements and marketing requirements. Coupled with the data provided by Roderick *et al.* (2004), these data will be used by the Organic Studies Centre to form the basis of a strategy for organic farming research and development in Cornwall. These data are also relevant to those involved in more direct market development.

Some of the problem areas identified, such as mastitis in cattle, helminth control in sheep and weed control in grass and crops are common to many organic systems throughout the UK. This represents an opportunity for collaborative effort to address these issues, develop guidelines and disseminate technical information through demonstration, training and publications. There are also regional issues, such as the suitability of crop varieties, which require more specific local effort and resources. The study has revealed a diverse range of enterprises within the county, offering opportunities for greater within and between farm integration and resource sharing.

REFERENCES

RODERICK S., BURKE J. and Le GRICE P. (2004) Organic farming in Cornwall. 2. Technical and socio-economic opportunities and constraints. In: Hopkins A. (ed.) *Organic farming: science and practice for profitable livestock and cropping*. Occasional Symposium of the British Grassland Society, No. 37, pp 47.50.

Organic Farming in Cornwall. 2. Technical and Socio-Economic Opportunities and Constraints

S. RODERICK, J. BURKE and P. LeGRICE
Organic Studies Centre, Duchy College, Rosewarne, Camborne, Cornwall, TR14 0AB

ABSTRACT
A survey of Cornish organic farmers (2002-2003) allowed investigation of the social, economic and environmental benefits of organic farming. Most farms were Soil Association certified and the average farm size of these was half that of farms registered with the Organic Farmers and Growers. There were considerable differences in enterprise type between the two certifying bodies. There was widespread adoption of the Organic Farming Scheme, with 49% of farms completing conversion during 2001. There were indications, based on farmers' perception, of improved biodiversity post-conversion. Income on most organic farms had either increased or remained static, although 30% of farms had suffered an income decline after becoming organic. Organic conversion also resulted in an increase in direct sales to the public, while marketing of meat products appeared to be unsatisfactory. Although mainly family farms, there was evidence of a slight decline in full-time employment and an increase in part-time and seasonal labour post-conversion. 72% said that they would continue to farm organically even with no price premiums. 61% said they enjoyed farming more since conversion, with 5% enjoying farming less. Many farmers had found difficulty in obtaining organic seeds, and paperwork was the main area of concern regarding application of organic standards.

INTRODUCTION
The potential social, economic and environmental benefits of organic farming have been recognized (Department for Environment and Rural Affairs, 2002). Whilst south-west England has the highest density of organic farms in the UK (Soil Association, 2002) there is an absence of detailed information describing the technical and socio-economic issues on these farms. In order to rectify this deficiency, a survey of producers was conducted during the latter part of 2002 and early 2003. The survey has also enabled the Organic Studies Centre to develop a training and research strategy that meets the needs of organic farmers and potential converters in Cornwall and the Isles of Scilly.
As well as providing detail of individual enterprise management and output, described by Burke *et al.* (2004), the survey provided details of a range of social, economic and environmental issues and technical constraints. These are summarized in this paper. The survey methodology is as described elsewhere in these proceedings (Burke *et al.,* 2004).

RESULTS
Organic certification
Seventy-six percent of producers were registered with the Soil Association (SA), although these accounted for 59.0% of the organic and in-conversion land area. The remainder were certified with Organic Farmers and Growers (OF&G). The average size of farms registered with the OF&G was approximately 113 ha, which is almost exactly double the average area of SA farms. There were also major differences in the prevalence of enterprise type between the two certifying bodies. The most obvious difference was in

the organic dairy sector, where 60.7% of all OF&G farms were involved in milk production, compared with only 15.4% of SA farms. The converse trend was found when the prevalence of horticultural enterprises was examined. For example, all of those organic producers involved in protected cropping were registered with the SA. They represented 14.3% of SA producers.

Only seven (5.9%) organic farms in Cornwall were fully converted to organic production before 1998. It was only after this period that there were new farms converting each year. Many of the organic farms (48.7%) in the county completed conversion during 2001. Ten of the surveyed farmers said that they still had unconverted land, of which seven said they were unlikely to convert any of this within the next five years.

Social and economic considerations

Land tenure. Less than half (47.1%) of holdings were at least partly tenanted, with 19.3% of all holdings consisting entirely of rented land. The total area of tenanted organic land was equivalent to 44.5% of organic and in-conversion land. 16% of farms were at least in-part in Less-Favoured Areas and 8% were, at least in part, designated as Environmentally Sensitive Areas. Eight of the organic farms in the county were National Trust properties.

Employment of farm labour. A series of questions were asked that referred to the farm economy, labour use and economic prosperity post-conversion to organic production. 6.8%, 13.2% and 29.2% of those who responded to this question said that they employed more full-time, part-time and seasonal labour, respectively, since conversion. Conversely, the responses regarding reduced labour requirement since conversion were 12.3%, 6.6% and 10.4% for full-time, part-time and seasonal labour, respectively. However, these data need to be treated with caution, as only 76 farmers (63.8%) responded to the question, with only 43 farmers responding specifically to the seasonal labour part of the question. Forty-seven farmers (39.5%) specifically stated that they did not employ staff, although the question did not reveal whether family labour was included in this response. Although 50.8% of employers said that they had had difficulty in recruiting farm staff, only 12.7% said that they found staff recruitment more difficult since conversion.

Farm income and marketing. Regarding income from organic farming in the county, there were more farmers who felt that their income had increased (41.2%) since conversion compared to those that felt income had decreased (29.4%). However, it should be noted that 11.8% of those who answered this question felt that the decrease had been significant. Some farmers stated that their income increased in the years immediately after conversion, but since then it had declined.

Questions relating to direct marketing of produce revealed that 31.9% of Cornish organic farmers viewed the amount of direct sales to have increased since conversion, whilst 5.3% said that these sales had declined. A similar response was received with regard to local sales of farm produce. Eleven organic farmers indicated that they operated a box scheme. All of these had either always farmed organically or had started the box scheme since conversion. Other notable results from specific enterprises was the lack of knowledge of carcass quality on many beef and sheep farms, and the general feeling by producers that the market for their products was chaotic.

A large percentage of farmers (72.2%) said that they would continue to farm organically even if there were no price premiums on organic produce. 62.8% of respondents said that they enjoyed farming more since conversion, whereas only 5.2% enjoyed farming less.

Business diversification. Eighty-seven percent of those interviewed said that organic farming was their main farming activity, i.e. provides more income than from conventional farming. However, 59.8% said that they had other on-farm income outside of farming. By far the most important source of other on-farm income was accommodation, with 39% of all farms surveyed having some form of income-generating accommodation on their farms. Farm shops (5.8%), horses (5.8%), food processing (5.0%), agricultural contracting (4.2%), craft/skills (4.2%) and leisure/tourism (1.7%) were also mentioned. 34.7% of organic farmers said that they were diversifying outside of farming more at the time of the survey, than they were before converting to organic production.

Conservation and the environment

Opinions on whether or not conversion to organic farming had improved the biodiversity on farms showed clearly that many felt that this was the case, with 58.9% of producers claiming that an increase in wild plant populations had occurred since conversion to organic production. The equivalent responses for other biodiversity indicators were: increased wild mammals (40.2% of responses), increased soil insects (52.7%) and improved soil structure (36.1%). All other responses indicated no observed changes in these parameters, apart from one farmer who observed a decline in wild plant populations.

Fifty-four percent of farmers (from 113) said that they had a whole farm conservation plan, although only 24.2% had been drawn up through a DEFRA-funded scheme. There was evidence of heavy subscription to the Organic Farming Scheme, with 72.0% of interviewed farmers indicating that they had successfully obtained the scheme (these data include farms that were organic before the scheme was launched). 26.1% of organic farmers had or were participating in the Countryside Stewardship Scheme, with 18.3% of all respondents applying after they had converted their farms. The application of other Government schemes was also in evidence with 9.6% of farmers having made successful application to the Woodland Grant Scheme and 6.1%, 5.2% and 4.4% having successfully applied for the Hill Farm Allowance, Environmentally Sensitive Area Scheme and the Farm Woodland Premium Scheme, respectively.

Technical constraints

Manure management. Open-ended questions referring to organic crop inputs provoked a number of responses referring to the use of farm yard manure. Many of these were concerned with the need to have more manure than the farm can produce, the cost and importance of storage and the importance of timing of manure applications. Specific questions were asked on the use and storage of farm yard manure. 14.0% of farms were bringing at lease some farmyard manure from other farms onto their own farms and 8.4% of farms were totally dependent on other farms for manure. Presumably most of this was coming from conventional farms, as only one respondent claimed that the farm supplied other farms with manure. With regard to storage, most farmers were either storing manure in a field pile (46.9%) or heaped on a concrete yard (23.5%), whereas only 16.3% had a specially designed storage area or used windrows (11.2%).

Organic seeds. Questions were asked regarding the availability of organic seeds. 74.8% replied that they had tried to buy organic seeds, and of these 62.3% had had difficulty in obtaining it. A quarter of farmers had not tried to buy organic seeds at the time of the interviews. Sixty-four farmers made comments on the availability of organic seeds and of these forty-seven specifically named a crop or crops that they had had difficulty obtaining seed for. By far the most frequently mentioned problem crops were cereals, with half of all comments referring to difficulties in obtaining organic cereal seeds for the varieties they wanted to grow. Fodder crops such as kale, stubble turnips and fodder beet were also frequently mentioned, as were brassica seeds, and in particular cauliflower seed, but also broccoli / calabrese, spring greens and cabbage. Other crops were also mentioned.

Organic Standards. The survey included an assessment of which organic standards the farmers of Cornwall found the most difficult to implement. Ninety-three (78.1%) of respondents made a total of 152 comments. By far the most widespread problem area was viewed as the quantity of paperwork, including record keeping that farmers have to complete in order to comply with regulations. Issues regarding animal health and welfare standards were also commonly expressed (19.7% of responses), although the specific issues raised were varied, and included the general impact on health and welfare of reduced conventional medicine use, whilst others were having difficulty controlling specific conditions, such as worm infestation in sheep.

DISCUSSION

In summary, this survey has provided important information regarding the relative strengths and weaknesses of organic farming in Cornwall, and has revealed a considerable variation across the sector in terms of income and marketing. The results do give a general impression of the social, economic and environmental benefits. However, this is not consistent across all farms, and perhaps future strategies for development of the sector need to be more system or circumstance specific. It appears that, with the increase in direct sales, a decline in full-time labour on some farms and more off-farm activity, organic farmers are having to work harder to obtain an income from farming. However, this hypothesis must be considered in the light of a general decline in farm incomes. Differences in enterprise spread between the two certifying bodies are probably related to the differences in stimulus for entry into organic farming. Further data analysis will reveal a more detailed picture of the production and structural factors that influence the benefits that may be accrued from organic farming.

REFERENCES

BURKE J., RODERICK S., and LeGRICE P. (2004) Organic farming in Cornwall. 1. A description of the systems and enterprises In: Hopkins A. (ed.) *Organic farming: science and practice for profitable livestock and cropping.* Occasional Symposium of the British Grassland Society, No. 37, pp 43-46.

DEPARTMENT FOR ENVIRONMENT, FOOD AND RURAL AFFAIRS (2002) *Action plan to develop organic food and farming in England.* London: DEFRA Publications.

SOIL ASSOCIATION (2002) *Organic Food and Farming Report.* Bristol: The Soil Association.

SESSION 2

PRACTICAL FORAGE AND LIVESTOCK PRODUCTION

Dock Management: a Review of Science and Farmer Approaches

R.J. TURNER, W. BOND and G. DAVIES
IOR-HDRA, Ryton Organic Gardens, Coventry, CV8 3LG, UK

ABSTRACT

This participatory project began in 2002 and has adopted a new approach to weed management in which farmers, researchers and other organic stakeholders identify, prioritise, trial and develop solutions to weed problems. Organic farmers were asked 'What are your main weed management problems?' and over 60% (n=152) responded that docks caused them the greatest concern. The aim is to collate published literature and other information on dock control and to document current farmer management practice. Over fifty farmer weed-management interviews have been undertaken and written into case study information from different farming systems. This paper gives a summary of the scientific knowledge and outlines current farming practice from surveyed farmers in relation to dock management.

SUMMARY OF SCIENCE KNOWLEDGE

The two main dock species are the broad-leaved dock (*Rumex obtusifolius*) and the curled dock (*Rumex crispus*). They are common throughout the UK both as the true species and as hybrids. The hybrids produce less seed but may be more vigorous than the parents. Docks reproduce from seed and by vegetative regeneration of the underground organs. Dock seedlings are poor competitors and in standing vegetation can only establish in open or disturbed patches. In grassland, the presence of docks is associated with uneven application of slurry or manure that leaves bare patches. The openness of a sward after cutting for silage is also linked with dock establishment. Poor grass management leading to overgrazing and poaching also allow dock seedlings to emerge and grow. Soils high in nitrogen or low in potassium are also said to favour docks.

It is reported that a single plant can produce up to 60,000 seeds that may become viable from the milk stage onwards. Viable seeds can develop on stems cut down just a few days after flowering. Intact plants can shed seed from late summer through to winter but the seeds may require a short after-ripening period before being ready to germinate.

Dock seed numbers in soil have been estimated at over 12 million per hectare. The seeds are capable of surviving in undisturbed soil for 50+ years. The seeds can germinate any time that conditions are favourable but the main flushes of emergence are in March-April and July-October. The seeds vary in size, seed-coat thickness and dormancy status and therefore respond differently to external factors, contributing to the opportunist ability of docks. The seeds are often shed around the parent plant but may be carried by animals, on machinery and in water. The main method of long distance dispersal is as a contaminant in seed, animal feed, straw and manure. The seeds can pass through cattle unharmed and will survive for several weeks in manure. They can also survive long periods of immersion in slurry that is not aerated. Treatment temperatures for sewage sludge may not be high enough to kill dock seeds. Seed viability is low in silage where additives are used to aid fermentation.

The underground parts of a dock consist of a vertical stem and a branched tap-root with a transition zone between them. The underground stem may reach 5 cm in length and is kept below ground by root contraction. There is considerable confusion about the ability of docks to regenerate from their underground organs. Some authors maintain that true roots do not regenerate and only the stem and transition zone can regenerate. Others insist that all parts will form new shoots if detached from the parent. At present it is generally agreed that only the upper 9 cm of the underground parts of broad-leaved dock and upper 4 cm of those of curled dock will regenerate. Uprooted dock plants can regenerate if left on the soil surface even following a period of dry weather. A dock seedling takes 40 days from emergence to develop a rootstock that will regenerate after decapitation.

Docks are grazed off by cattle, sheep, goats and deer, but not by horses. It may be that cattle or sheep should be put to graze with horses to prevent a build-up of docks. In pasture, plants of broad-leaved dock can be very long lived, forming compound crowns with multiple tap-roots. Curled dock often dies after flowering but will persist if repeatedly cut down. Mowing has little effect on established docks but will prevent seeding. In a pasture heavily infested with docks the best option may be to plough and reseed with grass but not immediately. The docks are likely to regenerate both vegetatively and from seed and a period of fallowing or arable cropping may help to reduce re-establishment of the docks. In arable land and elsewhere it is important to prevent the introduction of dock seed in straw, seed, manure, slurry and on machinery. In combinable crops the aim should be to collect dock seed shed during the harvesting operation and denature this before disposal. Biological control of docks is being investigated. A number of native insects and fungi attack docks but none is likely to have a dramatic effect on dock populations at the levels at which these are found in nature.

SUMMARY OF FARMER INTERVIEWS

The following information summarizes comments from 52 farmers in England and Wales interviewed between April and September 2003.

Attitude to docks

A few farmers were very pragmatic about docks, felt they did not need to take any specific action and just accepted that they had to live with this weed in their system. Most had very strong negative feelings towards docks and, although very aware of the benefits of a diverse weed flora, thought of this weed as an exception and would prefer a zero tolerance policy. Many farmers were very impressed with the dock as a weed and its amazing resilience. Some commented on the positive benefits it could offer in herbage, e.g. enhancing selenium or zinc supply to livestock.

What encourages docks in your system?

Farmers have found when reseeding grassland that docks have germinated in quantity. Any form of poaching which opens up the land has provided sites for dock germination and growth; cutting up of the ground by horses has also been problematic. Dock patches have developed along hedge lines, fences, gateways, below trees (where it is difficult to mow) and where cattle shelter. Patches have also been noted to spread on compacted ground. Some farmers have correlated spreading dirty water on to fields has increased the problem. Open stored water has been a potential receptacle for wind blown dock seed. Some farmers have noticed patches around cattle feeder units and water troughs.

Many feel their farms have historic dock problems that have been suppressed during conventional farming and are now re-appearing. They also feel that seeds blown in from surrounding fields and roadsides have increased the seedbank. Inappropriate or untimely cultivations were thought to have spread established docks.

What is your strategy for dock management?

Take no action

Several farmers had been surprised that in fields which had been covered with seeding docks, in the following year there had not been a real problem. Their conclusion was that if docks were left undisturbed there was a natural population fluctuation and not an exponential increase in population. In grassland it was felt that as the ley established and became more competitive the dock problem would diminish as the grass out-competed them.

Cultural controls

The policy of most farmers was to stop docks seeding and reduce the vigour of the plants they already had. Thus, attempts would be made to harvest arable crops before dock seed matured. In weedy fields a silage cut would be taken rather than making hay. If hay was taken from a weedy field then attempts were made to feed the hay back to stock only on those fields that already had a problem. The necessity to ensure that well cleaned seed was used was commented on, some farmers saving and cleaning their own seed.

If possible, established leys were not ploughed up, to avoid germinating new docks. Many farmers felt docks had spread from hedges and headland. Some now employed strict policies to keep field boundaries clean by mowing next to hedges or even establishing Stewardship strips which could be easily managed (although some commented on problems with Countryside Stewardship Schemes in organic systems: mowing field margins only after 15th July was too late to prevent weeds seeding).

Many farmers thought about their rotation and crops specifically in terms of dock management and were constantly thinking about when opportunities could be created to control docks, whether this was with suppressive cover crops, use of stock or direct mechanical action.

Direct action

In grassland systems the main control policy was integrated topping and grazing. These systems had the least options for control, particularly permanent systems where there was no cropping break in which to employ mechanical cultivation.

All farmers used some method of topping in grassland, typically to 10 cm sward height at least once in the growing season. It was commented that topping just encouraged a lower level of dock seeding down the stem. Some farmers felt topping was gradually reducing the vigour of their dock patches, whilst others thought they were just maintaining the same level. Some farmers specifically cut the dock patches in a field to give the grass a chance to out-compete the weed.

Sheep were felt to be more useful for dock management than cattle. Some farmers commented they used dry cows for mopping up docks. There was a range of methods to integrate topping and grazing, and grazing with different stock, e.g. fields grazed alternate years with cattle and sheep, or grazed one year then cut for hay the next. With cattle and sheep, cattle would graze first and sheep would be grazed tighter afterwards. Some farmers thought intensive grazing when docks were young stopped the spread. Some swards have been tightly grazed in spring with non-productive sheep to keep docks down.

Farmers with an arable phase used below-ground cultivations for dock management, often using bastard fallows and cultivation to desiccate the dock roots. The broad strategies are outlined below;

- On newly established docks: plough, cultivate several times with decreasing size of spring tines to bring roots up to desiccate, rake off with a heavy duty harrow. Drill a suppressive crop, e.g. red clover.
- Break ley early e.g. June, use heavy ducksfoot wide bladed tines (e.g. Terradisc) to 7-10 cm depth (needs to be done at end of ley so docks are held in position and machine can slice cleanly through). Could use a heavy duty rotovator. Leave roots to desiccate, follow up with a cultivator to rake and dry roots out. Plough everything down and sow a cereal. If still a problem take a Terradisc through again, or a heavy duty cultivator, then plough down and maybe another pass before sowing suppressive a crop, e.g. field beans.
- On well established docks, rotovate to 10 cm to cut off crowns, disk and harrow.
- Plough in summer, rotovate at 2-week intervals to gradually deeper depths 5 cm, 8 cm, 12 cm. Sow winter cereal, e.g. triticale.
- After a winter cereal rotovate the top 5 cm before sowing a catch crop. Then plough in the spring.
- Leave overwinter cereal stubble, cultivate for 6 weeks in spring with rotovator (2/3 passes) with tines rather than blades. Sow quick growing suppressive crop e.g. fodder rape or stubble turnip in late spring. In autumn graze, reseed, then tightly graze.
- Use sub-soiler after ploughing in compacted areas in spring which aids dock control.

A farmer with heavy land did comment that rotovation had not worked on their clay soil, and regardless of how much cultivation was undertaken dock roots never fully dried out and were able to regenerate.

Many farmers were still using some form of manual dock removal with spades, adapted forks or specialist hand-held dock removal tools. This was done when the ground was soft, e.g. in establishing leys or in cereal crops. This was felt to be the most cost-effective technique but time consuming, expensive and damaging to the backs of staff. One farmer used the forks of his lifter close together to dig out docks to a depth of about 60 cm. If there is not time for complete removal farmers will strim docks before seeding as an intermediate measure.

One biodynamic farmer was using a weed pepper to treat infested fields. Dock seed ash was potentized in a biodynamic preparation with water, this was then spread over the entire field in a thin covering applied with a hand held sprayer or spread with a dairy brush. The intention was to reduce the vigour of existing docks with repeated application. Some farmers commented on seeing rust on leaves and also beetles eating the dock leaves. Few felt this was reducing dock vigour significantly and none knew how to encourage the rust or insects, but all were keen to investigate the potential biocontrol agents.

No farmer thought they had all the answers to dock management and all were very keen to support and participate in the research project.

(A fully referenced science review is available on request from the authors and further information from www.organicweeds.org.uk).

Forage Production and Persistency of Lotus-based Swards under Organic Management

A. HOPKINS and R. H. JOHNSON
IGER, North Wyke, Okehampton, Devon, EX20 2SB, UK

ABSTRACT

A small-plot field experiment was sown to grass (*Phleum pratense* + *Festuca pratensis)* with lotus (*Lotus corniculatus* + *L. pedunculatus*) and the effects of cutting frequency and farmyard manure (FYM) on total dry matter (DM) yield and lotus persistency were investigated. Total DM yield was highest in year 1 (*c.* 12 t/ha) and lowest in year 5. The proportion of lotus in the harvested DM varied with treatment: up to 0.65, 0.65, 0.40, 0.35 in years 1, 2, 4 and 5 respectively. By year 5 most of the lotus was *L. corniculatus.* Treatments mown 3 times per year retained a higher proportion of lotus than those mown 4 times per year. FYM addition also decreased the proportion of lotus in some years. In a subsidiary experiment, chicory was sown as a companion species with grass-lotus and contributed 0.45 of total DM yield in year 1 and 0.32 in year 5. Results suggest that grass-lotus and grass-lotus-chicory have potential for use as high yielding medium-term forage ley mixtures suitable for organic and other low-input situations.

INTRODUCTION

Lotus (trefoil) species currently occupy a relatively minor role in UK and European agriculture, although they are sometimes present in permanent pastures. Cultivars of both the common lotus (bird's-foot trefoil, *Lotus corniculatus*) and greater lotus (*L. pedunculatus*) are widely used in many temperate grassland areas of the world, and they have attributes which are useful in organic livestock systems. The presence in lotus herbage of proanthocyanidins, also known as condensed tannins (CT) not only confers non-bloating properties, but can also improve dietary protein metabolism, with a consequent potential for reduced N losses to the environment (Aerts *et al.*, 1999). Anthelmintic properties associated with CT in lotus have also been widely reported (Molan *et al.*, 1999). Compared with other N-fixing legumes such as white clover, lotus species are also better adapted to less-fertile or moderate nutrient-input situations (Bologna *et al.*, 1996). However, lotus is not highly competitive in mixed-species swards, and in a previous experiment in the UK there was poor persistency after two harvest years (Hopkins *et al.*, 1996), though long-term lotus persistence in grazed New Zealand pastures was observed and found to be influenced by slope, soil fertility and genotype (Hopkins *et al.*, 1993). An improved understanding of lotus agronomy is certainly required before its use is likely to be more widely adopted. In 1998 we commenced a programme of small-plot experiments, and associated mesocosm studies, to identify factors affecting lotus establishment, and the management affecting herbage yield and persistence of lotus in sown swards with grasses and other companion species, using sites in several low-input situations. This paper focuses on the results obtained from one site, managed under organic conditions.

METHODS AND MATERIALS

Small-plot experimentation was carried out on organic farmland near Cirencester, UK (51°44'N/ 1°57'W, 140 m a.s.l.). The soil was a calcareous stony clay (Sherborne

Association), a type widely distributed in central-southern England and supporting both grassland and arable cultivation. At the start of the experiment soil pH was 7.8, P index was 2 and K index was 2. Mean long-term annual rainfall was 825mm. Following complete cultivation and seedbed preparation, a mixture of equal seed numbers of *L. corniculatus* (cv. Rocco) and *L. pedunculatus* (cv. Maku) giving a total lotus seed rate of 13.5 kg/ha, was broadcast-sown with grass (timothy, *Phleum pratense* cv. Promesse at 2 kg/ha plus meadow fescue, *Festuca pratensis* cv. Kasper at 2 kg/ha) on 18 June 1998. The seedbed was lightly harrowed then rolled to a target sowing depth of 5-10mm. Prepared cultures of lotus-specific rhizobia (NZP 2037 for *L. corniculatus* and RCR 3012 for *L. pedunculatus*) were applied to the seedbed immediately after sowing, using a watering can. The main experiment was a 2 x 2 factorial in randomized blocks, replicated three times, the four treatments comprising two cutting frequencies (three cuts and four cuts per year) and two fertilization rates (nil inputs, or composted FYM from within the farm, applied at 2.8 litres/m^2 in February each year).

The effects of chicory (*Cichorium intybus* cv. Puna, seed rate 8 kg/ha) or Caucasian clover (*Trifolium ambiguum*, cv. Endura, seed rate 8 kg/ha) as companion species with the same lotus-grass mixtures were investigated in a subsidiary experiment. The comparison was made with the grass-lotus mixture for the treatment cut three times per year and receiving FYM.

No herbage yield assessments were made in the establishment year of either experiment, but the plots were mown in September then grazed lightly with sheep. In 1999, 2000, 2002 and 2003 plots were harvested between mid-May and mid-September, using an autoscythe mower, and mown either three or four times each year according to the cutting frequency treatment. Assessments of herbage dry matter (DM) yield were made at each harvest, and additional sub-samples were hand sorted to determine the proportion of lotus and other species. The site was grazed by sheep each winter but no yield assessments under grazing were made, and during summer 2001 the site was also grazed only, but not assessed. Additional botanical assessments were made in July each year, and in autumn of the establishment year, to determine the per cent frequency values for the two lotus species (based on their presence/ absence in 10cm^2 cells within three 25cm^2 ground-level quadrats per plot).

RESULTS

By September of the establishment year there was vigorous growth of lotus and a high density of lotus plants in all plots; mean per cent frequency values were 92% for *L. corniculatus* and 33% for *L. pedunculatus*. Similar lotus frequency values were recorded for all treatments in years 1 and 2, but they decreased on all treatments in years 4 and 5, and varied according to treatment. In July of year 5, *L. corniculatus* was present at a frequency value of 40% for treatments cut three times per year, and 21% for treatments cut four times per year, while *L. pedunculatus*, though present in all plots, was a minor sward component. Applications of FYM had little effect on lotus frequency. The sown grasses also established well.

Total DM yield harvested was highest in year 1 and lowest in year 5 for all four main treatments, e.g. for the treatment receiving three cuts per year with FYM applied, these were 12.3, 8.8, 7.1 and 4.9 t DM/ha, respectively, for years 1, 2, 4 and 5. April-September rainfall data for these four years was 407, 468, 368 and 223 mm, respectively, and may explain the lower yields in year 5.

In year 1 lotus DM yield was particularly high, and it contributed 6.9 – 7.7 t DM/ha (0.59-0.69 of the total harvested yield; differences between treatments NS at P=0.05). In year 2 lotus again contributed about 0.65 of the total DM yield, but there were significant treatment effects: treatments receiving FYM had a lower lotus yield than the nil-FYM treatments (P<0.05), and the 3-cut treatments had a higher lotus yield than 4-cut treatments (P<0.01). In years 4 and 5 the lotus yield and its proportion of total DM yield were lower than in the previous years, but again varied according to treatment, particularly cutting frequency (P<0.01 in year 4 and P<0.05 in year 5). The highest lotus proportions were from the 3 cuts/year with no FYM treatment (Figure 1).

In the subsidiary 'companion species' experiment, chicory established well and contributed 0.45, 0.35, 0.41 and 0.32 of the total DM yield in years 1, 2, 4 and 5, respectively. In years 1 and 2 the addition of chicory to the seed mix resulted in reduced total DM yield and reduced grass and lotus DM yields, but in years 4 and 5 this effect had disappeared, and lotus DM yield and total DM yield were then similar to that of the equivalent treatment without chicory. Caucasian clover established poorly, and contributed only a negligible proportion of the herbage throughout.

Figure 1 Total DM yield, and yield of lotus ( ) and grass plus unsown species ( ) in successive years, for treatments (3/nil) cut three times, receiving no farmyard manure, and (4/fym) cut four times each year and receiving farmyard manure each spring.

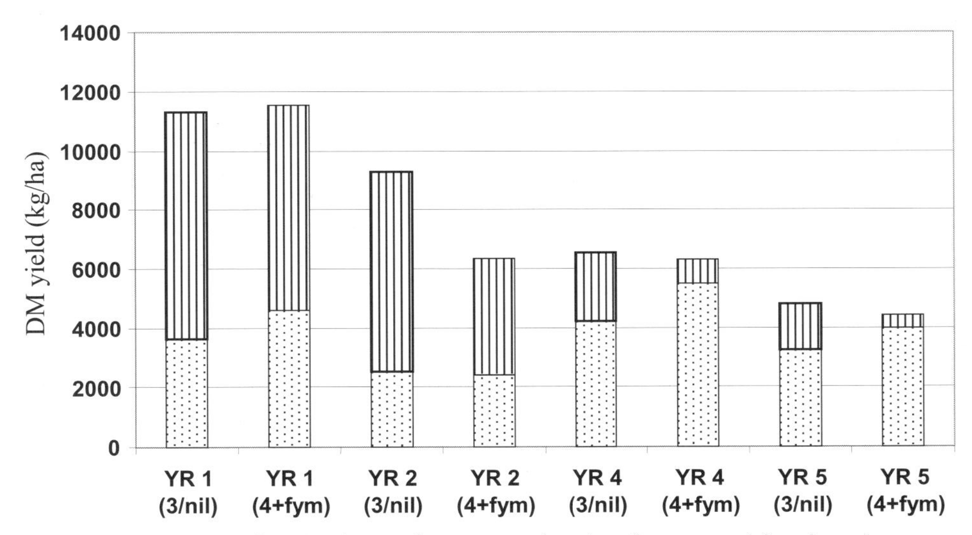

(N.B. plots were not mown in year 3)

DISCUSSION

Total DM yield, and the proportion contributed by lotus, were considered to be high, especially in years 1 and 2, and the results compare favourably with previous and other recent experiments using lotus, and with forage production yields in general. This, taken with the persistence of lotus as a worthwhile sward component into years 4 and 5, was extremely encouraging. Good seedbed preparation, a generous seed rate and favourable weather conditions after sowing were considered to have contributed to the excellent establishment of lotus obtained here. The choice of relatively non-competitive companion grasses was based on the suitability of timothy and meadow fescue in previous lotus experiments (Hopkins *et al.*, 1996). Persistence of lotus is often thought to require defoliation intervals that enable seed ripening and recruitment of new seedlings (Bologna *et al.*, 1996). The relative advantage of a 3-cut, compared with a 4-cut, interval in this experiment supports this, and it is suggested that whether lotus swards are mown or grazed that the intervals be suitably spaced. The effect of applying FYM was to slightly change the grass : lotus balance in favour of grass, but on soils with a lower PK status this might not necessarily occur. The decision to use a mix of two lotus species enabled the more successful species to develop according to the site (the relatively free-draining soil here favouring *L. corniculatus*). In the subsidiary companion species experiment, chicory was found to be a suitable additional species for this site, with good persistence and productivity. The failure of Caucasian clover was attributed to competition from the high grass-lotus seed rate and the absence of its specific rhizobia, and subsequent work with this alternative legume has been more successful.

ACKNOWLEDGEMENTS
The experimental site was provided by Abbey Home Farms whose cooperation is gratefully acknowledged. This research formed part of DEFRA funded contract, LS1307.

REFERENCES

AERTS R.J., BARRY T.N. and MCNABB W.C. (1999) Polyphenols and agriculture: beneficial effects of proanthocyanidins in forages. *Agriculture, Ecosystems and Environment*, 75, 1-12.
BOLOGNA J.J., ROWARTH J.S., FRASER T.J. and HILL G.D. (1996) Management of birdsfoot trefoil (*Lotus corniculatus* L.) pastures for productivity and persistence. *Proceedings Agronomy Society of New Zealand*, 26, 17-21.
HOPKINS A., SCOTT A.G., COSTALL D.A. LAMBERT M.G. and CAMPBELL B.D. (1993) Distribution of diploid and tetraploid *Lotus pedunculatus* in moist, North Island hill country. *New Zealand Journal of Agricultural Research,* 36, 429-434.
HOPKINS A., MARTYN T.M., JOHNSON R.H., SHELDRICK R.D. and LAVENDER R.L. (1996) Forage production by two *Lotus* species as influenced by companion grass species. *Grass and Forage Science*, 51, 343-349.
MOLAN A.L., WAGHORN G.C. and MCNABB W.C. (1999) Condensed tannins and gastro-intestinal parasites in sheep. *Proceedings of the New Zealand Grassland Association*, 61, 57-61.

Integrated Forage and Livestock Production

JOHN E. HERMANSEN and TROELS KRISTENSEN
Danish Institute of Agricultural Sciences, Dept. of Agroecology,
Research Centre Foulum, PO Box 50, DK-8830 Tjele, Denmark

ABSTRACT

Integrated forage and livestock production can be considered at the farm level and at the herd or animal level. At the farm level it is relevant to consider the overall utilization of N in the system in relation to crops and livestock. It is demonstrated that in organic dairy production a high transformation efficiency of N from input to edible products can be achieved, compared with conventional production. In addition, combining dairy and pig production allows an even higher N utilization. At the herd level, the quality of grass or clover-grass based forage is extremely important. This holds for the overall intake and milk production in dairy cows, and for the intake of clover-grass by grazing sows. In addition, the composition of the sward should be considered in relation to the influence of specific plant species on the development of endoparasitic infections in ruminants and on the wear strength in relation to free-range pig production.

For dairy production it is proposed that a strategy including only 20% concentrates (or cereals) of the dry matter in a total diet based on clover-grass and clover-grass silage represents an efficient milk production without impairing the health of the cows.

INTRODUCTION

The way feed and forage production takes place is a key element in organic livestock farming. It is specifically mentioned in the IFOAM principle aims of organic production that a harmonious balance between crop production and animal husbandry should be established and that the biological cycles within the farming system should be encouraged. This is also reflected in ECC-regulation of organic farming, stating that the farm management should mainly rely on internal farm resources rather than on external inputs. In this way, the environmental impact of intensive animal production is expected to be diminished. The detailed regulation includes limits for the maximum animal density per unit of land and on the minimum use of forage. However, it is also stated that the nutritional requirements of the livestock at the various stages of their development should be met to ensure a "quality" production and support the health of the animals.

The fact that all feed from 2005 should be organically grown will – from a economic point of view – make it even more important to rely on home-produced feed and thereby to focus on how the forage and livestock production can best be optimized.

THE INTEGRATED APPROACH

An important element in the integrated approach is to consider the nutrient cycle within the farm. In Table 1 we present figures on N-cycling measured on conventional and organic pilot dairy farms in Denmark. The conventional farms typically import more feed and fertilizer and have a higher milk yield and cash crop export, which is reflected in the N-turnover. The organic farms rely more on clover-grass yielding N to the crop rotation through its capacity for bio-fixation and on home-grown feed. If we consider the transformation efficiency for N in the animal component (N in edible products compared with N in feed) a slightly lower efficiency is observed in the organic system. This is probably a consequence of a higher milk yield per cow, obtained partly through a more balanced N-feeding in the conventional production system. However, the figures also show that the marginal efficiency of the imported feed is only around 22%.

Table 1. N-balances on organic and conventional Danish dairy farms in kg N/ha (Kristensen *et al.*, 2003).

	Conventional (1.5 LU[1]/ha)	**Organic (1.4 LU/ha)**
Animal component		
Input		
Purchased feed	97	49
Home-grown feed	110	123
Output		
Milk	36	30
Meat	10	6
Transformation efficiency	22%	21%
Crop/soil component		
Input		
Fertilizer	89	-
Fixation	23	70
Home produced manure	149	136
Precipitation	16	16
Total	277	223
Output		
Feed	110	123
Cash crop	9	3
Total	119	126
Transformation efficiency	43%	56%
Whole farm		
Transformation efficiency, %	24	30
Surplus, kg	172	108

[1] LU =livestock unit

If we consider the crop component it is clear that the organic system, through its use and choice of crops, is able to produce the same or more nitrogenous output even with less N input, compared with the conventional practice. This means a much higher transformation efficiency, which more than balances the reduction in the animal component and leads to a considerably higher whole-farm N-efficiency. These results underline the prospects of considering an increasing integration of livestock and crop production, even if such a practice may result in a lower output per animal.

This idea has been followed further in a simulation study (Kristensen and Kristensen, 1997). The exercise was carried out in relation to the planning of a Danish organic research station which was required to include both pig and dairy production. The overall hypothesis was based on the fact that in dairy production there was a huge amount of N available due to the type of feed produced, whereas in pig production, which has to rely on grain crops to a significant degree, there was a lack of N. Thus, combining the two enterprises should make it possible to increase the overall efficiency. Three systems were compared: a "normal" organic dairy production, a specialized organic pig production, and a mixed production, where in principle the pigs were given the major part of the cereal produced, and the dairy cattle had to rely to a very high extent on forage.

The results relating to level of production and cycling of N are given in Table 2.

Table 2. Results of model calculated production and N-balances of different farming systems (after Kristensen and Kristensen, 1997).

System	**Dairy**	**Pig**	**Dairy/Pig (mixed)**
Grass/clover, %	60	20	40
Import of animal manure, kg N/ha	0	45	0
Herd - cows/sows per ha	0.81/0	0/0.71	0.44/0.45
Feed			
- SFU/ha produced[1]	5.298	3.201	4.836
- SFU/ha imported	517	1.075	714
Production			
Milk, kg cow	7.357	-	6.286
Kg/ha	5.980		2.772
Meat kg/ha	239	1.239	120/784 (904)
N-balance, kg/ha			
Input			
Purchased feed	28	52	2/33 (35)
Atmospheric deposition	21	21	21
Fixation	89	30	54/5 (60)
Imported manure	0	45	0
Total	138	148	114
Output			
Milk	32	0	15
Meat	6	33	3/21 (24)
Total	38	33	39
Input - output	101	115	76
Efficiency			
N output/input	0.27	0.22	0.34

[1] SFU = Scandinavian Feed Unit (Barley net energy equivalent)

The output is given as kg of product, as milk and/or meat per animal or per ha. In addition, the output from the system is given as N in animal products per hectare, representing the amount of animal protein produced for human consumption. The model calculations indicate that the mixed dairy and pig system makes it possible to produce a higher amount of protein in animal products per ha (39 kg N), compared with the specialized dairy (38 kg N) or pig system (33 kg N). Moreover, it seems that this system results in the smallest difference between N-input and N-output. The better overall N-efficiency in the mixed system is primarily related to the better ability of this crop rotation to take advantage of the N left in the pre-crops, which means a lower risk of leaching/denitrification.

This approach is, of course, one of several approaches to be considered in an integrated production. In the following, strategies for a high self-supply in dairy production and systems of pig production are discussed.

RESULTS FROM DANISH PILOT DAIRY FARMS AIMING FOR A HIGH SELF-SUPPLY OF FEED

Results from commercial organic dairy farms that focus on a high self-supply of feed are given in Table 3. All feed given to the herd was organically produced since this was a prerequisite for obtaining a premium price for the organic milk at the dairy.

Table 3. Feeding, milk yield and land use in four commercial Danish organic dairy herds 2001-2002.

Herd	**I**	**II**	**III**	**IV**
Cows/herd	160	76	83	131
Intake per cow				
- DM, kg	6860	6790	6480	6640
- Energy, SFU	5780	5840	5930	6060
% of DM				
- Pasture	20	16	20	32
- Silage	54	51	45	44
- Cereal	11	22	16	12
- Others	15	11	19	12
% SFU self produced	87	88	93	84
Production per cow				
- kg ECM	8180	7720	7920	7600
- Fat %	4.21	4.21	3.95	4.35
- Protein %	3.29	3.30	3.17	3.27
Efficiency ECM/kg DM	1.19	1.14	1.22	1.15
Land use, ha/cow	1.34	1.24	1.22	1.39
- Clover-grass	0.62	0.66	0.77	0.90
- Whole crop silage	0.44	0.13	0.19	0.36
- Cereal	0.22	0.45	0.26	0.13
- Others	0.06			
Crop production				
- Clover-grass, kg DM/ha	6000	7000	6300	7700
- Cereals, kg/ha	4700	3900	4400	4400

The results are from the period May 2001 to October 2002 and illustrate that a high productivity is possible at herd and field level when between 84% and 93% of the energy intake of the cows is produced on the farm. The remaining part of the intake was primarily low protein concentrates based on Danish organically produced crops such as cereals, peas and rape seed cake.

The key component in these systems is the clover-grass. It occupies between 46% and 65% of the land use and more than half of the DM intake. The herd production was between 7600 and 8180 kg energy corrected milk (ECM) with a high efficiency: 1.44 to 1.22 kg ECM per kg DM intake.

OPTIMIZED USE OF FORAGE AND COMPLEMENTARY FEED IN DAIRY PRODUCTION

Several approaches can be taken. One approach is, that for a given group of cows, to decide how best to produce and use the forage in the short term. Much research can support decisions on the short-term response. A very important result in this respect is the finding that the digestibility of grass and clover-grass silage has a remarkable influence on the intake of silage both in dry matter intake and, in particular, in energy intake when fed *ad libitum*. Kristensen and Nørgaard (1987) showed in their classic experiments with increased level of concentrates to dairy cows in combination with silage differing in organic matter digestibility, that it was not possible through higher allowance of concentrates to compensate for a low digestibility of silage when striving towards a high milk yield (Figure 1).

A difference in DOM of 16 percentage units (66% versus 82%) caused a difference in FCM production of approximately 6 kg at 3.5 kg concentrate DM. This difference was reduced to just below 4 kg FCM at 9.5 kg concentrate DM. In order to obtain the same yield with grass silage with 70% digestibility of OM as with 82% digestibility of OM, the amount of concentrates should be increased from 3.5 to 9.5 kg DM.

Although these results were obtained using silage produced by use of fertilizers, there is no reason to believe that the principles will not be valid in organic conditions also. Therefore, the results deserve considerable attention when planning feed supply based on a high proportion of home-grown feed in dairy production.

Figure 1. Effect of silage digestibility (DOM) and amount of concentrate on milk production, per cow daily.

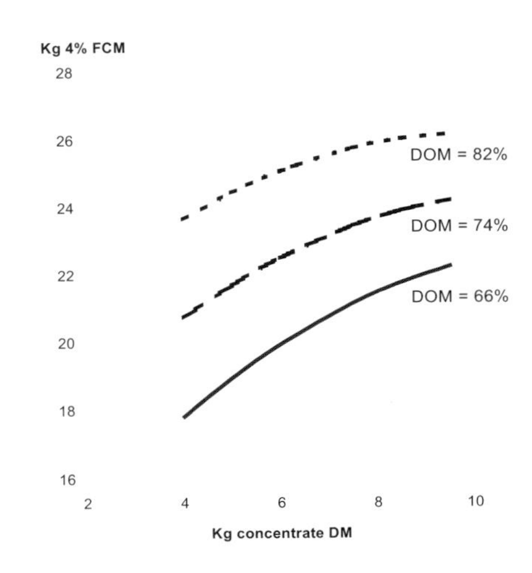

Mogensen *et al.* (2003) used a different approach. Based on the yields of cereals, rape-seed and clover-grass that could be produced on a given area of land for feed, a feeding experiment was conducted on each of two organic dairy farms. The rations were formulated so that the expected feed consumption could be produced at the same area per cow, assuming crop yields typical for organic conditions in Denmark, i.e. 3,700 kg DM/ha of cereal, 2,200 kg DM/ha of rape seed, and 6200 kg DM/ha of clover-grass (as silage) calculated as net yield (feed available at the fodder board).

The corresponding diets are shown in Table 4. For technical reasons the roughage could not be entirely clover-grass silage, but included in addition barley and pea whole-crop silage as well as grass pellets.

The results indicate that although cereal can be given as the largest amount of feed and feed energy produced, it is perhaps not always the most efficient supplement when the feeding is based on home-grown crops. Inclusion of crushed rapeseed contributes especially to dietary fat in the diet, which tends to increase milk yield but at the same time to reduce fat and protein concentration in milk, the overall response in fat-corrected milk not being influenced. This pattern of response is typical when adding unsaturated fatty acids to the diet of dairy cows. In conclusion, the income from milk in the two situations should not be different and there were no indications that the different feeding regimens changed the risk of diseases of the cows. Consequently, the growing conditions for the complementary feeds in terms of economy and risk might be the most important factor in the choice of feed production.

Table 4. Feed intake and milk yield comparing different diets representing feed from the same area per cow, per cow daily (after Mogensen, 2003[x]).

Diet	Cereal	Cereal and rape seed
Intake	(N=73)	(N =74)
Cereals, kg DM	4.0	1.2
Rape seed, kg DM	-	1.2
Roughage (estimated), DM	16.0	16.7
Total, DM	20.0	19.2
ME, MJ	227	222
Yield		
Milk, kg	25.7	26.8
FCM, kg	25.4	25.4
Fat, %	4.14	3.85
Protein, %	3.20	3.06

x) combined results of two experiments (on two organic dairy farms)

Another approach is to investigate the consequences of long term strategies for feed production and feeding. However, there is only a limited number of long-term studies, particularly with regard to organic production. Sehested *et al.* (2003) reported the results of one such long-term investigation of the means to rely as much as possible on home-grown feed, carried out on the Danish Organic Research Station, (Rugballegård). Three strategies were compared:

- No supplementation to a clover-grass mixture (grazed or given as silage).
- Low level of supplementation to the above diet (3 kg concentrates mixtures in the first 24 weeks of lactation).

- Normal level of supplementation to the above diet (8 kg concentrates mixtures in the first 24 weeks of lactation).

The grazing sward was composed of perennial ryegrass (*Lolium perenne*) and white clover (*Trifolium repens*). The average content of white clover above grazing height was 36%, on a dry matter basis.

Feed intake, milk production, and feed conversion on a per cow and year basis are shown in Table 5. Increasing the level of concentrates from 100 kg DM to 2,400 kg DM per cow reduced intake of clover-grass and silage by 800 kg DM resulting in an increase in total DM intake of 1,400 kg and in energy intake of 41%. As a result, milk yield was increased by 32%. However, looking at the group with only a small supplement (1,000 kg), which increased total intake of energy by 13%, the milk yield increased by 22%. In fact, the overall feed conversion rate was considerably increased at the small supplement diet compared with both a higher and a lower concentrate supplementation. This really illustrates the law of diminishing returns.

Table 5. Feed intake and milk yield per cow and year (after Sehested *et al.*, 2003).

	Supplementation					
Strategy	**No**		**Small**		**Normal**	
Feed intake	**kg DM***	**MJ NE***	**kg DM***	**MJ NE***	**kg DM***	**MJ NE***
Silage	2.9	18	2.7	16	2.4	14
Grazing	1.5	11	1.4	10	1.2	9
Fodder beets	0.2	2	0.2	2	0.2	2
Concentrates	0.1	1	1.0	8	2.4	19
Total	4.8	32	5.3	36	6.2	45
Milk yield						
Kg	5030		6027		6646	
(Energy corr.)	5090		6230		6723	
"Feed efficiency" (Winter)						
Feed utilization, %	98		97		84	
kg ECM per DM	097		1.08		1.04	

x 1,000

The non-supplemented group had a lower incidence of clinical diseases compared with the other two groups, which did not differ significantly. The main differences were in limb and metabolic diseases. No difference in mastitis-related diseases occurred and no differences in somatic cell counts were found. In terms of effects on reproduction, numbers of days to first insemination after calving, and the calving interval, decreased with increased feed supplementation. At no supplementation the milk quality, in terms of the content of free fatty acids, was impaired compared with normal level of supplementation.

The group with moderate supplementation showed results very much comparable to the modelled results for combined pig and dairy production as detailed in Table 2. This strategy appears to represent a good balance between feed and dairy production with respect to overall efficiency, animal health, and product quality. This means that the system should include cereal or comparable concentrate feed.

MAINTAINING LIVESTOCK HEALTH AT GRASS

Helminth infection in young stock is the most common health problem in organic livestock (Younie and Hermansen, 2000). Grazing management is a main component of worm control strategy and the system must be designed primarily to minimize parasite infection. This can be achieved by preventive, evasive or diluting strategies. Preventive strategies involve the movement of uninfected animals to swards uncontaminated with worm larvae, e.g. by alternating cattle, sheep, other livestock species and/or conservation cuts from year to year on any one field. In an evasive strategy, stock are moved from a contaminated area to a clean sward, e.g. a silage aftermath. Dilution involves restricting the stocking rate of susceptible animals, for example by mixing young stock with another species (e.g. heifers and sows), or mixing young and adult stock of the same species, or simply be reducing stocking rate *per se* (Roepstorff *et al.*, 2000).

The overall forage production strategy needs to take these considerations into account. In some situations this may be difficult and there is also a need to consider other approaches such as breeding for resistance to parasites in the sheep flock (Eady *et al.*, 2003) and exploiting the effect of different pasture species and herbs on the development of endoparasite infections in ruminants. Although there is no full understanding of the mechanisms, it has been shown that plant species with a high content of condensed tannins (e.g. *Lotus pedunculatus*, *Lotus corniculatus*, and *Cichorium intybus*) can in some cases reduce the parasite burden of the livestock. In the light of these results there is a need to focus more on the species composition of the ley.

OUTDOOR PIG PRODUCTION

Outdoor pig production – at least for the sow part – fits very well some of the aims of organic farming, but also represents some challenges that need to be met.

Typically, in Denmark sows are kept in outdoor systems all year round, and pigs are moved to an indoor pig unit with an outdoor yard when they are weaned at seven weeks of age. In this way the sows automatically have access to grazing in the summer period, and the farmers have only one production system for their sow herd instead of having both a system for summer housing and a system for winter housing. The layout of the paddocks depends on soil type and the available land on the individual farm. The paddocks are normally moved to a new field every spring, often in a two-year crop rotation - one year with barley with an under-sown grass-ley and one year with sows on pasture. The stocking rate is adjusted to an excretion of 140 kg N in pig manure per ha each year (often practised as 280 kg N/ha every second year).

One of the major concerns in keeping sows on grass in intensively managed production has been the potential environmental impact due to high excretions of plant nutrients, especially N and P in the manure. The environmental impact of outdoor pig production is related to the amount of nutrients in the supplementary feed and the stocking density. Recent investigations have shown a surplus of 330-650 kg N per ha of land used for grazing sows on organic farms (Larsen *et al.*, 2000). Although this level is lower than that of average conventional outdoor sow herds, this nutrient surplus definitely represents an environmental risk, as it has proved difficult to obtain optimal efficiency of the nutrients deposited during grazing. The adverse consequences of this include considerable nutrient losses from grazed pastures and undesirably low nutrient availability in the rest of the crop rotation.

Another concern for outdoor production is the maintenance of the grass sward. A well-maintained grass sward serves several important purposes. The uptake of nitrogen and water by the grass decreases the risk of nitrogen loss by leaching (Watson and Edwards, 1997). In paddocks for lactating sows, a high level of grass cover is one of the factors which seems to decrease piglet mortality (Kongsted and Larsen, 1999) probably related to the ability of the sow to keep the hut dry and clean. In addition, for pregnant sows grass can constitute a significant part of their daily energy requirement (Sehested *et al.*, 2003).

Larsen and Kongsted (2001) investigated the importance of grass seed mixture on the amount of grass cover in a paddock with lactating sows under Danish conditions. The traditional mixture of perennial ryegrass, Kentucky bluegrass, red fescue, and white clover (WC) was compared with a mixture of meadow fescue + WC, a mixture of red fescue + WC, and a short-turf ryegrass +WC. The effect of grass-mixture on the grass cover in the paddocks was very little, compared with other management factors such as time of moving huts etc. It was concluded that the mini-turf mixture was a good choice for farrowing paddocks since this mixture had less disposition to culm-formation and had at least the same wear strength as the other mixtures.

As regards growing-pigs at pasture, several investigations indicate that growth rate obtained in outdoor systems can be comparable to the growth rate for indoor production. Although the growing pig can consume grass and other herbage to meet up to 20% of its daily day matter intake (Carlson *et al.*, 1999), the overall contribution to the energy supply of the pig when fed *ad libitum* than with concentrate mixtures is normally much lower, ranging from 2-8%. This means that most of the feed needs to be supplied as concentrates given to the pigs at pasture, and consequently a high risk of environmental impact can be expected unless measures are taken to counteract this.

We are investigating strategies for combining grazing and rearing in barns from the perspective of reducing the risk of environmental impact, and at the same time allowing the growing pigs to have plenty of space when they are young and most active. In the experiment piglets are moved indoors at weaning, or at 40 kg liveweight, or at 80 kg liveweight, or stay at pasture until slaughter.

The preliminary results show a normal growth rate (approximately 750 g daily gain) and no marked differences between the pigs fed *ad libitum* outdoors or *ad libitum* indoors. However, the feed intake per kg gain was increased by 13% when fed *ad libitum* outdoors. On the other hand, outdoor pigs, which were restricted in energy intake, had the same feed conversion rate as the indoor pigs and, in addition, a significantly higher lean content (approximately 4 units), but growth rate was of course reduced (by 16%). A very interesting finding occurred in the strategy where pigs were kept outdoors until 80 kg live weight followed by *ad libitum* feeding indoors. This strategy resulted in a feed conversion rate comparable to indoor feeding, and the overall daily gain was reduced by only 10-15% compared with *ad libitum* feeding indoors. These results indicate that there are options that can be used in order to obtain very good production results from outdoor-kept finishers.

However, with the stocking rate applied (100 m^2 per outdoor pig kept from 20 kg to 100 kg live-weight), all vegetation was destroyed. Complementary measurements on risk of N-leaching will elucidate the environmental risks in the systems, but these data are not yet available. However, it seems as if a choice has to be made: i.e. using a considerably lower stocking rate than used in this experiment in order to keep a good vegetation cover, or to accept the rooting and try to take advantage of it.

Table 6. Liveweight gain and estimated grass intake for grazing heifers and pregnant sows grazing separately or mixed (average of two experiments; after Sehested *et al.*, 2003).

Grazing system:	Separately	Mixed
Heifers (per heifer and day)		
Live weight gain, g	866	1063
Grass intake, NE, MJ	41.1	52.5
Sows (per sow and day)		
Daily live weight gain, g	512	557
Supplementary concs, NE, MJ	11.0	11.0
Grass intake, NE, MJ	10.3	10.8

Figure 2. Numbers of infective *O. ostertagi* larvae per kg dry grass on two pastures grazed by heifers only or by a mixed herd of pregnant sows and heifers (after Roepstorff *et al.*, 2000).

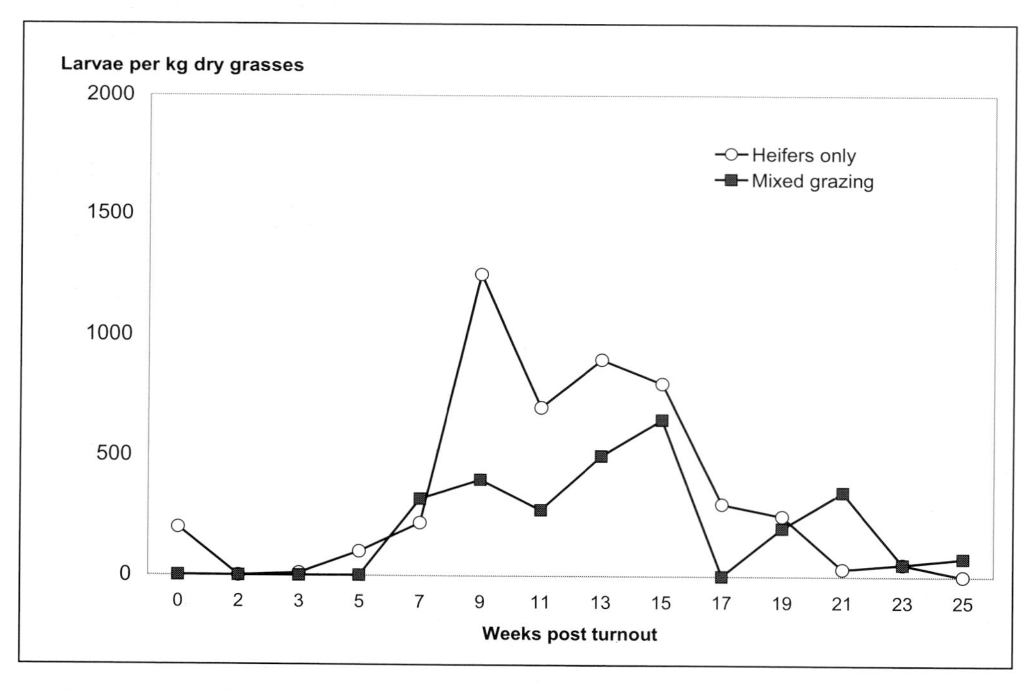

Several ways for better integration of pig production within the land use need to be considered. In the case of pregnant sows, which can be handled in relatively large herds, one perspective could be to base feed intake on forage. There is no doubt that forage can constitute a very large part of the nutrient requirement for pregnant sows. In addition, it has been shown that co-grazing sows and heifers reduces the parasite burden of the heifers and results in an overall better sward quality, compared with grazing separately (Roepstorff *et al.*, 2000; Sehested *et al.*, 2003). The liveweight gain and the estimated grass intake for heifers and pregnant sows grazing together or separately are summarized in Table 6, and in Figure 2 shows the larvae infection in the grass sward.

It appears that both sows and heifers had a higher daily gain when grazed in the mixed systems, although only the different growth rate for heifers was significant in each experiment. It can also be observed that the sows' grass-intake corresponded to half of the energy requirement. The peak of larvae infection of importance for the heifers per kg grass DM was in the mixed system, only half of the infection in the separately grazed systems. *Serum pepsinogen* levels in blood samples of the heifers confirmed the lower infection rate in the mixed grazing systems. No differences in parasite burden in the sows were observed.

These results were obtained from sows fitted with a nose ring, but since this strategy seems suitable in combination with a low stocking rate for the pigs, one may expect a lower overall incidence of rooting and, consequently, that similar results could be obtained with unringed sows.

Following the results on grass-sward compositions one proposal can be that for pregnant sows the sward should have a considerable growth of highly digestible species which can support the nutrient supply of the pig. In the case of lactating sows the intake of energy from the grass will in any case be modest, and the main emphasis could be put on the wear strength of the sward.

CONCLUSIONS

Integrated forage and livestock production should be considered at farm level and herd/animal level, respectively. At the farm level an appropriate balance of different crops or different livestock species makes it possible to obtain a high production efficiency, expressed as N transformation (input into edible products) and milk yield per kg DM consumed. This supports a livestock production with a lower environmental impact than conventional high-input systems.

At the herd or animal level there is a need to be very much aware of the "quality" of the forage. This relates to intakes of grass and clover-grass in ruminants as well as in sows, especially non-lactating sows. The inclusion of clover is particularly important in this respect. However, additional features of the sward species related to their impact in endo-parasitic infections in ruminants need to be considered, although there is a need for much more knowledge in this field.

REFERENCES

CARLSON D., LÆRKE H.N., POULSEN H.D. and JØRGENSEN H. (1999) Roughages for growing pigs, with emphasis on chemical composition, ingestion and faecal digestibility. *Acta Agriculturae Scandinavica, Section A - Animal Science,* 49, 129-136.

EADY S.J., WOOLASTON R.R. and BARGER I.A. (2003) Comparison of genetic and nongenetic strategies for control of gastrointestinal nematodes of sheep. *Livestock Production Science*, 81, 11-24.

KONGSTED A.G. and LARSEN V.A. (1999) Pattegrisedødelighed i frilandssohold [Piglet mortality in outdoor sow herds] *DJF- rapport (Husdyrbrug)*. 11. 56 pp.

KRISTENSEN I.S. and KRISTENSEN T. (1997) Animal production and nutrient balances of organic farming systems; prototypes. Presentation 3rd ENOF Workshop 'Resource use in organic farming', Ancona, July 1997.
KRISTENSEN I.S., HALBERG N. NIELSEN A.H., DALGAARD R. and HUTCHINGS N. (2003) N-turnover on Danish mixed dairy farms. Paper presented at the workshop: *Nutrient management on farm scale: how to attain European and national policy objectives in regions with intensive dairy farming?* Quimper, France, 21pp.
KRISTENSEN V.F. and NØRGAARD P. (1987) Effect of roughage quality and physical structure of the diet on fed intake and milk yield of the dairy cow. In: *Research in Cattle Production: Danish Status and Perspectives*, Landhusholdningsselskabet, Copenhagen, pp. 79-91.
LARSEN V.A., KONGSTED A.G. and KRISTENSEN I.S. (2000) Udendørs sohold - Balancer på mark- og bedriftsniveau. [Outdoor sow production - Nutrient balances on field and farm level] In: SOMMER S.G. and ERIKSEN J. (Eds.), Husdyrgødning og kompost. Næringsstofudnyttelse fra stald til mark i økologisk jordbrug, *FØJO-rapport*, 7, pp. 67-76.
MOGENSEN L., INGVARTSEN K.L., KRISTENSEN T., SEESTED S. and THAMSBORG S.M. (2003) Organic dairy production based on 100% organic feed grown at equal area per cow – Rape seed, rape seed cake or cereal as supplement to silage ad libitum. (*Acta Agriculturae Scandinavica, Section A - Animal Science*, Submitted).
ROEPSTORFF A., MONRAD J., SEHESTED J. and NANSEN P. (2000) Mixed grazing with sows and heifers: Parasitological aspects. In: Hermansen, J.E., Lund, V., Thuen, E. (Eds.). *Ecological Animal Husbandry in the Nordic Countries*, pp 41-44, DARCOF Report No. 2, Tjele, Denmark.
SEHESTED J., SØEGAARD K., DANIELSEN V., ROEPSTORFF A. and MONRAD J. (2003) Mixed grazing with heifers and sows: herbage quality, sward structure and animal weight gain. *Livestock Production Science,* (in press).
SEHESTED J., KRISTENSEN T. and SØEGAARD K. (2003) Effect of concentrate supplementation level on production, health and efficiency in an organic dairy herd. *Livestock Production Science*, 80, 153-165.
WATSON C. and EDWARDS S.A. (1997) Outdoor pig production: What are the environmental costs? In: *Environmental & Food Sciences, Research Report, Scottish Agricultural College*, pp.12-14.
YOUNIE D. and HERMANSEN J.E. (2000) The role of grassland in organic livestock farming. In: Grassland Farming, Balancing Environmental and Economic Demands (K. Søegaard *et al.*, eds.) *Proceedings of 18th General Meeting of the European Grassland Federation*, Aalborg, Denmark, 22-25 May 2000, *Grassland Science in Europe,* 5, 493-509.

Controlling Internal Parasites in Organic Sheep and Cattle

R. KEATINGE[1], R.F. JACKSON[2] , I. KYRIAZAKIS[3] and J. DEANE[4]
[1]ADAS Redesdale, Rochester, Otterburn, Newcastle upon Tyne NE19 1SB, UK
[2]Moredun Research Institute, Bush Loan, Penicuik, Edinburgh, EH26 0PZ, UK
[3]Scottish Agricultural College, Bush Estate, Penicuik, Edinburgh EH26 0PZ, UK
[4] Institute of Rural Sciences, University of Wales, Aberystwyth SY23 3AL, UK

ABSTRACT
Internal parasites are a potentially serious threat to the health, welfare and productivity of organically managed livestock. The standards for organic production emphasize preventive control strategies based on appropriate management, breeding and nutrition. The ultimate goal is to eliminate dependence on antiparasitic drugs; however, this is rarely achieved in practice. Focusing on management, nutrition and the use of novel crops, a Defra-funded study (OF0185) combines epidemiological studies with replicated experiments in order to develop better systems of control applicable to UK organic farms. Results from the study are used to support an overview of parasite control in organic sheep and cattle systems, as well as potential application for conventional farmers wishing to reduce dependence on anthelmintics.

INTRODUCTION
EU regulation 1804/99, governing the production of organic livestock, requires disease control to be based on appropriate animal husbandry, rather than routine prophylactic treatment. Homeopathic or phytotherapeutic intervention should be used in preference to allopathic treatments. Anthelmintics may only be used by derogation from the sector body, as part of an agreed disease reduction plan, or to safeguard animal welfare.

At the same time, anthelmintic resistance is becoming increasingly prevalent, with triple-resistant nematodes now confirmed in sheep (Yue *et al.,* 2003). For sheep and goats, complete control of parasites through the use of anthelmintic appears to be no longer sustainable. There is increasing recognition of the importance of maintaining a proportion of the parasite population *in refugia,* either within relatively resilient host animals or on the pasture, in order to delay the proliferation of anthelmintic resistant roundworm parasites (Stubbings, 2003). More integrated strategies are required, not only for organic producers, but also to prolong the life of drenches currently used in conventional farming.

The main categories of internal parasite relevant to UK sheep and cattle are tapeworms, liver fluke, gastro-intestinal roundworms and lungworm. Each is considered below, in the context of organic livestock production.

TAPEWORMS (*CESTODES*)
In the absence of anthelmintic, tapeworms sometimes give rise to concern because segments are often seen in the faeces. While heavy infestation of young animals has been associated with a failure to thrive (Radostits *et al.,* 1994), and vague digestive disturbances such as constipation, mild diarrhoea or dysentery, there is little evidence that tapeworms significantly affect animal health on organic farms in the UK.

LIVER FLUKE (*FASCIOLA HEPATICA*)

Liver fluke has a relatively complex lifecycle, an essential component in which is the presence of a mud snail (*Lymnaea truncatula)* as intermediate host. A warm damp environment is required for both snail and the developmental stages of the parasite, but very acid soils do not suit the intermediate host. Fluke has tended to be associated with wetter areas in the north and west (Eales, 1990). However, in recent years fluke appears to be spreading geographically (Logue, 2003), possibly due to climatic changes, exacerbated by movement of infected stock to areas where fluke is not normally seen.

The likelihood of fascioliasis can be predicted from climatic conditions during the season (Dunn, 1978). A wet spring and early summer predisposes to a rapid increase in the snail population and a good hatch of fluke eggs. Dry weather over this critical period will prevent the snails breeding, and will kill many fluke eggs. Acute disease can occur in the late summer autumn, from large numbers of immature flukes migrating to the liver. Chronic liver damage results when the population becomes established in the bile duct. Tissue damage caused by migrating flukes can also provide opportunities for clostridial disease to develop (Henderson, 1990), and is a risk factor in assessing the need to vaccinate an organic flock. Cattle appear to have a higher tolerance to fluke than sheep.

Fluke is a recognized problem on specific organic farms. Where it is endemic, it is difficult to control in sheep other than by chemical means (Eales, 1990). In the most severe infestations, strategic drenching in March and May has been proposed for conventional systems, to reduce pasture contamination with fluke eggs (Taylor, 1995). The occurrence of triclabendazole-resistant flukes is suspected in the UK, but is now well recognized in the West of Ireland, where fluke is endemic. A measure of cultural control can be achieved by drainage (Dunn, 1978), or by excluding grazing livestock from the wettest areas infested by the snail (Taylor, 1987). However, it can be difficult to identify all such habitats, and would be impractical on many upland farms due to the large areas affected. Furthermore, many farmers are now involved in agri-environmental schemes which limit drainage. Ducks will eat the mud snail, but such biological control is not practical on any scale.

GASTRO-INTESTINAL NEMATODES (STOMACH WORMS)

In the UK, over 20 different roundworm species infest the abomasum, small intestine and large intestine of sheep and cattle. The lifecycle of the main strongyle (*Ostertagia, Trichostrongylus, Haemonchus*) group follows a typical pattern (Dunn, 1978). From excretion in the faeces, the egg develops on the pasture to form the L1 larva which hatches, grows and moults into L2 and then L3 phase. The L1 and L2 larvae are free-living, feeding on micro-organisms within the faecal pat. The infective L3 larva is protected by a sheath and must survive on its own food reserves until it enters a host. Summer rainfall favours the emergence of larvae from the dung pat, and can result in a rapid increase in infective larvae on the herbage. The egg can resist dry conditions, and both egg and L3 larvae can survive moderate freezing. However, larvae are very susceptible to desiccation, and in very hot, dry periods (e.g. summer 2003) the pasture larval population may be decreased to zero. The L3 larvae move to the top of the pasture herbage in a film of moisture where they may be ingested by the host. Once ingested, further development occurs within the mucosa until the fifth stage larvae or young adults emerge into the lumen of the gut. The period from ingestion of the parasite to the appearance of eggs in the faeces i.e. the prepatent period, typically varies from 14 to 24 days. Some producers are still wedded to the concept of 3-week grazing rotation as a means of controlling stomach worms. However, development from the egg through to infective larvae can take 3 - 25 weeks, depending on temperature and moisture. Eggs passed later in the season develop faster because of the higher ambient temperature, with the result that the bulk of eggs passed on the pasture through spring and early summer tend to complete their development around the same time.

Although young cattle can harbour *Nematodirus spp*, clinical disease is confined to lambs, with the main transmission from one lamb crop to the next. Infection tends to increase progressively on contaminated pasture which has been continually grazed by ewes and young lambs. The most notable feature of *Nematodirus* is its ability to survive climatic extremes. The larvae develop from L1 to L3 entirely within a tough egg-shell, and are capable of surviving for up to two years, even if fields are ploughed and reseeded. The eggs hatch in response to specific conditions (classically, a period of chill followed by warmer conditions), resulting in the abrupt release of many infective larvae onto the pasture. However, there is evidence that development may also be retarded by drought, so that *Nematodirus* is increasingly seen during the summer and autumn. The occurrence of disease depends on susceptible lambs being present on the pasture to coincide with large numbers of infective larvae. If the hatch occurs early in the spring, susceptible lambs may not be consuming sufficient herbage to succumb to disease. If the hatch occurs in late April or early May, there is a high risk to young March-born lambs (Taylor, 1987). When the hatch occurs late, the lambs may be sufficiently resistant to withstand challenge.

Control strategies

Grazing management. Clean grazing strategies for controlling stomach worms were widely promoted during the 1970s and 1980s (MAFF, 1982). These were based on rotating sheep and cattle grazing on an annual basis, and where possible integrating with arable production, or the conservation of forage. These plans relied on the strategic use of anthelmintic, given before animals were put onto clean pasture, in order to preserve the cleanliness of the grazing. Despite proven efficacy in commercial practice (Emmerson, 1983), these systems were not widely taken up (Davies *et al.,* 1996) because of perceived inflexibility in grazing management, constraints in farm infrastructure and the ready availability of effective anthelmintics. The demise of new leys in arable farming, and generally decreasing proportions of cattle to sheep, have been additional limiting factors.

Organic farmers have more incentive to adopt preventive management, and possibly better sheep:cattle ratios in their enterprise mix. The principles behind clean grazing systems have equal application in organic farming, the major difference being ready access to anthelmintic. Evidence from Defra-funded project OF0185 (*Controlling internal parasites without the use of pharmaceutical anthelmintics*) indicates the extent to which some organic farmers are willing to go to adjust their enterprises and re-organize their farming systems in pursuit of better parasite control. However, there is often a financial penalty from such system adjustment. Overall levels of parasite burden can be reduced through diversity of enterprises, moderate stocking rates, and system changes e.g. in lambing date, away-wintering of hoggs, or the sale of store lambs. Many use agri-environmental schemes to facilitate lower stocking rates. Without strategic use of anthelmintic, exposure must be regulated during the grazing season by more frequent movement of animals e.g. a change of pasture 4-6 weeks after lambing, or movement of weaned lambs to aftermath grazing, stubble turnips or undersown reseeds.

Role of nutrition. Undernutrition in energy, protein and mineral/trace elements have all been implicated, directly or indirectly, in predisposing animals to parasitic infection (Holmes, 1993). This provides a basis for the preventive nutritional approach advocated in the organic standards. The "spring" or "periparturient rise" in ewe faecal egg output which occurs close to lambing and during early lactation (Dunn, 1978) features strongly in the epidemiology of roundworm parasites in sheep. There is good evidence from controlled studies (Houdijk *et al.,* 2001) for the positive role of protein and other nutrients in maintaining immune responses, particularly during the periparturient period. This is based on the prioritisation of protein for lactation, over immune responses, at a time of high demand or nutrient scarcity (Coop and Kyriazakis, 1999). The principle was tested in a replicated experiment (OF0185) using organically managed ewes, carrying a mixed naturally-acquired infection (Keatinge *et al.,* 2003). Ewes were lambed at a body condition score of 2-2¼ and grazed at a sward height of 4cm-6cm. Control ewes received no energy or protein supplement. Despite an increase of 50% in calculated metabolisable protein supplied to supplemented ewes, a statistically significant reduction in faecal egg output was seen only during weeks 4-6 of lactation i.e. the period of assumed peak metabolic requirement. While the results support the theory that nutritional manipulation can improve immunity to gastrointestinal parasites, there is a limit in the extent to which increased protein can offset parasitic challenge. Excess dietary protein will also be uneconomic, and detrimental to the environment. The result underlines the importance of meeting nutrient requirements in early lactation, as well as prioritising inputs towards more vulnerable classes of animal e.g. lambed gimmers, and ewes rearing twin lambs.

Crop type and pasture ecology. The potential for crop type to influence parasite burden it is not surprising, given the complex interactions which occur between host, parasite and the grazing environment (Thamsborg *et al.,* 1999). The crop itself will combine characteristics related to chemical composition, nutritional value, plant morphology, and microclimate within the sward canopy, which will affect how infective larvae are presented to the host, and how the host copes physiologically with subsequent infection.

Plants contain many phytoactive compounds, some of which may have activity against internal parasites. However, as a direct replacement for pharmaceutical anthelmintics, there is no convincing evidence for the efficacy of herbs such as garlic and wormwood in ruminant animals. Trials in the UK with tanniferous crops such as *Lotus spp* have shown a reduction in parasite burden in weaned lambs (Marley *et al.,* 2003), consistent with data from earlier New Zealand studies (Niezen *et al.,* 1998). These results may arise through a direct effect of tannin on certain parasite species (Athanasiadou *et al.,* 2000), or through the protective effect of tannin on plant proteins within the abomasum, resulting in improved supply of protein to the hindgut. The current commercially available varieties of *Lotus* in the UK do not compete well with weeds, are not very winter hardy, and tend to be sensitive to heavy grazing by sheep. Agronomic performance is therefore likely to remain a barrier until better and more persistent varieties become more widely available.

Chicory has much better agronomic performance (Deane *et al.,* 2003). The crop promotes good lamb growth rates through its nutritional value, and/or improved trace element status (Younie *et al.,* 2001). Lambs grazing chicory were shown to have lower faecal egg output (Deane *et al.,* 2003). However, data on total worm burden were less convincing, suggesting an effect on fecundity. Fewer L3 larvae were recovered above the 5cm horizon on chicory plants compared to perennial ryegrass/white clover. This suggests a possible inhibiting effect of leaf morphology on larval migration. The range of crops currently being evaluated for anti-parasitic effects at research institutes in the UK includes chicory, lucerne, sainfoin, red clover and sulla (*Hedysarium coronarium*).

Investigations into the role of pasture ecology in plant parasite interactions have also been conducted within Project OF0185. Sward structure will influence microclimate, which in turn may affect conditions (e.g. temperature, moisture, the density of coprophytic and nematophagus invertebrates) for larval development, survival and migration. Deane *et al* (2003) reported the slowest rates of faecal degradation, and lowest numbers of invertebrates in chicory compared with a perennial ryegrass/white clover sward. Ecological effects are likely to be multifactorial, for example, in addition to any influence of crop type the number of soil dwelling organisms on the pasture may be affected by faecal density i.e. stocking level. For biological control of free-living stages, delivery methods for nematophagus fungi are also being investigated (Waller *et al.*, 2001).

Breeding for resistance. The heritability of resistance to roundworms is thought to be 0.2-0.4. The value of this to the farmer is twofold. Firstly, progress can be made in selecting more resistant individuals. Secondly, it has been estimated that in lambs 50% of pasture contamination arises from 10% of individuals within the group, so by removing the most susceptible individuals overall levels of challenge can be reduced.

More precise genetic selection methods are now commercially available, based on faecal egg counts. An index which includes resistance to roundworms is available through Signet for Suffolk, Charolais and Texel breeds, which could be readily applied for selecting terminal sires on lowland organic sheep farms. Some organic farmers select against dagginess, but dirty animals may in fact be mounting an aggressive immune response (Jackson, 2003). There is a debate as to whether selection should be for animals which are genetically resistant i.e. shed fewer eggs, or for animals which are more resilient i.e. may continue to shed significant numbers of eggs, whilst being tolerant of the effects of infection. Some evidence exists for a negative correlation between resistance/resilience to parasites and productive traits, possibly because a higher proportion of total nutrient resources are being deployed to meet immune responses. Furthermore, resilient stock will continue to contaminate the environment with potentially detrimental effects on the less resilient faction within the flock. The ideal animal for selection may well be one which performs well physically, but in a parasitized environment. Within the next 10-15 years, continuing advances in DNA technology hold the promise of more accurate, marker-assisted selection techniques.

LUNGWORM (*DICTYOCAULUS VIVIPAROUS*)

While dictycaulosis can occur in cattle, sheep and deer, husk is generally seen as a more severe problem in cattle. In conventional cattle production, the use of the lungworm vaccine declined during the 1980's, in favour of routine anthelmintic, given as a drench or ruminal bolus. It is thought that this almost wholly suppressive approach may have reduced the opportunity for animals to develop a natural immunity and led to a resurgence in lungworm cases, even in adult dairy cattle.

The mature worms live in the bronchi, and are prolific egg producers. Moisture is essential for larval survival and development. The infective third stage larvae are relatively inactive, and few migrate from the faecal pat. Dispersion relies on external factors such as rain, earthworms and the activities of a fungus (*Pilobolus spp.).* Grazing strategies alone are not wholly effective, because of the unpredictable nature and climatic dependency of the disease (Taylor, 1995). Under optimum conditions larvae can survive on the pasture for over a year (Radostits *et al.,* 1994), and although highly susceptible to desiccation, are capable of over-wintering through freezing conditions. Adult worms can live for up to 6 months, and their presence within a symptomless carrier can contaminate pasture for more susceptible stock. Manure from sheds in which affected animals have been wintered has also been implicated in the spread of infection onto pasture.

A significant proportion of organic dairy farmers vaccinate cattle against lungworm (Source: Soil Association). Suckled calves could be expected to be less susceptible than young dairy-bred stock. Concentrate supplementation of youngstock after turnout may assist control, by improving general nutrition as well as reducing herbage intake from the pasture. Where vaccine is used some background level of infection is required, in addition to the vaccine, in order to generate a full immune response. An issue which often arises in organic practice is whether (and at what stage), vaccination can be stopped. An ELIZA test carried out on a representative number of blood samples may help determine antibody levels, and indicate whether background levels of the naturally occurring parasite are still present. A homeopathic nosode is available (McLeod, 1981), but no reliable information is available on its performance relative to the commercial attenuated live vaccine.

EPIDEMIOLOGICAL DATA FROM UK ORGANIC FARMS

Within project OF0185, five commercial organic farms, reflecting a range of production systems, were selected for detailed epidemiological study. The objective was to determine the extent to which parasites could be controlled on commercial organic farms, identify limiting factors for each representative system, and where possible facilitate development towards better systems of control. Comprehensive epidemiological data (including faecal egg output, pasture larval counts, and total worm burdens) were collected from sentinel groups of animals, over two grazing seasons. Epidemiological measurements were supported by management, veterinary, meteorological and animal performance data.

Sheep systems

An overview of sheep performance data is given in Table 1. Apart from the extensively managed flock, specific groups of animals were drenched each year in the other three flocks. Nevertheless, the overall amount of anthelmintic used was substantially less than levels typically used in conventionally managed flocks (Source: Moredun Foundation).

Ewes showed a typical periparturient rise in faecal egg output, which subsequently declined to low levels by weaning. Generally, ewes were not drenched apart from the specialist sheep flock, where anthelmintic was given on veterinary advice to reduce pasture contamination at lambing.

Nematodirus was a particular problem early in the season on two of the four sheep farms. *Nematodirus* had previously been detected on the upland farm, but levels in 2002 and 2003 were sufficient to visibly affect lamb performance, and drenching was required. On the specialist lowland sheep farm, *Nematodirus* was a recognized problem from the outset of the study. Lambing date, a high ratio of sheep to cattle, and little or no clean grazing were clear predisposing factors. During the early years of conversion, veterinary advice was to drench ewe and lambs twice in order to try and break the cycle. However, this proved unsuccessful. An expansion in the size of suckler herd, and the provision for

Table 1. Summary epidemiological and performance data 2003 (2002 in brackets).

	Extensive hill	Upland beef & sheep	Lowland mixed	Specialist sheep
No. of routine drenches: Ewes	0(0)	0(0)	0(1)	2(2)
Lambs	0(0)	2(2)	1(1)	2(2)
Lamb liveweight at weaning (kg)	33(29)	34(32)	32(27)	30(32)
Mean ewe FEC: 1st mth lactation	522(486)	171(153)	504(373)	505(797)
Weaning	279(212)	25(35)	117(252)	183(55)
Mean lamb FEC: 8 weeks	2(4)	233(1030)	791(1457)	505(759)
12 weeks	670(783)	*2(*77)	*12(690)	301(*126)
Weaning	575(451)	81(*29)	699(*129)	189(272)

(* lambs drenched previously)

more alternate grazing by sheep and cattle, was designed to improve parasite control. However, the provision of a one year break from sheep on some of the land, was not sufficient to break the cycle, and there were indications that young cattle were acting as a 'biological bridge' to carry infection over to the third grazing season after sheep.

The mixed lowland farm showed the best prospects of eliminating anthelmintic through moderate stocking levels and the integration of sheep, cattle, and arable enterprises. Lambing date had been put back to late April/early May to avoid the main *Nematodirus* period. Prior to the study, no routine anthelmintic was used in ewes or lambs, and individual treatments were typically 5-6 treatments per 100 lambs. In 2002, the clean grazing system broke down for the sentinel group of animals, due to inadvertent contamination of the previous years grazing by rams, and ewes and lambs were drenched in July. Subsequent performance was acceptable, and lambs finished well on stubble turnips despite high faecal egg counts (>1000 e.p.g.).

Pasture larval counts showed considerable variation, but in addition to the other parameters measured, were a useful additional surveillance tool for *Nematodirus.* From the tracer and permanent lamb data, *Teladorsagia* tended to predominate in mid-season. *Haemonchus* was more prevalent than expected, consistent with the view that milder winters may be enabling more *Haemonchus* to over-winter on the pasture, rather than within infected animals. Also noteworthy was the presence of large bowel worms such *Chabertia* and *Oesophagostomum*, species not normally seen in significant numbers in conventionally reared lambs subject to frequent anthelmintic drenching. The greater incidence of these species may be significant. While they can lay substantial numbers of eggs, they are generally less pathologic, which could have implications for the interpretation of raw FEC data in organically managed sheep.

Cattle system

During the grazing season, faecal egg counts in cattle were consistently low (less than 200 e.p.g.) and no anthelmintic was used for stomach worms at any stage. Even on the organic dairy farm, integration with conservation and frequent movement of youngstock, kept parasites to a low level. However, during 2002 lungworm was detected in one of the two suckler herds. The dairy farm routinely vaccinated for lungworm.

The incidence of winter scours has virtually disappeared from conventional systems, due to the efficacy of modern anthelmintics (Taylor, 1987) against immature stages of the parasite (*Ostertagia ostertagi*). However, the sector bodies continue to receive a number of applications for derogation to treat organic cattle at housing on the basis of perceived risk from ('Type II') winter parasitism (Source: SOPA). As a separate study (within OF0185), four additional organic herds (two dairy and two beef) were used to assess the risk of winter scour in housed cattle. Animals were paired on the basis of previous grazing history, breed, sex, and liveweight. One of each pair was drenched in Dec 02/Jan 03, and performance measured at monthly intervals until turnout. Individual pairs (n=64) were maintained in common pens, and under the same management regime. Faecal egg output was low (less than 100 e.p.g.) throughout the experiment. As expected, drenching significantly reduced FEC. For individual farms, there were no statistically significant treatment differences in daily liveweight gain (DLWG). Three of the four herds had significantly lower blood pepsinogen levels in treated animals, although levels were at all stages within the standard reference range of 0-2 IU/l.

Table 2. Effect of winter drenching on liveweight gain and blood pepsinogen at turnout.

		Suckler herds		Dairy herds		All herds combined
		O	R	M	H	
DLWG (kg)	Drench	0.575	1.11	-0.01	0.63	0.65
	No drench	0.426	1.13	0.10	0.63	0.57
	Significance	NS	NS	NS	NS	P<0.05
Pepsinogen (IU/l)	Drench	0.431	0.298	0.320	0.982	0.579
	No drench	0.604	0.535	0.690	1.854	1.049
	Significance	P<0.05	P<0.05	NS	P<0.001	P<0.001

CONTROLLING INTERNAL PARASITES - A WHOLE FARM APPROACH

Organic livestock producers frequently face difficult judgements, balancing a desire to adhere to the spirit of organic production, whilst still ensuring high standards of health and welfare for their stock. Success in controlling internal parasites requires an integrated whole-farm approach, in which three main components can be identified.

Planning

From the start of conversion, UK certifying bodies require the production of an animal health plan. However, the quality of the plans produced varies enormously (Hovi, 2003). The value of a flock/herd health plan should be to enable control principles to be used in context, identify key system-limiting parasites and combine the best options and approaches for the particular circumstances. In effect, it should form the basis of a covenant between farmer, vet and sector body to develop an acceptable parasite control strategy, pre-empt problems and agree circumstances in which intervention is appropriate. It should allow for (where possible measure) progressive improvements in disease status.

Monitoring

The value of monitoring is to provide regular information. Assessments done at single points in time are difficult to interpret, and are of limited value. Faecal egg counts (FECs) are the most common tool used - through a vet, commercial lab or, with appropriate training, using on-farm systems such as FECPAK (Butler, 2004). However, FECs are an indirect measure, an overall consequence of the total burden, physiological state and relative fecundity of the parasite species present. To arrive at a more robust conclusion, information should be drawn from a range of sources (current FEC, status of current and future grazings, weather conditions, age and animal performance) to provide a more rounded disease picture. Post mortem results, abattoir data, and any information fed back from the purchaser of livestock are all relevant. Another useful concept is 'contamination mapping' - a risk assessment based on likely previous contamination of the pasture by monitored animals. Over time, data and experience will build up to support some of the subjective judgements which have to be made. Following any treatment of organic sheep, a faecal egg reduction test is recommended to check the drug for continuing efficacy.

Biosecurity

Specific guidelines have been developed to counter the potential introduction of anthelmintic resistant nematodes (Stubbings, 2003), based on the quarantine of newly-introduced stock and sequential drenching with levamisole and avermectin-based products. These support earlier recommendations on drenching practice (e.g. dose rates, gun calibration, fasting and standing regimes) designed to increase the effectiveness of treatments. The organic sector bodies need to take a pragmatic view when considering these inputs, on the basis of the benefit to the wider herd/flock from strategically treating a small number of individual animals. Additional benefits of a quarantine period and strategic treatment, may be realized for other important conditions such as foot-rot and sheep scab, a preventive approach entirely consistent with organic ideology.

ACKNOWLEDGEMENTS

Financial support from Defra (OF0185) and support and collaboration of participating farmers.

REFERENCES

ATHANASIADOU S., KYRIAZAKIS I., JACKSON F. and COOP R.L. (2000) Consequences of long-term feeding with condensed tannins on sheep parasitized with *Trichostrongylus colubriformis*. *International Journal for Parasitology*, 30, 1025-1033.

BUTLER G. (2004) DIY Faecal Egg Counts in Sheep. In: Hopkins A. (ed.) *Organic farming: science and practice for profitable livestock and cropping*. Occasional Symposium of the British Grassland Society, No. 37, pp 121-124.

COOP R.L. and KYRIAZAKIS I. (1999) Nutrition-parasite interaction. *Veterinary Parasitology*, 84,187-204.

DAVIES M., PEEL S., HARLEY R., ELLIS S., MERRELL B. and POWELL L.N. (1996) Study to assess the uptake of results generated from grassland R and D. *MAFF Review Project MS1413*.

DEANE J., CORK S., HARESIGN W., LAMPKIN N., MARLEY C., WARREN J. and

WILLIAMS B. (2003) *Control of internal parasites without the use of pharmaceutical anthelmintics: Parasite pasture ecology*. Final report 2000-2003. University of Wales.
DUNN A.M. (1978) *Veterinary Helminthology,* 2nd Edition, William Heinemann Medical Books.
EALES F.A. (1990) Welfare in hill sheep in extensive and organic livestock systems. *Proceedings of a Symposium by the Universities Federation for Animal Welfare and the Humane Slaughter Association,* RAC, Cirencester, September 1990, pp. 83 - 89.
EMMERSON S.A. (1983) Clean grazing for sheep and cattle. *Advisory Project Report (PW/W & M/81/5)* ADAS Midlands and Western Region, December 1983.
HENDERSON D.C. (1990) Internal parasites. *The Veterinary Book for Sheep Farmers*. Ipswich: The Farming Press, pp 477-534.
HOLMES P.H. (1993) Interactions between parasites and animal nutrition: the veterinary consequences. *Proceedings of the Nutrition Society*, 52, 113-120.
HOUDIJK J.G.M., JESSOP N.S. and KYRIAZAKIS I. (2001) Nutrient partitioning between reproductive and immune functions in animals. *Proceedings of the Nutrition Society*, 60, 515-525.
HOVI M. (2003) Devising and implementing animal health plans. Workshop proceedings: *Positive Health and Welfare in Organic Beef and Sheep Systems*, ADAS Redesdale, April 2003.
JACKSON F. (2003) Breeding for resistance – the epidemiological impact. Workshop proceedings: *Positive Health and Welfare in Organic Beef and Sheep Systems*, ADAS Redesdale.
KEATINGE R., KYRIAZAKIS I., HOUDIJK J. and JACKSON F. (2003) The effect of maternal body reserves at lambing on nematode faecal egg output in lactating, organically managed ewes. *Proceedings of the British Society of Animal Science*, 2003, p.29.
LOGUE D. (2003) Fluke facts. *Grass farmer*, No. 76, Autumn 2003, p8.
MAFF (1982). Grazing plans for the control of stomach and intestinal worms in sheep and cattle. *Booklet 2154.* London: Ministry of Agriculture Fisheries and Food.
MACLEOD G. (1981) *The treatment of cattle by homeopathy*. Published by C.W. Daniel Co.
MARLEY C.L., COOK R., KEATINGE R., BARRETT J. and LAMPKIN N.H. (2003) The effect of birdsfoot trefoil (*Lotus corniculatus*) and chicory (*Cichorium intybus*) on parasite intensities and performance of lambs naturally-infected with helminth parasites. *Vet Parasitology*, 112, 147-155.
NIEZEN J.H., CHARLESTON W.A.G., HODGSON J., MILLER C.M., WAGHORN T. S. and ROBERTSON H.A. (1998) Effect of plant species on the larvae of gastrointestinal nematodes which parasitise sheep. *International Journal for Parasitology,* 28, 791-803.
RADOSTITS O.M., BLOOD D.C. and GAY C.C. (1994) Eds. *Veterinary Medicine*, 8th edition. Bailliere Tindall.
STUBBINGS L. (2003) Internal Parasite Control in Sheep. Short Term Strategies to Slow the Development of Anthelmintic Resistance in Internal Parasites of Sheep in the UK. Proceedings of a Defra-funded workshop, 11-12 March, 2003, London.
TAYLOR M.A. (1995) Strategies to control worms in grazing cattle. *Farmers Weekly Supplement.*
TAYLOR M.A. (1987) Parasitic worms in ruminants and their control. *Seminar on Parasite Control in Ruminants for British Organic Farmes, July 1987. British Organic Farmers, Bristol.*
THAMSBORG S.M., ROEPSTORFF A. and LARSEN M. (1999) Integrated and biological control of parasites in organic and conventional production systems. *VetParasitology*, 84, 169-186.
WALLER P.J., KNOX M.R. and FAEDO M. (2001) The potential of nematophagus fungi to control the free-living stages of nematode parasites of sheep: feeding and block studies with *Duddingtonia flagrans*. *Veterinary Parasitology*, 102, 321-330.
YOUNIE D., UMRANI A.P., GRAY D. and COUTTS M. (2001) Effect of chicory or perennial ryegrass diets on mineral status of lambs. In (J. Isselstein, G. Spatz and M. Hoffman, eds.) *Organic Grassland Farming; Volume 6 Grassland Science in Europe*, European Grassland Federation, Witzenhausen, Germany, pp. 278-280.
YUE C, COLES G and BLAKE N (2003) Multiresistant nematodes on a Devon farm. *The Veterinary Record*, November 8, p 604.

Suitability of Purebred and Crossbred Dairy Cows for Organic Systems

S. BROTHERSTONE, J. SANTAROSSA and M.P. COFFEY
Scottish Agricultural College, West Mains Road, Edinburgh EH9 3JG, UK

ABSTRACT

Organic livestock producers are required to choose breeds and crosses that are most suited to organic systems, in order to minimize the incidence of disease and other health and welfare problems. To investigate any benefit of crossbreeding to organic farmers, data on milk production, somatic cell count (SCC), calving interval and survival from first to second lactation were extracted for a variety of pure and first-cross cows. No useful heterosis was found for any of the production traits, but useful heterosis was found for SCC in all first-cross animals for all breeds. Similar genes control SCC and mastitis, so by crossbreeding, advantage can be taken of heterosis for SCC to reduce the incidence of mastitis in the herd. Heterosis for calving interval in the Ayrshire-Jersey cross indicates that fertility is better in this cross than in either purebred Ayrshires or purebred Jerseys. Economic values for traits of interest were calculated twice, using assumptions applicable to conventional and organic systems. From an economic perspective, and irrespective of system, Holsteins performed best but Shorthorns ranked last. If maximizing profitability is their primary objective, it is not evident that farmers running organic systems should consider crossbreeding as a first option.

INTRODUCTION

In all production systems, it is important for milk producers to choose breeds and crosses that are most suited to their systems and the markets that their milk is sold into. This is especially true for organic milk producers since they have particular difficulties to address, e.g. the need to minimize the use of drugs to treat or prevent disease. This is important as the prophylactic use of antibiotics and some other drugs on a herd basis is prohibited in organic systems. There has been little research on disease resistance in organic systems, but it has been shown that crossbred animals generally show greater disease resistance and overall fitness than their purebred counterparts, due to heterosis, or hybrid vigour, for these traits (Harris *et al.*, 1996; Lopez-Villalobos, 1998). This suggests that, from a disease perspective, crossbred animals may be better suited to organic systems than purebred animals.

Useful heterosis is found when the crossbred mean is more advantageous then either of the purebred means, e.g. when the crossbred mean for milk yield is higher than either of the purebred means or the crossbred mean for SCC is lower than either of the purebred means. Heterosis is exploited regularly in crop production and also pig and poultry production and is associated with good fertility and overall fitness.

In the UK, bull and cow predicted genetic merit for milk, fat, protein, lifespan, SCC (to predict mastitis) and locomotion (to predict lameness) are each weighted by their economic value and combined in a profit index (Brotherstone *et al.*, 1998). These economic values are based on many fixed assumptions including milk price, yield level and concentrate use. These assumptions are appropriate for the majority conventional farming systems but may not be appropriate for organic systems.

The objective of this work was to quantify the benefits of using crossbred animals in an organic system, by (1) estimating pure and crossbred differences for breeds in the UK national herd and for traits of interest to the organic sector, and (2) comparing the financial performance of these breeds and crosses in both conventional and organic systems.

METHODS AND MATERIALS

Pedigree data were obtained from National Milk Records (NMR). We calculated the proportion of each breed in every animal with a milk record, and stored the information in a relational database (8 million females). Animals were than assigned to the appropriate breed group or cross group for analysis.

Herds identified as participating in crossbreeding (by having some crossbred animals) were chosen, and only animals which are a mix of, at most, two genotypes were selected. Data on milk production, survival from first to second lactation, calving interval (where present) and somatic cell count were extracted for all heifers in the chosen herds and calving after 1990. Breeds and crosses analysed were: Holstein, Ayrshire, Jersey, Guernsey, Shorthorn, Holstein-Friesian, Ayrshire-Jersey, Ayrshire-Shorthorn and Jersey-Guernsey.

An animal model with pedigree was used for all analyses apart from the analyses of the Holstein-Friesians, which utilized a simple sire model due to the reduced size of the data set. Age at calving was included in the model as a linear and a quadratic covariate, and month of calving was fitted as a fixed effect. The proportion of one of the breeds in the cross and the heterosis coefficient were included in the model as covariates. Both animal and herd-year of calving were defined as random effects.

Economic values were estimated using a new method which takes land use and soil characteristics into consideration, and results in the values being, in part, dependent on the long-term profitability of the land. Using this system, economic values for production, longevity, mastitis and lameness were derived under both organic and conventional assumptions.

In order to gain an insight into differences between the systems (conventional versus organic) and between breeds, a comparative analysis was conducted which used risk measurement techniques. Based on variation in income and costs, the amount of risk involved in stocking a farm with a particular breed or cross was calculated under both organic and conventional farming systems, and breeds were ordered from 1 (least financial risk) to 9 (most risky).

RESULTS

Heterosis

Irrespective of the breeds involved, no useful heterosis was identified for the production traits (milk, fat, protein, fat%, protein%), although a significant amount of heterosis was found for fat% in Ayrshire-Jerseys and Jersey-Guernseys. Useful heterosis was found for SCC in all first-cross animals. As similar genes control both somatic cell count and mastitis (Kadarmideen and Pryce, 2001), SCC is a useful predictor of mastitis. By crossbreeding, advantage can be taken of useful heterosis for SCC, so reducing the incidence of mastitis in the herd. This is particularly useful to organic producers who are prevented from using antibiotics on a regular basis.

Evidence of hybrid vigour for survival from first to second lactation was found in the Ayrshire-Jerseys and the Ayrshire-Shorthorns, and the presence of heterosis for calving interval in the Ayrshire-Jersey cross indicates that fertility is better in this cross than in either purebred Ayrshires or purebred Jerseys.

Economic values

Purebred and crossbred economic values for production traits, longevity, mastitis, fertility and lameness under organic farming assumptions were similar to those derived using conventional farming assumptions. For example, for the Holstein-Friesian the economic value for an additional lactation under conventional farming assumptions was £24, whereas under organic farming assumptions it was £26. The similarity in economic values across systems may be a consequence of limited information on organic systems. For example, information on the incidence of mastitis and lameness in organic dairy herds in not available, so incidences in conventional farming systems had to be used.

Comparative analysis

The scores for each of the breeds and crosses, ranked from 1 (least financial risk) to 9 (most risky) are given in Table 1.

Table 1. Breeds and crosses ranked from least risk (=1) to most risky (=9) in organic and conventional production systems

	Organic	Conventional
Holstein	1	1
Ayrshire	3	3
Ayrshire-Jersey	4	4
Ayrshire-Shorthorn	5	5
Guernsey	7	7
Holstein-Friesian	2	2
Jersey	8	8
Jersey-Guernsey	6	6
Shorthorn	9	9

The comparative analysis shows clearly that Holsteins outperform all other breeds in both systems, whilst Shorthorns rank last. Ranking was identical across systems.

DISCUSSION

It is important for organic milk producers to choose breeds, crosses and within-breed selection goals which are most suited to their production systems. A particular requirement is the need to reduce the use of drugs to treat or prevent disease. Mastitis is an important disease for the dairy cattle industry as it is a major reason for culling (in conventional herds approximately 10% of culling in each lactation is due to mastitis) and treatment costs are high. This work has shown that somatic cell count, which is a good predictor of mastitis, is significantly lower in all crossbreds investigated than in their purebred parents. By cross breeding, advantage can be taken of useful heterosis for somatic cell count, so reducing the incidence of mastitis in the herd. This advantage may remain even if there is selection pressure for milk production applied to the parent populations.

Economically, Holsteins and Holstein-Friesian crosses outperformed all other breeds and crosses in both organic and conventional systems, suggesting that they are appropriate for both sectors of the market. Whilst being a popular suggestion amongst those who discuss organic milk production, it is not evident from these results that farmers running organic systems should consider crossbreeding as a first option to improve profitability.

This investigation has revealed that insufficient data is currently recorded on dairy cattle to allow accurate statistics to be compiled for the organic sector. We need to know the incidence of all major diseases, per lactation, in organic dairy herds. This will allow us to refine our economic values for traits currently in the UK profit index, and will allow us to expand the profit index to include other costly diseases such as fertility. Identification of cows in the organic dairy sector would allow us to investigate their energy balance profiles and compare these with energy balance profiles of cows in conventional systems.

ACKNOWLEDGEMENTS

We would like to thank the Participatory Group who collectively guided this project and National Milk Records for providing the data. We also thank colleagues, in particular Geoff Simm and John Woolliams, for helpful and constructive comments.

REFERENCES

BROTHERSTONE S., VEERKAMP R.F. and HILL W.G. (1998) Predicting breeding values for herdlife of Holstein Friesian dairy cattle from lifespan and type. *Animal Science,* 67, 405-412.

HARRIS B.L., CLARK J.M. and JACKSON R.G. (1996) Across breed evaluation of dairy cattle. *Proeedings of the New Zealand Society of Animal Production,* 56, 12-15.

KARDAMIDEEN and PRYCE (2001) Genetic and economic relationships between somatic cell count and clinical mastitis and their use in selection for mastitis resistance in dairy cattle. *Animal Science,* 73, 19-28.

LOPEZ-VILLALOBOS N. (1998) *Effects of crossbreeding and selection on the productivity and profitability of the New Zealand dairy industry.* PhD Thesis. Massey University, NZ.

Combining Ethological Thinking and Epidemiological Knowledge to Enhance the Naturalness of Organic Livestock Systems

M. VAARST[1], S. RODERICK[2], V. LUND[3] and W. LOCKERETZ[4]
[1] Danish Institute of Agricultural Sciences, PO Box 50, DK – 8830, Tjele, Denmark.
[2] Organic Studies Centre, Duchy College, Rosewarne, Camborne Cornwall, TR14 0AB UK
[3] National Veterinary Institute, Pb. 8156 Dep., N-0033 Oslo, Norway
[4] Friedman School of Nutrition Science and Policy, Tufts University, Boston MA02111, USA

ABSTRACT
Organic livestock farming places strong emphasis on conditions that allow animals to exhibit behavioural needs. This involves the provision of a natural environment and, in particular, outdoor conditions and a reliance on natural forages. Such environments also allow animals to be effectively integrated into crop production. However, there are potential disease risks associated with these conditions, with control options being partly limited by restrictions on chemoprophylactic measures. Examples from dairy and poultry production demonstrate how a basic understanding of ethology and a knowledge of disease epidemiology can enhance the welfare of animals whilst satisfying the ecological objectives of organic farming. Existing epidemiological models and published data can be used to examine the potential ensuing health hazards and control possibilities and to suggest alternatives.

INTRODUCTION
The concept of naturalness underpins the special philosophy of organic livestock farming. This includes the idea that farm animals should live in an environment corresponding to that which they are adapted to through evolution, and that they should be allowed to perform species-specific behaviour and be fed according to their physiological needs. In Northern Europe, animals have increasingly been confined in artificial environments that do not allow them to exhibit their physiological and behavioural needs. Also, the time the farmer has available for each animal has dramatically decreased. As a result, many farmers have lost a good understanding of animal behaviour, e.g. herding, controlling natural breeding, and identifying sick animals for treatment. This becomes particularly detrimental in organic farming, where management should be based on understanding of animal behaviour and where preventive health care is vital. This involves a certain freedom of choice. 'Care' is understood as the counterpart of naturalness, expressing humans' special responsibility towards domestic animals (Alroe *et al.*, 2001).

Organic farming puts greater emphasis on animal needs and integration with the environment. However, naturalness has not always been embodied in practice. For example, organic dairy production tends to differ from conventional systems in provision of feeds and disease controls. If naturalness is taken seriously, management will be based more on insights into animal behaviour, would involve more consideration of the animal as a part of a herd and would offer greater provision of a natural environment.

Since the overall goal in organic farming is to create sustainable systems, this philosophical framework can create several dilemmas, particularly in relation to integrating naturalness into the systems of production as well as the other animal health, economic and environmental objectives of organic farming. In this paper, we focus on the basic understanding of ethology and knowledge of disease epidemiology, and how the combination of knowledge in these two areas can help deal with this problem whilst satisfying the ecological objectives of organic farming.

APPLYING ETHOLOGICAL THINKING

Improved knowledge of animal behaviour can be viewed as important in organic farming for at least three reasons: 1) to solve immediate challenges in animal care and production; 2) to understand what is required for animals to express their natural behaviour; and 3) to improve the animals' welfare in the long term.

With few exceptions, all farm animal species are social. In the wild, group living can provide many advantages for the individual, such as increased efficiency in detecting and acquiring food and better protection against predators, but there can also be disadvantages, such as increased exposure to parasites and pathogens. Thus, natural populations have optimal group sizes, adapted to their communication and recognition abilities, season and developmental stage. When these more natural group structures are introduced on the farm, animals are allowed a greater opportunity to express their social behaviour. However, the structure that best suits the animal may be uneconomical. For example, under wild conditions groups of mixed sexes and ages are most likely, yet under commercial conditions would increase costs and cause management difficulties. Studies of animals living under wild or semi-wild conditions can be used when developing alternative and more 'natural' systems. In organic farming this idea is being applied more to pig and poultry production than to dairy farming.

DISEASE EPIDEMIOLOGY AND 'NATURAL LIVING'

Conversion to organic production may result in changes in disease epidemiology, possibly as a consequence of changing farmers' attitudes and perception, treatment thresholds or prohibition of preventive medication, changes in cost-benefit relationships, new feeding strategies or change in disease factors associated with greater outdoor access. Although focusing on natural living potentially gives many welfare benefits and reduces many of the behavioural and bacterial disease problems in crowded and poorly ventilated indoor systems, free-range organic animal husbandry has raised the question of whether the animals' welfare is at risk. There may be specific diseases that justify this concern, and the few studies published so far indicate that the major health concerns are related to parasitic and other infections connected with outdoor rearing (Thamsborg *et al.*, 2003).

Organic animal husbandry places particular emphasis on health promotion and disease prevention, including breed selection, animal husbandry practice, feeding natural forages, free-range conditions and appropriate stocking densities (CEC, 1999). Patterns of disease are influenced by biological, economic, cultural and environmental factors. Epidemiological studies that include such factors can enhance our understanding of how to promote health and manage disease organically.

The aim in organic farming is co-operation with nature, including improving the animals' ability to deal with disease challenges. One can debate whether outdoor and free-range systems enable a more balanced immune response or present an animal welfare issue. The animals need to build up their ability to handle infections, e.g. through low stress levels and low dosage exposure to infective agents in cases where immune response can develop. However, we also need to discuss what disease levels are 'tolerable' in organic farming, since the aim is not to eradicate diseases at any price. Ultimately, the organic farmer can intervene with treatment or other forms of care. It is not relevant to raise concern about 'suffering' of organic animals, and the emphasis on animal health and welfare in organic production should not be ignored. With correct feeding, stocking, breeding, and care, the risk of disease need not be a major concern.

COMBINING ORGANIC THINKING

The aim of organic farming is to develop viable agro-ecosystems whose parts support each other to create a better and more productive whole (Lund *et al.*, 2004). Integration of more than one livestock species and of livestock with cropping can be the basis of a balanced and sustainable farming system, allowing nutrient recycling and effective resource use. Due consideration must be taken of the whole system. For example, if not managed properly, natural animal behaviour can cause problems. Under natural conditions pigs and chickens spend most of the day looking for food. Restricted foraging (e.g. where animals are fed concentrated diets from a feeder) can lead to behavioural disturbances, and it has been suggested as a cause of feather pecking in chickens (Blokhuis, 1986). The organic requirement to provide animals with roughage and give them access to grass helps to mitigate some problems related to foraging behaviour, but a challenge remains for production systems to use animal behaviour as an asset, and not simply solve behavioural problems. This must be done as an integrated part of the whole organic system. However, some conflicts are difficult to avoid and care must be taken to solve these in a constructive way, still including the animal welfare objectives.

Example 1: Group living in dairy calves

Given the choice, most farm animals would live in groups for at least part of their lives, and this is a requirement in housed organic animals. This is contrary to how most young dairy calves are reared in North-west Europe, where individual housing is most common. If learning from how a 'natural' system works, there would be a calving season, cows would naturally bond with their own calves and there would be a small but stable grouping of calves of the same age within the herd. The advantages of being in a group like this is the learning element: the calves will gradually move from playing to forming a hierarchy, which will be supported by the stability of the group. How can this be brought into daily practice in large European dairy herds?

The argument for single-calf housing often refers to disease risk (e.g. pasteurellosis), suckling on other calves in the group, and that single penned calves are less demanding to manage, and disease is easier to monitor. The ease of disease surveillance may, in part, explain why disease incidence is higher in group-housed calves: they are more time-consuming to observe and manage. Yet, from an ethological and natural perspective group-housing would be most beneficial. Solutions are required that enable calves to live within groups and yet remain healthy. One key could be the emphasis on stability, allowing calves to build up a common immunity, as well as gradually forming a group structure and hierarchy without being disturbed by new animals and new infections. Stable groups, reared together, will assist in promoting good health and supporting the calves' needs for developing social behaviour. The next step could be groups of dairy cows with calves, thus facilitating the learning process through adult contact. Good examples of such systems exist, but are still more complex, e.g. in larger herds on limited space and with no distinct calving season.

Example 2: Integration of poultry in the crop rotation

Free-range systems are required in organic poultry production, and these systems do provide an opportunity for group living, albeit in single-age groups. However, this system could present increased parasite and predation risk. In large-scale poultry production, all-in / all-out systems will typically be introduced as a necessary means to prevent diseases, which implies single-age groups. The inclusion of cockerels in poultry flocks can, however, improve group dynamics. Group size is another feature that for economic reasons usually deviates greatly from what is considered the optimum in natural habitats. Mobile houses offer opportunities for smaller groups and are better suited to integrating poultry into the farming system.

A mobile poultry system that fits into a crop rotation builds soil fertility and uses farm resources more efficiently, including 'wastes'. There also are natural parasite control benefits from pasture rotation, which could extend to benefit ruminants within the crop rotation. However, annual cropping does not allow the establishment of trees and bushes that resemble the natural environment of the species (i.e. jungle shrub) and which satisfy the animals' basic instinct to escape predation. One way to grow annual crops and still provide a more natural environment for poultry could be to grow tall crops such as maize, although this may be difficult to combine with mobile systems. A trade-off must be made between the poultry's need for a natural environment, and the epidemiological requirements of parasite control. Integration of animals with orchards or fruit plantations appears to suit organic systems better than integration with annual crops; a good example of layer birds both benefiting from, and contributing to, a raspberry crop is provided by Reid (2002).

CONCLUSIONS

In organic farming, naturalness and a natural life are considered important for animal welfare. This paper has highlighted the significant difficulties associated with combining these with the other objectives of organic farming, particularly high standards of animal health and the requirement for economic efficiency. However, the paper has also illustrated that, with a basic understanding of behavioural needs and knowledge of disease risks and patterns, combined with an innovative approach, sustainable, functional and mutually beneficial animal and crop system are feasible. This knowledge must be supported by the practical implementation of appropriate housing facilities and grazing arrangements, disease surveillance and good human-animal relationships. Existing epidemiological models and published data can be used to examine the potential ensuing health hazards and control possibilities and to suggest alternatives. Finally, whilst the principles of naturalness are embedded within the legal framework for organic farming, it is imperative that producers not only fulfil these legal requirements, but also embrace the underlying principles when developing and managing organic livestock systems.

REFERENCES

ALROE H.F., VAARST M. and KRISTENSEN E.S. (2001) Does organic farming face distinctive livestock welfare issues? A conceptual analysis. *Journal of Agricultural and Environmental Ethics*, 14, 275-299.

BLOKHUIS H.J.(1986) Feather pecking in poultry: its relation with ground pecking. *Applied Animal Behaviour Science,* 16, 63-67.

CEC (1999) Council Regulation No. 1804/1999 supplementing Regulation No 2092/91 on organic production. *Official Journal of the European Communities*, 42, L222, 1-28.

LUND V., ANTHONY R. and RÖCKLINSBERG H. (2004) The ethical contract as a tool in organic animal husbandry. *Journal of Agricultural and Environmental Ethics*, 17, 23-49.

REID F. (2002) Integrating layer chickens into a certified organic raspberry and vegetable farm. In: *Proceedings of the 14th IFOAM Organic World Congress*, 21-24 August, Victoria, Canada, p.80.

THAMSBORG S.M., RODERICK S. and SUNDRUM A. (2003) Animal Health and Diseases in Organic Agriculture: an Overview. In: Vaarst M., Roderick S., Lund V. and Lockeretz W. (eds.): *Animal Health and Welfare in Organic Animal Husbandry*, pp 227-252. Wallingford UK: CABI.

Vegetation Change on an Upland Organic Livestock Unit in the North East of England from 1992-2001

H.F. ADAMSON, C.N.R. CRITCHLEY and A.E. MOON
ADAS Redesdale, Rochester, Otterburn, Newcastle upon Tyne NE19 1SB

ABSTRACT

The effects of organic farming on livestock performance and botanical change have been investigated on an upland sheep and beef farm in the north-east of England since its conversion to organic management in 1991. Vegetation change was monitored from 1992 to 2001 in one compartment (heft) that was divided into comparable organic and conventional sub-hefts across three land types (unimproved hill, improved hill and intensive inbye grassland). Botanical monitoring was based on fixed quadrats. On the hill land, stocked at 2.1 ewes/ha, wet heath and acid grassland communities became increasingly dominated by rough grasses, while heather declined. Decline of white clover on improved hill grassland under conventional management was probably attributable to inorganic fertilizer N application. On inbye grasslands, naturally occurring species colonized after the initial establishment of sown species. The most important effects were related to the intensity of management, irrespective of organic or conventional status. The biodiversity benefit of organic management will be better realized if it incorporates positive conservation measures.

INTRODUCTION

Organic farming is generally perceived to be beneficial to biodiversity. Most evidence for this emanates from studies on lowland farms, although the greatest area of organic land in the UK is upland pasture (Shepherd *et al.*, 2003). Moreover, there have been few long-term studies at the system scale. In 1991, part of a progressive upland livestock farm in Northumberland, UK was entered into organic conversion. One compartment (heft) was divided into two new hefts of comparable size and vegetation type, with one being entered into organic conversion, and the second managed conventionally, to compare livestock performance in the two systems. The original experimental objective was to test the extent to which production could be pushed under an organic system, and similar sheep stocking rates were maintained on organic and conventional sub-hefts. Conversion was completed in 1993. Changes to the vegetation over a nine-year period under the two systems are described below.

MATERIALS AND METHODS

Sampling design and field methods

There were three land types on each heft: unimproved hill (open hill, with stocking density 2.1 ewes/ha), improved enclosed hill (limed and sown with a perennial ryegrass/clover mix in the 1970s) and inbye land (agriculturally improved, intensively managed enclosed grassland, close to the farmstead). The organic and conventional hefts each comprised *c.* 65 ha of unimproved hill, *c.* 20 ha of improved hill, and inbye areas of approximately 5 ha and 4 ha, respectively.

Fixed 1m x 1m quadrats were located in a grid layout on each land type. Each quadrat was divided into 100 10cm x 10cm cells and the dominant plant species (i.e. that with greatest cover) in each cell recorded. Species dominance was recorded at intervals of between one and three years in the period 1992-2001, from 126 quadrats on the unimproved hill, 23 on the improved hill and 19 on the inbye. Although the different land types were not all assessed in the same years, it did allow an overall comparison of temporal trends to be made.

Analysis

Variation in species composition in the first year of survey was assessed for each land type using Detrended Correspondence Analysis (DCA). A fuzzy clustering analysis was then applied to the ordinated data (Equihua, 1990) and the resulting vegetation types described with reference to the National Vegetation Classification (Rodwell, 1991; *et seq.*). Species data from subsequent years were then added to the ordination as supplementary (passive) variables to identify temporal change in the composition of the different vegetation types within the ordination space.

Changes over time of the percentage dominance of the main species within each vegetation type were analysed on the organic and conventional hefts separately. Ranked Friedman Analysis of Variance (ANOVA) was used on the unimproved hill data (to account for the different vegetation types identified) and one-way ANOVA on the improved hill and inbye.

RESULTS

Unimproved hill

Five vegetation types were identified. Three of these were wet heath (equivalent to NVC community M15 *Scirpus cespitosus - Erica tetralix*) and dominated by, respectively, *Calluna vulgaris* (heather) only (n=26), heather and *Molinia caerulea* (purple moor-grass) (n=22) or purple moor-grass only (n=32). The others were *Nardus stricta* (mat-grass) dominated grassland (NVC community U5 *N. stricta – Galium saxatile*) (n=14) and acid grassland (U4b *Festuca ovina – Agrostis capillaris – G. saxatile, Holcus lanatus - Trifolium repens* sub-community) (n=8).

Over time, all vegetation types on both the organic and conventional hefts moved in ordination space away from the heather dominated wet heath or the acid grassland towards those dominated by rough grasses such as mat-grass and purple moor-grass (Figure 1).

The most notable changes in individual species dominance occurred in the wet heath where heather decreased significantly over time on both the organic and conventional hefts respectively ($H_{12,3}$ = 21.92, P<0.001; $H_{14,3}$ = 33.13, P<0.001). There was a corresponding increase in *Carex nigra* (common sedge) on the conventional heft ($H_{14,3}$ = 1.55, P<0.01) and a similar trend on the organic heft that was not statistically significant. For acid grassland, *T. repens* (white clover) decreased on the conventional heft ($H_{5,3}$ = 14.04, P<0.01).

Unimproved hill

Two grassland vegetation types were identified, which were separated along axis 1 of the DCA ordination. The first was dominated by *Holcus lanatus* (Yorkshire fog) and white clover (NVC community MG6 *Lolium perenne-Cynosurus cristatus*) (n=15) and the second by *Lolium perenne* (perennial ryegrass) and white clover (MG7b *L. perenne, L. perenne-Poa trivialis* sub-community) (n=8).

On both organic and conventional hefts the DCA showed that, over time, the vegetation showed a gradual transformation away from the more fertility-adapted perennial ryegrass type towards the low fertility-adapted Yorkshire fog type. No change was detected in individual species dominance on the organic heft. In the two conventional fields, however, there was a decline in white clover dominance between 1995 and 2001 ($F_{1,4}$ = 13.18, $P<0.05$ and $F_{1,12}$ = 26.67, $P<0.001$).

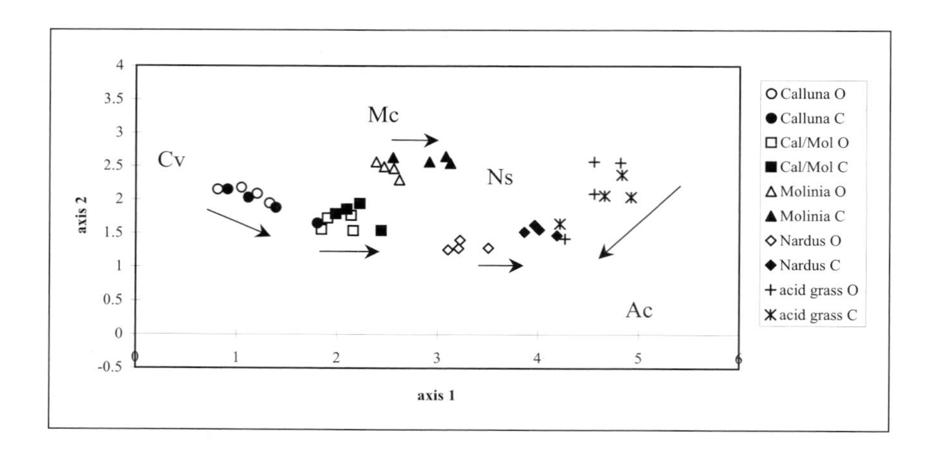

Figure 1. DCA ordination of first two axes of variation of the hill quadrats. Each point represents the mean position of a vegetation type for either the organic (O) or conventional (C) heft in each year. The arrows indicate the approximate direction of change for each vegetation type over time. The positions of the key species, *Calluna vulgaris* (Cv), *Molinia caerulea* (Mc), *Nardus stricta* (Ns), and *Agrostis capillaris* (Ac) have been plotted to assist interpretation.

Inbye land

The organic area was reseeded in 1992, and the conventional area in 1991 and 1997. No distinct vegetation types were identified by the fuzzy clustering analysis, which indicated that there was little variation in species composition across the inbye fields. However, the DCA of species data from 1995 revealed that the sown species (perennial ryegrass and white clover) were more closely associated with the organic field (n=10).

In contrast, the naturally regenerating Yorkshire fog was more closely associated with the conventional field (n=9). This reflected the elapsed time since each field had been reseeded.

On the organic field, from 1995 there was a decrease in the dominance of both ryegrass ($F_{4,40}$ = 19.69, $P<0.001$) and white clover ($F_{4,40}$ = 8.15, $P<0.001$) from 45-50% to approximately 30%. There were corresponding increases in *Ranunculus repens* (creeping buttercup) ($F_{4,40}$ = 4.06, $P<0.01$), Yorkshire fog ($F_{4,40}$ = 4.98, $P<0.01$), *Phleum pratense* (timothy) ($F_{4,40}$ = 6.34, $P<0.001$) and *Poa trivialis* (rough meadow grass) ($F_{4,40}$ = 14.70, $P<0.001$), although of a smaller magnitude (from 0-5% to 10-15%).

Following reseeding of the conventional area in 1997 there was an increase in the dominance of sown perennial ryegrass ($F_{5,48} = 10.93$, P<0.001) and a decrease in that of Yorkshire fog ($F_{5,48} = 4.50$, $P<0.01$). In the years following reseeding, ryegrass remained at approximately 50% dominance, then decreased to 18% in 2001, with corresponding increases between 1998 and 2001 of rough meadow grass ($F_{5,48} = 11.15$, $P<0.001$), creeping buttercup ($F_{5,48} = 2.78$, $P<0.05$) and timothy ($F_{5,48} = 10.86$, $P<0.001$).

DISCUSSION

On the hill land, on both the organic and conventional hefts, wet heath and acid grassland communities became increasingly dominated by rough grasses, while heather declined. The loss of heather was probably attributable to the relatively high stocking rate applied on both hefts. These changes represent a decline in the condition of these important upland plant communities, and a reduction in their value for other upland biodiversity. On other areas of the organic farm, where stocking rates were lower, the hill vegetation was maintained. The changes detected in the acid grassland reflected a decline in its agricultural value, but also a reduction in the diversity of vegetation types across the hill land.

On the improved hill, the overall trends indicated a continuing reversion of the reseeded grassland towards a semi-improved state under both organic and conventional management. The reduction in white clover on the conventional fields was probably attributable to the application of inorganic nitrogen fertilizer.

On the inbye, naturally occurring species colonized after the initial establishment of sown species but there were no major differences that might be attributable to the different systems. In organic systems, these species might be expected to establish more readily than under conventional management that includes herbicide application. However, the value of short-term leys to biodiversity is probably limited unless annual weeds are available to farmland birds as a source of seed and invertebrate prey.

Many of the vegetation changes seen here were the consequence of a relatively high intensity of management. In upland organic systems, stocking densities are often lower because of reduced forage production and stock carrying capacity (Shepherd *et al*., 2003). At this site, stocking rates were reduced on the organic heft for this reason. In organic upland livestock systems, benefits to biodiversity are only likely to accrue when they include extensive practices such as reduced stocking rates, removal or reduction of sheep in winter, and incorporation of longer-term leys into the inbye management. Enhancement of biodiversity is central to the philosophy of organic farming, but positive conservation measures will also need to be taken to fulfil its potential benefits.

REFERENCES

EQUIHUA M. (1990) Fuzzy clustering of ecological data. *Journal of Ecology*, 78, 519-534.

RODWELL J.S. (1991 *et seq.*) *British Plant Communities, Volumes 1-5*. Cambridge University Press, Cambridge.

SHEPHERD M., PEARCE B., CORMACK B., PHILIPPS L., CUTTLE S., BHOGAL A., COSTIGAN P. and UNWIN R. (2003) *An Assessment of the Environmental Impacts of Organic Farming*. London: DEFRA.

Some Possibilities for Sustainable Organic Production from Clover-rich Pastures in the Hills and Uplands

G.D. ANDERSON
Edgerston Tofts, Camptown, Jedburgh, TD8 6NF, UK

ABSTRACT

The paper reviews results of experiments in the 1970s in which white clover was oversown at 4 kg/ha on to moorland swards in Yorkshire, without cultivation or herbicide. Within 3-4 years, mean yields of *Molinia* and *Nardus* swards almost doubled, to 4t DM/ha/annum, and on *Calluna* heath increased five-fold to 1.45t /ha, associated with 51% and 35% clover cover, respectively. This was achieved by one application of 12t lime/ha with 90kg P_2O_5/ha as triple superphosphate or basic slag, but after 4 years rock phosphate was catching these up. N fertilizer reduced clover, while clover without N in *Molinia* and *Nardus* swards gave DM yield increases exceeding 1.5t /ha/annum. These findings were subsequently put into practice on a Scottish farm where application of 7.5 t/ha magnesium lime, followed after grazing by 2.5 kg/ha each of clover, timothy and ryegrass with 90kg P_2O_5/ha, enabled doubling of the livestock on *Molinia* and *Nardus* swards within 4 years. A further 6.0-7.5t lime/ha increased clover spread and sward palatability. In rejuvenated in-bye and hill swards, S184 clover has persisted for over 20 years.

INTRODUCTION

Improvement of swards on acidic hill soils by oversowing white clover with lime and phosphate has the potential to increase their capability for sustainable production and organic management. Unimproved *Molinia, Nardus* and *Calluna* swards produce below 2300kg DM/ha annually, much of which is not utilized. Improvement of hill pastures has often been by ploughing or rotavating and complete reseeding, but costs, steepness, stoniness, or wetness have limited the adoption of such methods. Though some benefits of oversowing have been known for many years, there have been few investigations of the nutrient requirements for establishment, productivity and persistence of oversown white clover on moorlands. This paper reviews results of hill land improvement experiments from the 1970s, when the author was employed by ADAS, and describes how the findings have been put into practice to improve organic production on a hill farm in the Scottish Borders.

EXPERIMENTAL

Oversowing clover on to *Molinia, Nardus* and *Calluna* swards

In 1976-79 the effects of various lime and fertilizer inputs on the establishment and productivity of oversown clover were investigated on six sites in the Yorkshire Pennines, two on each of *Molinia, Nardus* and *Calluna* sward types, all with surface pHs below 3.8. Table 1 outlines the treatments, which were applied in a 2 x 3 x 2 x 3 lime PKN factorial of 36 plots on each site. *Rhizobia*-inoculated S184 white clover was oversown at 4 kg/ha. Contiguous 'All minus One' experiments had 'minus clover', 'minus inoculant', 'minus lime' and 'control' treatments and a comparison of triple superphophate (TSP), basic slag and Gafsa rock phosphate (RP) in a seven-plot layout without N, having two replicates giving four per sward type. All plots were scored for clover cover prior to harvesting cuts, usually in June and August each year.

RESULTS OF EXPERIMENTAL WORK

An early appraisal of effects on clover establishment was reported in Anderson (1979) and an outline of changes in sward productivity in Anderson (2001).

Dry Matter Responses. With added lime and phosphate, DM yields increased annually, as clover spread for at least 4 years after sowing. Table 1 shows responses 3-4 years after fertilizer application.

Table 1. Effect of nutrient application at sowing on the DM yield (t/ha) 3-4 years after over-sowing on different sward types (mean % increases over controls in brackets).

Treatment	Plots	*Molinia*	*Nardus/Juncus*	*Calluna*	Mean
6t Lime (basal P, K&N)	36	0.45(19)	1.07(58)	0.96(342)	0.83(55)
12t Lime (basal P,K&N)	36	0.60(25)	1.18(64)	1.10(393)	0.96(64)
12t Lime(basal P*K no N)	8	1.21(51)	1.93(104)	0.94(335)	1.36(91)
P30 (basal L, K&N)	24	0.29(12)	0.38(15)	0.74(102)	0.47(25)
P90 (basal L, K&N)	24	1.04(42)	0.82(32)	1.04(144)	0.96(51)
P90*(basal L, K no N)	8	1.34(60)	1.77(88)	1.00(455)	1.37(92)
K90 (basal L, P&N)	36	0.31(11)	0.12(4)	0.42(38)	0.28(12)
N30/yr. (basal L,P&K)	24	0.33(13)	0.39#(15)	0.65(102)	0.46(24)
N90/yr. (basal L,P&K)	24	0.82(33)	0.56#(21)	1.43(223)	0.94(49)
Clover(basal L,P*K no N)	8	1.65(86)	1.52(67)	0.60(97)	1.26(79)

* mean of TSP& slag, #part of response only residual from previous N application.

There was a good response to 6t/ha lime on all sites and additional response at 12 t/ha. *Molinia* swards responded least to lime. Added P gave large responses on all sward types ($P<0.05$). Phosphorus, either as TSP or as slag, virtually doubled DM yield of *Molinia* and *Nardus/Juncus* swards and increased it over five-fold on *Calluna* (Table 2).

Table 2. Effect of different phosphates at sowing on mean yields after 3-4 years (t DM/ha/annum). (Basal lime, potassium and clover but no N) (percentage increases over controls in brackets).

Phosphate source	*Molinia*	*Nardus/Juncus*	*Calluna*	Overall
TSP	3.80 (96.9)	3.53 (75.6)	0.99 (350)	2.77 (85.9)
Basic Slag	3.36 (74.0)	4.02 (100)	1.45 (559)	2.94 (97.3)
Rock Phosphate	2.83 (46.6)	3.16 (57.2)	1.10 (400)	2.36 (58.4)

Potassium had little effect initially, and nitrogen depressed clover on all sites. In the absence of fertilizer N, clover produced an additional 1.65t DM/ha on *Molinia* and 1.52 t/ha on *Nardus/Juncus* swards. TSP gave similar overall responses to slag, TSP being slightly better on *Molinia* and slag on *Nardus,* both of these yielding around 4t/ha/annum after 3-4 years. Rock phosphate was inferior in the early years, but by year 4 it had surpassed TSP on *Calluna.*

Table 3. Mean Percentage clover covers with different phosphates after 3-4 years. (Basal lime, K and clover, but no N- percentage increases over controls in brackets)

Phosphate source	*Molinia*	*Nardus/Juncus*	*Calluna*	Overall
TSP	51.7(45.0)	51.6(51.6)	35.0(35.0)	46.1(43.9)
Basic Slag	45.5(38.8)	57.7(57.7)	35.9(35.9)	46.3(44.1)
Rock Phosphate	30.9(24.3)	34.2(34.2)	21.7(21.7)	28.9(26.7)

White Clover. Mean clover covers for TSP and slag were similar overall at 46%, but rock phosphate resulted in only 29% clover overall (Table 3). Because of the depressive effect of N on clover, the responses to lime and phosphate were much greater in terms of both yield and % clover cover, when N fertilizer was absent (Tables 1 and 4). Clover was still present when the sites were revisited 26 years after treatments were applied, and 20 years after abandonment to sheep grazing. On some treatments clover still contributed over 20% of cover, but some reversion to the original vegetation had occurred, this being least and clover persistence greatest with the higher rates of lime and P applied in 1975/76.

Table 4. Mean effects of nutrients at sowing on percentage clover cover 3-4 years after oversowing

Establishment fertilizer	*Molinia*	*Nardus/Juncus*	*Calluna*	Overall
12t Lime (basal P, K&N)	+10.8	+25.0	+13.7	+16.5
12t Lime (basal P*K no N)	+31.9	+52.1	+35.5	+39.8
P90 (basal L, K& N)	+23.9	+19.6	+19.2	+21.1
P90*(basal L, K, no N)	+42.0	+54.6	+35.5	+44.0
N90 annually(basal LKP)	-11.0	-13.2	- 6.1	- 10.1
K90 (basal L, P & N)	+ 3.7	+ 2.2	+ 3.1	+ 3.0

*Means of TSP and slag (8 plots). Other effects are means of 24 or 36 plots.

FARM APPLICATIONS FOR SUSTAINABLE ORGANIC PRODUCTION

The above findings have been applied on a Scottish Borders farm, Edgerston Tofts, Jedburgh, over the past 20 years.

Nutrient requirements for oversowing clover

Lime rate. As 12 t lime/ha gave greater DM, clover cover and clover yield than 6t, 7.5t /ha has been used initially in practice, followed 3-4 years later by a further 6.0-7.5t /ha to improve clover spread. Magnesian lime has been used, this reducing deaths of cattle and sheep due to magnesium deficiency in herbage.

Phosphate. TSP is prohibited in organic systems. Basic slag is unavailable and Steelworks slag has not been appraised in this situation. However RP, though not generally as effective as TSP or slag, can still be beneficial, even where lime is essential for clover establishment. In other situations, where a small amount of soluble phosphate, say 30kg P_2O_5 /ha could be applied initially, RP would probably be as effective in the longer term. One application of 90kg P_2O_5/ha as TSP or slag was adequate for the first 4 years, but double this rate of RP may enable a similar early effect. Enhanced microbial activity with liming can result in greater availability of soil P, as shown by Floate *et al.* (1981).

Nitrogen and Potassium. Fertilizer N, even in an N deficient hill situation, is unnecessary to establish productive clover-rich swards. Indeed, N90 was clearly detrimental to clover establishment and productivity, as Younie *et al.* (1984) also found. Its value in hill swards where enough clover has been sown with adequate lime, P and K, can only be to encourage grasses and then only before clover is established. Potassium application has generally proved unnecessary for clover establishment on grazed hill and upland swards, except on heather dominant podzols (Table 1) and on deep peat (Reith *et al.,*1973).

Pasture species

Seeds. Sowing sufficient clover is important in establishing clover-rich swards. With the 4kg/ha sown in Yorkshire, the small-leaved cultivar S184 persisted for over 20 years. Also, in each limed plot there was a spread of indigenous grasses like Yorkshire fog. In practice, 2.5kg/ha has proved satisfactory in establishing swards with 20-30% clover, but 2.5kg/ha of timothy and perennial ryegrass in addition has helped to replace some *Molinia* and *Nardus*.

Weeds. Liming hill ground encourages thistles and, on peaty gleys, rushes also. Rushes can thrive in clover-rich swards and if not topped regularly may become prevalent. In the longer term, liming may also encourage earthworms and moles so that drainage is improved and rushes reduced (Crompton, 1960). However, some form of control of thistles may need to be developed, especially where topping is difficult.

DISCUSSION

Oversowing white clover on hill swards

In situations where agricultural improvement of hill swards is permitted, the following procedures for oversowing white clover are suggested. When land is fairly dry, apply 7.5t lime /ha. Graze the sward hard, preferably with cattle. The following April/May or mid July/August, apply 180kg P_2O_5/ha as rock phosphate. Broadcast this mixed with 2.5kg/ha each of S184 clover, timothy and ryegrass or oversow the seeds separately, ideally when the soil is moist and more rain forecast. This package costs about £180/ha in 2003 After 3-4 years a further 6.0-7.5t lime/ha is advised, to encourage clover spread.

Establishing clover-rich meadows for organic silage or hay

The amount of fodder conserved is a major factor determining the livestock carried on hill and upland farms. A few years after implementing the above improvement programme at Edgerston Tofts, Jedburgh, some less steep areas of *Molinia* and *Nardus* were cut for big-bale silage or hay. While these gave fibrous fodder, it often contained about 20% of DM as clover and was eaten readily by suckler cows. Overseeding clover-deficient in-bye meadows, usually with just 2.5kg /ha S184 clover but similar levels of lime, P and K, has also increased silage production.

After establishing clover-rich hill swards, some areas have been ploughed or rotavated, fenced and a rape crop grown. The second application of 6.0-7.5t lime/ha was applied after ploughing, followed by 90kg P_2O_5/ha as TSP, 90kgK_2O/ha and 120kgN/ha. Now, under organic production, the N is omitted, the phosphate given as RP and K is applied only when deficient. The key to good organic rape crops is ploughing out clover-rich swards, but liming the seedbed also favours N release. After grazing such crops with lambs, the subsequent ploughing is given a further 6.0-7.5t lime/ha and 90 kg P_2O_5/ha. This means that 22.5t lime/ha may be applied over about 6 years. The seed mixture should contain 4 kg/ha of clover, with at least 1kg small-leaved white clover and the rest medium-leafed varieties. About 20kg of mainly late perennial ryegrasses and 6 kg/ha timothy should be sown. Ploughing facilitates deeper nutrient-enrichment, thus encouraging better rooting. Such clover-rich swards can sustain good silage crops for many years without reseeding, provided lime, P and K are maintained and N fertilizer avoided.

REFERENCES

ANDERSON G.D. (1979) Effects of lime and fertilizers on establishment of white clover on 3 types of Pennine moorland. *Journal of Science Food and Agriculture*, 30, 336.

ANDERSON G.D. (2001)Changes in the productivity of 3 types of Pennine moorland effected by oversowing lime, NPK and clover. *Proceedings British Society of Soil Science, Soils and the Uplands*. Durham. (Abstract).

CROMPTON E. (1960) Soil features associated with the development of good permanent pasture in moorland areas. *Proceedings of the 8th International Grassland Congress, Reading*. 257-260.

FLOATE M.J.S., HETHRINGTON R.A., COMMON T.G., EADIE J. and IRONSIDE A.D. (1981) Long-term responses of a range of grazed hill pasture types to improvement procedures. 2. Nutrient cycling and soil changes. In: Frame J. (ed) *The Effective Use of Forage and Animal Resources in the Hills and Uplands*. British Grassland Society Occasional Symposium No. 12, 147-149.

REITH J.W.S., ROBERTSON R.A. and INKSON R.H.E. (1973) Nutrient requirements of herbage on deep acid peat. *Journal of Agricultural Science, Cambridge*, 80, 425-434.

YOUNIE D., WILSON J.F., CARR G. and WATT C.W. (1984) The effect of undersowing, N and date of sowing on white clover establishment. In: D.J. Thomson (ed.) *Forage Legumes*. British Grassland Society Occasional Symposium No.16, 182-183.

Towards an Organic System for Cattle Fly Control: a Push and a Pull?

MICHAEL A. BIRKETT
Biological Chemistry Division, Rothamsted Research, Harpenden, Herts., AL5 2JQ, UK

ABSTRACT

The role of volatile semiochemicals in mediating the location and selection of Holstein-Friesian heifers by nuisance and disease-transmitting cattle flies was investigated. Using volatile extracts collected by air entrainment from heifers, a number of active peaks were located by coupled GC-electrophysiology (GC-EAG) for *Musca autumnalis* and *Haematobia irritans*. In total, 18 compounds were identified by coupled GC-MS. Of these, 6-methyl-5-hepten-2-one, when applied at physiologically relevant levels to low and high fly-loading heifers in a small herd, reduced fly-loads on these individuals and the difference in fly-load within the whole herds. This study is the first report on the identification and use of volatile semiochemicals to reduce fly-loads on individuals in the field, and provides, for the first time, evidence for the hypothesis that differential attractiveness within a host species is, in part, due to volatile semiochemicals emitted from the host.

INTRODUCTION

The behaviour of nuisance and disease-transmitting cattle flies (Diptera) that settle or feed on grazing hosts leads to increased disease incidence, reproductive failure and also reduced meat and milk yields, with significant economic losses arising as a consequence (Fraser and Broom, 1990). Current cattle fly control is mainly obtained through the use of broad-spectrum toxicant insecticides, in the form of ear-tags, pour-ons, dips, sprays, etc. However, concerns over the build-up of insecticide resistance, and public concern over the possible environmental impact of such chemicals requires alternative control methods to be sought. One such approach is to employ volatile semiochemicals (behaviour-modifying chemicals) e.g. attractants or repellents. It has been established that herbivorous insects avoid energy-wasting visits to feed on, or colonize plants on which they cannot successfully develop, by responding to volatile semiochemicals emitted by unsuitable hosts (Pickett *et al.*, 1998). For haematophagous insects, it has been demonstrated for cattle flies that individual hosts vary in their attractiveness (Thomas *et al.*, 1987). The hypothesis that has been developed to explain this differential attractiveness is that all heifers release volatile attractants, but those less-attractive individuals release additional compounds that interfere with the activity of the attractants. These compounds at higher levels act as true repellents but at lower, or at the natural physiological levels, effectively mask the normal attractiveness of the host species (Pickett et al., 1998). Such compounds, once identified, could be used in novel fly control strategies, e.g. the SDDS, or "push-pull" strategy (Miller and Cowles, 1990). In this study, volatile semiochemcials were collected from heifers known to differ in their attractiveness, and compared by coupled gas chromatography-electrophysiology (GC-EAG), using the face fly, *Musca autumnalis*, and the horn fly, *Haematobia irritans,* as model insects.

MATERIALS AND METHODS

Volatile collection. Volatiles from Holstein-Friesian cattle were collected by drawing the air surrounding the animals through Porapak tubes (50 mg per tube) over a period of several hours. The animals selected were based in the Netherlands and Denmark. The animals selected in the Netherlands displayed significantly higher fly-loads when compared to other animals in the same herd. In Denmark, animals shown in a preceding study (Jensen *et al.*, 2004) to have very high and low fly-loads (heifers 1257 and 1270 respectively) were selected. Air entrainments of clean empty stalls were carried out as controls. Porapak tubes containing trapped volatiles were eluted with diethyl ether and samples stored at -20°C until further use.

Insects. The face fly, *M. autumnalis*, and the horn fly, *Ha. irritans,* were reared at the Danish Pest Infestation Laboratory (DPIL), Kgs. Lyngby, Denmark, and sent to the United Kingdom as pupae for coupled GC-EAG studies. Flies were kept in controlled environment cabinets at 20°C, L16:D8 until required. The emerged female flies were fed with water prior to use.

Coupled GC-electrophysiology (GC-EAG). The coupled GC-electrophysiology system, in which the effluent from the GC capillary column is delivered simultaneously to the antennal preparation and the GC detector has been described elsewhere (Wadhams, 1990). Separation of air entrainment samples was achieved on an AI93 GC (AI, Cambridge, UK) equipped with a cold on-column injector and a flame-ionization detector (FID). Two columns were used. The non-polar column (50 m x 0.32 mm i.d., HP-1) was maintained at 40°C for 1 minute, then 5°/minute to 100°C, then 10°/minute to 250°C. The polar column (30 m x 0.3 mm i.d., BP-20) was maintained at 40°C for 2 minutes, then 10°/minute to 225°C. The carrier gas was hydrogen. The outputs from the EAG amplifier and the FID were monitored simultaneously on a chart recorder.

Coupled GC-mass spectrometry (GC-MS). A Hewlett-Packard 5890GC equipped with a non-polar GC column (50 m x 0.32 mm i.d., HP-1), and a cold on-column injector, was connected to a VG Autospec mass spectrometer (Fisons, Manchester, UK). Ionization was by electron impact at 70eV, 250°C. The GC was maintained at 30°C for 5 minutes, then programmed at 5°C/minute to 250°C. Compounds tentatively identified by GC-MS were confirmed by co-injection of authentic samples on GC using non-polar and polar columns.

Field study. A field trial was conducted in Denmark using a small herd (n = 7) of Holstein-Friesian heifers. 6-Methyl-5-hepten-2-one was released from slow release formulations (release rate = 14 mg/day) comprising polyethylene bag sachets placed in protective metal dispensers. On day 1, the flies on each animal were identified and the number counted every 0.5 hr, altogether 4-5 times, and the data used to rank the animals according to fly load. Dispensers, without slow release formulations, were hung around the necks of the two least and two most fly attractive individuals to allow the animals to become accustomed to the devices. On the morning of day 2, the metal dispensers on the two least attractive cattle were loaded with slow release formulations, and the cattle given an hour to settle before the counting protocol was initiated. At the end of day 2, the formulations were removed from the dispensers and returned to the freezer. On day 3, the same counting procedure was carried out but with the formulations applied to the metal dispensers on the two most attractive cattle.

RESULTS AND DISCUSSION

Coupled GC-EAG using recordings from *M. autumnalis* and *Ha. irritans* antennae located a number of electrophysiologically active compounds in the samples of cattle volatiles, which were identified tentatively by coupled GC-MS and confirmed by peak enhancement using authentic standards (Table 1).

Table 1. Electrophysiologically active compounds identified from cattle volatiles using recordings from *Musca autumnalis* and *Haematobia irritans* antennae.

Class of compound	Compound
Aromatic hydrocarbon	Propylbenzene Naphthalene Acenaphthene Styrene
Polar aromatic	Phenol *o*-Cresol *m*-Cresol *p*-Cresol 4-Methyl-2-nitrophenol
Fatty acid derivative	(*Z*)-3-Hexen-1-ol 1-Octen-3-ol[a] 2-Heptanone Propyl butanoate
Aliphatic hydrocarbon	Decane Undecane
Isoprenoid	α-Pinene Camphene[a]
Isoprenoid derivative	6-Methyl-5-hepten-2-one

[a]Stereochemistry undefined.

In a field study using a small herd of cattle in Denmark, 6-methyl-5-hepten-2-one appeared to act as a repellent (Figure 1). When applied to the two least attractive heifers (A and C) on day 2, the fly-loads on these animals were reduced even further. Application of the compound to the two most attractive heifers (F and G) on day 3 resulted in an almost instantaneous reduction in the fly-loads on these animals (individual F, reduction from 48% to 11%; individual G, reduction from 35% to 21%), and reduced the differences in fly-load amongst the whole herd.

This study is the first report on the identification and use of volatile semiochemicals to reduce fly-loads on individual heifers in the field, and provides, for the first time, evidence for the hypothesis that differential attractiveness within a host species is, in part, due to volatile semiochemicals emitted from the host. It also demonstrates the potential of developing a semiochemical-based push-pull approach for cattle fly control.

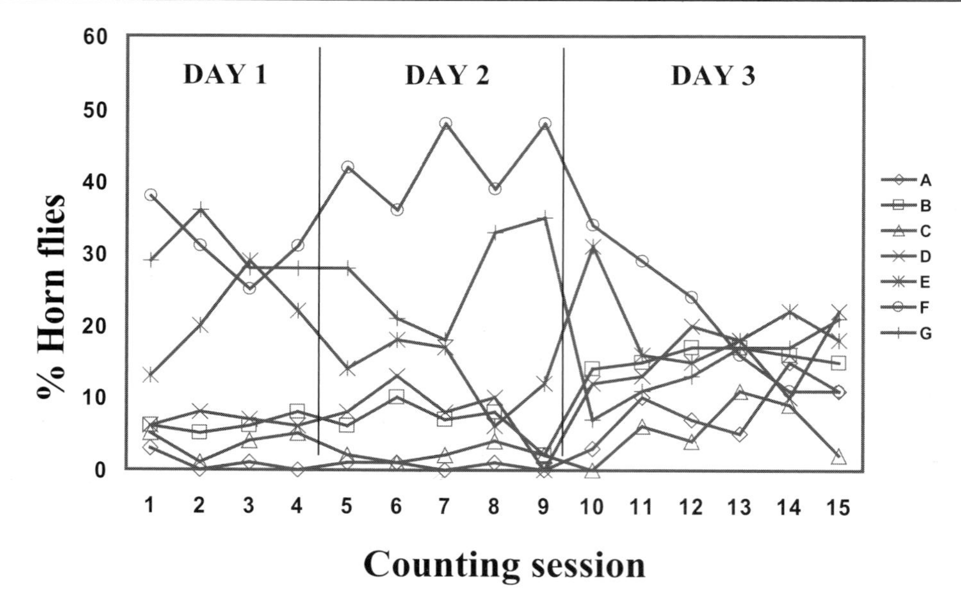

Figure 1. The percentage of horn flies distributed on a herd of Holstein-Friesian cattle (n = 7) in response to the addition of slow-release sachet formulations of 6-methyl-5-hepten-2-one (release rate = 14 mgs/day). Labels A-G denote individual heifers.

ACKNOWLEDGEMENTS

This work was supported by the European Commission (CEC Contract Mo. AIR3-CT 93-1445). Rothamsted Research receives grant-aided support from the UK BBSRC.

REFERENCES

FRASER A.F. and BROOM D.M. (1990) *Farm Animal Behaviour and Welfare*. London: Bailliere Tindall.

JENSEN K.-M.V., JESPERSEN J.B., BIRKETT M.A., PICKETT J.A., THOMAS G. and WADHAMS L. (2004) Variation in the load of the horn fly, *Haematobia irritans* (Diptera: Muscidae) in cattle herds is determined by the presence or absence of individuals. *Medical and Veterinary Entomology*, submitted.

MILLER J.R. and COWLES R.S. (1990). Stimulo-deterrent diversion: a concept and its possible application to onion maggot control. *Journal of Chemical Ecology*, 16, 1367-1382.

PICKETT J.A., WADHAMS L.J. and WOODCOCK C.M. (1998) Insect supersense: mate and host location by insects as model systems for exploiting olfactory interactions. *The Biochemist*, August, 8-13.

THOMAS G., PRIJS H.J. and TRAPMAN J.J. (1987) Factors contributing to differential risk between heifers in contracting summer mastitis. In: Thomas, G., Over, H.J., Vecht, U. and Nansen, P. (Eds). *Summer Mastitis* pp. 30-39. Dordrecht, Boston and Lancaster: Martinus Nijhof.

WADHAMS L.J. (1990) The use of coupled gas chromatography: electrophysiological techniques in the identification of insect pheromones. In: McCaffery A.R. & Wilson I.D. (Eds). *Chromatography and Isolation of Insect Hormones and Pheromones*, pp. 289-298. New York: Plenum Press.

Organic Forage Crop Production in Atlantic Zones: an Ecophysiological Approach

N. PEDROL and A. MARTÍNEZ
Servicio Regional de Investigación y Desarrollo Agroalimentario (SERIDA).
Estación Experimental 'La Mata'. Apdo 13. 33820 Grado (Asturias), Spain.

ABSTRACT

Within a schedule of long-term organic forage crop rotations, two mixed stands of ryegrass-clover were evaluated as following crops after winter cereal-legume mixtures under different sowing and tillage methods in Asturias (N. Spain). Significant differences were found in productivity, sward components, chemical composition and in growth dynamics. The crop sequence perennial ryegrass-white clover established under a triticale-field bean canopy produced an additional 6 t DM/ha, compared with oat-vetch followed by Italian ryegrass-red clover under regular post-harvest sowing, thus being a promising method for improving organic forage productions with minimum tillage. Changes in productivity and feeding values observed for both these ryegrass-clover swards along different growth periods were studied together with soil nutrient evolution and seasonal climatic fluctuations. Correlation analysis showed ecophysiological links among crop, soil and climate variables.

INTRODUCTION

With the disadvantages of small holding size, organic farmers in the Atlantic zones of Spain face a great challenge of self-sufficiency, in order to reduce as much as possible external fodder purchase. Thus, new strategies to maximize forage production are required that are not detrimental to sustainability. Research in organic farming (OF) is demanding more holistic approaches, by linking as many components of the agroecosystem as possible (Isart and Llerena, 1999). Useful agronomic results obtained from OF studies may be considered from a wider ecophysiological perspective, by taking into account the complex interactions among crop growth, soil and climate (Lambers *et al.*, 1998). In this work, we tested minimum tillage methods to improve organic forage productions in the Atlantic zones and propose an ecophysiological approach to the growth evolution of ryegrass-clover swards.

MATHERIALS AND METHODS

Two forage crop sequences were started on a mixed meadow site at Llanera, Asturias, N. Spain (46°26'N, 5°50'W) on 30 October 2001. Based on previous soil analysis, tillage and amendment was in accordance with the organic management rules (commercial organic fertilizer, N:P:K 42:56:105; dolomite, Ca:Mg 883:494). Sequence 1 was started with a mixed crop of oats (O) (cv. Prevision)/vetch (V) (cv. Acisreina) (OV 3:2, 265 seeds/m^2) followed by a mixture of Italian ryegrass (Ir) (cv. Serenade)/red clover (Rc) (cv. Marino) (IrRc 5:4, 45 kg/ha) sown on 16 May 2002 after harvesting the cereal/legume crop. For sequence 2, an open mixed stand of triticale (T) (cv. Senatrit)/field bean (B) (cv. Rutabon) (TB 6:1, 195 seeds/m^2) was sown simultaneously with a mixture of perennial ryegrass (Pr) (cv. Tove)/white clover (Wc) (cv. Huia) (PrWc 6:1, 40 kg/ha), so that the grass/clover sward was already established once the cereal/legume crops were harvested on 30 April 2002. Each rotation was replicated four times in plots of 300 m^2.

Different cuts corresponding to subsequent growth periods (GP) were obtained from each grass/clover sward until late April 2003. For IrRc the cutting dates were 9 November 2002 (GP2, summer), 11 December 2002 (GP3, autumn), and 29 April 2003 (GP4, winter-spring).

The PrWc was mown on the same dates plus an extra early cut on 8 July 2002 (GP1, late spring-early summer). Following soil nutrient-status analyses (3 May 2002 and 27 February 2003), organic fertilizer (N:P:K 25:33:63) was applied to both rotations on 4 March 2003, during GP4. Climatic variables (Tmin, Tmax, Tmed and rainfall) were measured daily along all growth periods. For each sward and cut, growth rate (GR), yield and its components (ryegrass, clover, other species, and dead matter) were obtained on a DM basis. Subsamples were analysed for chemical composition: crude protein (CP; TECATOR, 1995), water soluble carbohydrates (WSC; Hoffman, 1937), neutral-detergent fibre (NDF; Van Soest *et al.*, 1991), and acid-detergent fibre (ADF; Goering and Van Soest, 1970). Evolution of parameters within mixtures along growth periods were statistically analysed by one-way ANOVA and LSD-test. Total yields were compared by t-test. Effects of mixtures, common growth periods (GP2 to GP4), and their interactions were tested by two-way ANOVA. Correlations among all measured plant, soil and climate variables were analysed by Pearson's coefficient. Chemical composition and yields were also measured for initial cereal/legume mixed crops and compared by t-test.

RESULTS AND DISCUSSION

For cereal/legume mixed crops, TB stands were significantly more productive than OV (9600 vs. 8320 kg DM/ha; $P<0.05$). The success of TB mixtures in previous work has been based on the early growth and high competitive ability of field beans compared with other forage legumes, resulting in TB organic productions that compete in yield and quality with conventional ones (Pedrol and Martínez, 2003). Both mixed crops did not show significant differences in chemical composition (data not shown).

From the sum of yields taken at subsequent cuts from ryegrass/clover swards, total yields for PrWc were 1.7 times higher than for IrRc, (11113 *vs.* 6502 kg DM/ha; $P<0.001$). This was attributed to the contribution of a productive cut of PrWc in early summer (GP1, Table 1) obtained after an excellent establishment of the mixture under the open canopy of TB, which protects seedlings from chilling, imposing minimum shading. In this first growth period PrWc had the most intense growth rate, the highest ryegrass and fibre contents, and a good WSC balance (Table 1). Undersowing the organic sward with a winter cereal/legume crop is shown to be a good strategy to encourage grass growth in the forthcoming spring-summer.

The two-way ANOVA obtained for the subsequent common cuts (Table 1) illustrates the differences in yield parameters and chemical composition between PrWc and IrRc mixtures as a whole. Growth period (cGP) had highly significant effects on independent variables, but seasonal differences in T and rainfall did not covariate with all of them (Table 2). For example, the relative contribution to yield of ryegrass and dead matter was not correlated with the climatic conditions of each GP, but covariated significantly in different directions with changes in soil fertility. In both of the grass/clover swards, ryegrass percentage decreased with litter increase (Table 1) probably due to progressive depletion of N and K, whereas this tendency changed notably after fertilization (GP4). CP contents were, surprisingly, negatively correlated with soil N and K, and were not correlated with clover contents (Table 2). Regarding the highly significant positive correlation with litter percentage, the maintenance or even the rise in CP contents was probably due to resorption from litter. The nitrogen mainly invested in the photosynthetic apparatus is exported via phloem from senescent tissues (as amino acids or active protein pools) to growing tissues. Nutrient resorption is particularly efficient in grasses such as ryegrass, which have active growth of new leaves (Aerts, 1996), especially under limited nitrogen supply, e.g. some organic swards (Culleton *et al.,* 2002). Of course, soil N availability in each GP was also positively correlated with yield and GR (Table 2).

Table 1. *Above:* Evolution of yield, components, and chemical composition in subsequent growth periods (GP) of two organic swards: perennial ryegrass/white clover (PrWc) and Italian ryegrass/red clover (IrRc). For each mixture asterisks denote significant differences among growth periods (one-way ANOVA) and mean values (n=4) labelled with distinct letters are significantly different (P<0.05; LSD-Test). *Below:* Two-way ANOVA of the effects of mixtures and their common growth periods (cGP). ***P<0.001, **P<0.01, *P<0.05, NS (not significant) P>0.05. [All values refer to DM, except DM % (over FM)].

		DM	Yield	rye	clo	others	litter	GR	CP	WSC	NDF	ADF
sward	**GP**	%	kg/ha	%	%	%	%	kg/ha·day	%	%	%	%
PrWc	1	20.2^{a}	4352^{a}	67.0^{a}	3.8^{a}	26.3^{a}	3.3^{a}	63.1^{a}	11.0^{a}	12.7^{a}	58.6^{a}	31.8^{a}
	2	11.4^{b}	1151^{b}	37.8^{b}	13.5^{b}	34.3^{a}	14.5^{b}	17.7^{b}	17.5^{b}	6.3^{b}	53.1^{b}	31.7^{a}
	3	16.0^{c}	1629^{b}	24.5^{b}	1.8^{a}	49.5^{a}	23.5^{c}	17.9^{b}	17.7^{b}	14.0^{a}	46.3^{c}	27.0^{b}
	4	17.9^{c}	3981^{a}	36.5^{b}	15.0^{b}	39.5^{a}	8.8^{ab}	29.7^{c}	14.5^{c}	12.2^{a}	49.8^{bc}	26.9^{b}
		***	***	**	*	NS	***	***	***	**	***	**
IrRc	2	10.9^{a}	1000^{a}	39.8^{a}	9.0^{a}	38.8^{a}	12.5^{a}	8.5^{a}	16.7^{a}	6.5^{a}	51.4^{a}	31.1^{a}
	3	15.5^{b}	1361^{a}	17.0^{b}	9.3^{a}	40.5^{a}	32.8^{b}	15.0^{b}	17.8^{a}	11.7^{b}	44.7^{b}	24.7^{b}
	4	17.7^{c}	4142^{b}	29.3^{ab}	36.8^{b}	25.3^{a}	8.3^{a}	30.9^{c}	17.3^{a}	11.5^{b}	47.0^{b}	24.1^{b}
		***	***	*	***	NS	***	***	NS	***	**	***
M		NS	NS	NS	*	NS	NS	*	NS	NS	*	*
cGP		***	***	**	***	NS	***	***	*	***	***	***
M×cGP		NS	NS	NS	*	NS	*	*	*	NS	NS	NS

Legend: rye, ryegrass; clo, clover; others, other species; litter, dead matter; GR, growth rate; CP, crude protein; WSC, water-soluble carbohydrate; NDF, neutral-detergent fibre; ADF, acid-detergent fibre.

Clover contribution to total yield was significantly superior in IrRc, achieving values of optimal establishment of red clover in GP4, after fertilization (Table 1, above). This is a clear indicator of a well-developed tap root system during the preceding growth periods, which allows a fortified regrowth after winter (Culleton *et al.*, 2002). The significant loss of white clover in GP3 could be due to the reduction of soil K levels (Baars and Younie, 1998), but soil K contents and depletion (from 72 to 43 mg/l) were almost identical for both swards. The % of "others" was negatively correlated with ryegrass and clover contents (Table 2). Although total amount of weeds did not change significantly with soil fertility and seasonal climatic fluctuations (Tables 1 and 2), the relative contribution of different species differed between swards and changed along subsequent growth periods. White clover growth in GP3 (autumn) was restricted by the increased presence of the highly competitive Agrostis and Yorkshire fog (components of "others"), whereas this fraction in IrRc was mainly composed of the non-aggressive plantain.

Table 2. Correlation analysis of yield-, chemical composition-, climate- and soil fertility related variables measured in each subsequent growth periods of two organic ryegrass/clover swards. Asterisks denote the significance level of positive (+) or negative (–) bivariate correlations tested by Pearson's coefficient (n=28): *** $P<0.001$, ** $P<0.01$, * $P<0.05$, NS (not significant) at $P>0.05$.

Variable	DM	Yield	rye	clo	others	litter	GR	CP	WSC	NDF	ADF
Yield	***+										
rye	NS	*+									
clo	NS	NS	NS								
others	NS	*–	***–	*–							
litter	NS	***–	***–	NS	NS						
GR	***+	***+	***+	NS	*–	**–					
CP	***–	***–	***–	NS	NS	***+	***–				
WSC	***+	**+	NS	NS	NS	NS	*+	NS			
NDF	NS	NS	***+	NS	*–	*–	***+	***–	NS		
ADF	NS	NS	***+	*–	NS	NS	NS	*–	*–	***+	
T	***–	***–	NS	*–	NS	NS	NS	NS	***–	*+	***+
Rainfall	NS	*+	NS	*+	NS	NS	NS	NS	NS	NS	*–
N	NS	*+	*+	NS	NS	***–	*+	***–	NS	***+	***+
P	NS	NS	NS	NS	NS	NS	NS	NS	NS	NS	NS
K	NS	NS	***+	NS	NS	***–	NS	**–	NS	***+	***+

T and Rainfall, mean temperature (°C) and accumulated rainfall (mm) of each growth period. Others as listed in Table 1.

Cuts taken after summer growth (GP2, Table 1) were the least productive, having the lowest DM and WSC contents but high fibre proportions; all these parameters showed significant correlations with T (Table 2). As a response to high temperatures in summer, stomatal closure is induced to avoid excessive water loss by transpiration, thus leaf expansion and plant growth are reduced. Due to the compromise between carbon gain and water loss, diminished net photosynthetic rates lead to a decrease in WSC concentrations, whereas reduced growth leads to C storage and deposition in cell walls as reflected in higher fibre contents (Table 1).

From our results, we conclude that the establishment of the perennial ryegrass-white clover sward together with the starting crop triticale-field bean is a promising, low-cost method to improve organic forage production. In this trial we obtained an additional 6 t DM/ha, compared with oat-vetch followed by Italian ryegrass-red clover under regular post-harvest sowing. Besides being more productive, the system has faster work rates, requires minimum tillage and reduces crop establishment costs. While these are early results from a long-term trial, some of the ecophysiological processes that underlie our field agronomic observations can be explained. From consideration of correlations among some plant, soil and climate parameters, it is possible to achieve a better understanding of the crop performance in a given agroecosystem.

ACKNOWLEDGEMENT
Reseach work was supported by the INIA project RTA01-144-C5-2. The authors wish to thank the team from "La Mata" for their essential field work assistance.

REFERENCES
AERTS R. (1996) Nutrient resorption from senescing leaves of perennials: are there general patterns? *Journal of Ecology*, 84, 597-608.
BAARS T. and YOUNIE D. (1998) Grassland and choices for sustainability. *FAO/REUR meeting*. La Coruña, Spain.
CULLETON N., BARRY P., FOX R., SCHULTE R. and FINN J. (Eds) (2002) *Principles of Successful Organic Farming*. NDP- AFDA, Teagasc, Dublin, Ireland.
GOERING H.K. and VAN SOEST P.J. (1970) Forage fiber analysis. *USDA, ARS Agriculture Handbook* 379, pp 1-12. Washington DC, USA.
HOFFMAN W.S. (1937) A rapid photoelectric method for the determination of glucose in blood and urine. *Journal of Biological Chemistry*, 120, 51-55.
ISART J. and LLENERA J.J. (Eds) (1999) *Organic Farming Research in the UE, Towards 21st Century*. ENOF White Book . LEAAM-Agroecología (CID-CSIC). Barcelona, Spain.
LAMBERS H., CHAPIN III F.S., PONS T.L. (1998) *Plant Physiological Ecology*. Springer-Verlang, NY, USA.
PEDROL N. and MARTÍNEZ A. (2003) Asociaciones cereal-leguminosa en rotaciones ecológicas forrajeras de zonas húmedas. [Mixed cropping of cereals and legumes in organic forage crop rotations from Atlantic zones]. In: Robles A.B. *et al.* (Eds) *Proc. XLIII Reunión Científica de la SEEP*, pp 131-136. Spain.
TECATOR (1995) The determination of nitrogen according to Kjeldahl using block digestion and steam distillation. *Perstop Analytical*. Application note, AN 300.
VAN SOEST P.J., ROBERTSON J.B. and LEWIS B.A. (1991) Methods for dietary fiber, neutral detergent fiber and no starch polysaccharides in relation to animal nutrition. *Journal of Dairy Science*, 74, 3583-3597.

Vaccine Use in Organic Cattle and Sheep Systems – DESTVAC: a Decision Support Tool based on Qualitative Risk Assessment

M. HOVI[1], D. GRAY[2], S.M. RUSBRIDGE[2] and K. CHANNA[1]
[1]VEERU, Dept. of Agric., University of Reading. PO Box 237, Reading RG6 7AR, UK
[2]SAC Veterinary Services, Craibstone, Bucksburn, Aberdeen AB21 9TB, UK

ABSTRACT
A project is described for the development of a web-based decision support tool on vaccine use for organic dairy, beef and sheep farms. The project focused on 24 diseases which cattle and sheep in the UK are routinely vaccinated against. The tool allows for specific exploration of risk factors and risk management measures, information on vaccines, farm assessments and scenario building for submissions to certification bodies. The lack of evidence for national disease prevalence/incidence data was considered to be a major hindrance to the development of evidence-based decision making on vaccine use on organic farms.

INTRODUCTION
The UK national organic standards (UKROFS, 2001) state that statutory health plans for organic farms must "allow for the evolution of a farming system progressively less dependant on allopathic veterinary medicinal products". While vaccines are classified as allopathic veterinary medicinal products, the UKROFS standards permit their use "where there is a known disease risk". In practice, most of the organic sector bodies require justification of vaccine use either in a written health plan or by a separate application. The definition of a "known disease risk" has, however, caused problems both in situations where veterinary opinion has not been considered adequate by the certification body, or where a producer has decided to discontinue all vaccine use without implementing any other disease control measures. This has occasionally led to accusations of poor welfare on organic farms as a result of disease outbreaks that could potentially have been prevented by immunization (Anon, 2001). Certification bodies have struggled to find a balance between "routine" vaccine use and farm system redesign that would lead to reduced dependency on vaccines, without jeopardizing animal welfare (Chris Atkinson, personal communication). The certification bodies also find it difficult to assess vaccine-use decisions taken by producers in the absence of information on the potential risk factors relevant to each farm. This information is seldom provided by the farmer's veterinary practitioner when justifying the need for vaccine use.

The aim of the project reported here was to support the development of a consistent approach to decision making on vaccine use among organic dairy, beef and sheep farmers and their advisers, including their veterinarians, by developing a decision support tool on vaccine use.

MATERIALS AND METHODS
Scope of the project
The project focused on twenty-four diseases that are routinely vaccinated against in UK cattle herds and sheep flocks.

Risk assessment

Evidence-based risk assessment for vaccine use and immunization policies is often difficult at farm level. Local disease situations may not be well known; traditions and financial constraints are often the main determinants in shaping these policies (Barrett, 2001). Poor understanding and implementation of herd and flock health security (biosecurity) on British cattle and sheep farms, as recognized by the DEFRA SCG (Anon, 2000a; Anon 2000b), further complicates an objective risk assessment exercise on immunization. There are, however, existing models of risk assessment on farms in relation to vaccine use and disease control on cattle and sheep farms, in the form of commercial animal health accreditation schemes (e.g. Premium Cattle Health Scheme, Maedi Visna Accreditation Scheme). Quantifiable risk assessment and risk modelling are also commonly used to assess the need for statutory immunization or prophylaxis policies (Morris *et al.*, 2001).

The project was initiated by an assessment of the VIDA database in order to determine whether there was scope for quantifiable risk assessment, regionally or nationally, in the case of any of the diseases listed. It was concluded that the absence of denominator data on the VIDA database would make it impossible to introduce quantifiable risk assessment into the tool by using national data. Literature sources of quantifiable epidemiological data on the listed diseases in the UK were reviewed. It was concluded that due to the variable quality and nature of the existing data, this information would be presented as additional decision support information in the tool, rather than part of the risk assessment. A short review of existing information was created, and in cases where this was possible, this information page was linked to an existing, DEFRA-funded decision support tool on the web: Compendium for Animal Health and Welfare in Organic Farming (www.organic-vet.reading.ac.uk) and Economics of Livestock Diseases (www.reading.ac.uk/livestockdisea). Hazard characterization for the user was provided by a link to the existing DEFRA-funded support tool (www.organic-vet.reading.ac.uk).

In the absence of quantitative epidemiological data, it was decided to assess exposure at individual farm levels by qualitative farm-specific risk factor assessment. This was based on

- an extensive review of literature in relation to risk factors for each disease and creation of lists of risk factors for each disease;
- a review of disease specific lists of risk factors by animal health experts at the Scottish Agricultural College (SAC) for deletion/addition/modification and ranking (ranking from 1 to 3; 1 = serious risk, 2 = intermediate risk and 3 = minor risk);
- a workshop review of the risk factor lists and ranking by a group of experts, veterinary surgeons, farmers and representatives of vaccine manufacturers and certification bodies; and
- a final peer review of listings and rankings by the SAC experts.

Information for risk management at farm level was provided by a similar process whereby risk management measures were reviewed, ranked and peer reviewed.

Database creation

Using information from the risk factor and risk management measure assessment described above, an SQL Server -database was created. This database also included information on vaccines, hazard characterization and exposure.

Tool building
An interactive, web-based tool to query the SQL Server-database and to provide access to other additional information was built by using ASP.NET.

User assessment
User assessment was carried out during two workshops (November 2002 and April 2003). Feed-back from these was incorporated where appropriate.

RESULTS

The results have been built into a web-based (www.destvac.reading.ac.uk) decision support tool. This allows farm-specific exploration of risk factors and risk management measures, access to additional information on vaccines, exposure and financial impact of disease at farm level. The tool also allows the user to produce reports on individual farm assessments or exploratory scenario building for submission to certification bodies as part of a health plan.

CONCLUSIONS

It is evident that the lack of national disease prevalence/incidence data for all of the diseases included in this project is a major hindrance to the development of evidence-based decision making on vaccine use on organic farms. This absence of data obviously also reduces the value of any decision support tools offered. In the absence of quantitative analysis, the current tool has shifted the emphasis of decision making from assessment of exposure to risk reduction at farm level by system and husbandry redesign, in accordance with the organic principles. It is, therefore, believed that the tool will be useful both to organic farmers and their veterinary surgeons, and to the organic certification bodies, in moving the animal health management on organic farms from input substitution towards true system redesign.

ACKNOWLEDGEMENTS

The authors would like to recognize the contribution of all experts, workshop participants and the Scottish Organic Producers Association to this project. The project was funded by the Department of Environment Food and Rural Affairs (DEFRA).

REFERENCES

ANON (2000a) *Zoonotic infections in livestock and the risk to public health.* Proceedings of the first meeting organized by DEFRA, Edinburgh, 28 June 2000. (http://www.maff.gov.uk/animalh/diseases/vtec/index.htm)
ANON (2000b)*Zoonotic infections in livestock and the risk to public health.* Proceedings of the Second Meeting organized by DEFRA, London, 7 December 2000.
ANON (2001) *Veterinary Record*, 7 April 2001.
BARRETT D. C. (2001) Biosecurity and herd health – a challenge for the 21st century. *Cattle Practice*, 9:2, 97-103.
MORRIS R.S., WILESMITH J.W., STERN M.W., SANSON R. L. and STEVENSON M.A. (2001) Predictive spatial modelling of alternative control strategies for foot-and-mouth disease epidemic in Great Britain, 2001. *Veterinary Record* 149, 137-144.
UKROFS (2001) *Standards for organic livestock and organic livestock products*, p42.

Animal Welfare in Organic Systems – a Summary of Four Surveys on Standards Perception and Inspection Practice

M. HOVI[1], M. KOSSAIBATI[1], R. BENNETT[1], S. EDWARDS[2], J. ROBERTSON[3], S. RODERICK[4] and C. ATKINSON[5]

[1]The University of Reading, School of Agriculture, PO Box 234, Reading RG6 7AR, UK
[2]The University of Newcastle, Department of Agriculture
[3]The University of Aberdeen, Department of Agriculture
[4]Duchy College, Organic Studies Centre, Camborne, Cornwall
[5]The Scottish Organic Producers Association, Edinburgh

ABSTRACT

Potential animal welfare benefits and problems in organic livestock production were studied through surveys of inspectors, advisers and veterinarians, a workshop with certification bodies and interviews with producers. It was concluded that while existing organic standards did not require changes, there was a need for certification bodies to encourage better implementation of standards, for veterinary support on farms to be strengthened, and consideration to be given to formal welfare assessment at farm level.

INTRODUCTION

Several studies have confirmed that animal welfare is an important part of the consumer perception of organic livestock production (Harper and Henson, 1998; Holmberg, 2000). As organic livestock production struggles to widen its markets and secure further growth, it is becoming increasingly important to maintain consumer confidence in the high levels of animal welfare in organic production. There are several aspects of organic livestock production that have been recognized as potentially problematic with regard to animal welfare. A potential conflict has been seen between some of the other objectives of organic production and the welfare objective (Thamsborg *et al.*, 2000). The veterinary profession has been particularly concerned about the limits that organic standards set to the use of conventional veterinary medicinal products. There has also been a concern that organic farming, alongside other farm assurance schemes with input-based welfare standards, will not necessarily deliver good welfare on all farms simply by stipulating welfare inputs, rather than evaluating welfare outcomes (Main *et al.*, 2003).

A policy research project, aimed at identifying potential animal welfare benefits and problems in organic livestock production systems and developing recommendations for welfare assessment and promotion within the UK organic certification system, was carried out over a two-year period in 2001-2003. This paper summarizes the main findings of the study.

MATERIALS AND METHODS

A range of data collection methods were employed in the study. These consisted of the following:

- a questionnaire survey of organic inspectors, advisers and veterinarians, with 370 useful returns;
- a telephone interview survey of 50 organic producers;

- a field survey of inspection practices during 26 inspection visits by four different organic inspectors; and
- a workshop with representatives of organic certification bodies.

RESULTS

Questionnaire survey of inspectors, advisers and veterinarians

Overall response by all respondent groups indicated that organic standards were perceived primarily to have a positive impact on animal welfare. However, there were significant differences between the different respondent groups, specifically veterinarians gave more negative responses than inspectors and advisers regarding practices differing from conventional husbandry, e.g. a systems-based disease prevention rather than a medicinal approach, the ban on mutilations, the extensive nature of organic systems, etc.

The survey identified "good" and "bad" standards in terms of their perceived impact on animal welfare. The five "best" standards (i.e. >85% of respondents considered them to have a positive impact on welfare) were those requiring that:

- poultry buildings must be emptied of livestock between each batch of poultry reared;
- transit time between the farm and destination is kept to a minimum and that, where practical, the nearest appropriate approved abattoir is used;
- runs for poultry must be left empty for at least two months to allow vegetation to grow back;
- positive health and welfare must be provided by a plan drawn up by the farmer, preferably working together in partnership with a veterinary surgeon and agreed between them during and after conversion; and
- exercise areas for pigs must permit rooting and dunging by the animals.

The "worst" standards (i.e. >40% of respondents felt that the standards would have a negative impact on animal welfare) were those that:

- limit the feed ingredients allowed in organic systems by stipulating that only products listed in the standards can be used for animal feeding – this was with special reference to the fact that synthetic amino acids are not allowed in pig and poultry feeding in organic systems;
- allow physical castration in order to maintain the quality of products and traditional production practices;
- prohibit the use of chemically-synthesized allopathic veterinary medicinal products or antibiotics for preventive treatments; and
- limit the number of courses of treatment with chemically-synthesized allopathic veterinary medicinal products or antibiotics within one year to three, with the exception of vaccination, treatments for parasites and any compulsory eradication schemes.

None of the respondent groups consistently identified conditions or diseases that they perceived to be more or less prevalent in organic compared with conventional systems. Improved stockmanship, advice and support and animal health plans were identified by all respondent groups as the best ways to improve animal welfare. More intensive inspections and stricter or more lenient standards were viewed as less important factors in guaranteeing high welfare standards.

Telephone survey of organic livestock producers

A vast majority of the fifty respondents felt that organic standards had a positive impact on animal welfare, mainly due to lower production levels, i.e. less stress, and that there were no negative impacts on animal welfare of having to operate under organic standards. A minority felt that organic standards had a negative impact, specifying these as being mainly due to restrictions on medicine and vaccine use. When asked about more or less frequent welfare or health problems on farms since conversion to organic production, the responses were contradictory and confusing. With the exception of external and internal parasitism in sheep and overall better health in dairy herds, no other conditions were consistently perceived as being more or less prevalent in organic herds/flocks.

Fifty percent of respondents felt that welfare assessment would not be useful, citing increased bureaucracy and paperwork, and objectivity/reliability as being the main problems with assessments. Those that viewed welfare assessment as being potentially useful fell into two equally proportioned groups: assessments should be based on standards compliance according to environmental parameters; and record-based assessment. A majority of respondents expressed concern regarding the reliability and/or objectivity of welfare assessments. The only groups of respondents not expressing this concern were those advocating environment-based compliance assessment. Also, a majority of respondents felt that more and better advice was the best way to improve welfare on organic farms.

Field study of inspection practices regarding animal welfare assessment

It was apparent that the inspectors had poor access to health plans before and during the inspections. In cases where health plans were available, they tended to provoke an inspection concern. There was also poor availability of animals for inspection during the majority of the visits to beef and sheep farms. The inspectors suggested that the availability of animals could only be improved by special arrangements, due to the extensive nature of many sheep and beef enterprises.

All inspectors assessed the stockperson on all visits, using a combination of interview and observation. The inspectors also used a wide variety of both input and outcome related welfare inspection parameters. However, there was no consistent and formal approach to these parameters. As a result, it appeared that welfare concerns did not get adequately reported or noted. In a majority of cases, inspectors had a welfare concern and verified it during inspection, but only in two cases was this concern included in the inspection report. In all other cases, the issue was discussed with the stockperson and guidance was given.

Workshop with the organic certification bodies

- There appeared to be a general acceptance that, whilst organic standards are probably sufficient, animal welfare issues were not currently being adequately addressed by inspection and certification. However, the certification bodies' representatives felt that some of the problems arising from "poor" standards were being addressed through guidance notes produced by all certification bodies. The need to collate this guidance was acknowledged. Regarding welfare assessments, the certification bodies' representatives also recognized the need to train the inspectors and to standardize inspection procedures. They felt, however, that the introduction of pro-forma health plans and intensification of health planning requirements would be difficult as they would encounter farmer resistance. Similarly it was felt that statutory veterinary involvement was not appropriate as there was insufficient competence in organic farming available within the veterinary profession.

There was a strong feeling that animal behaviour should have a greater role in welfare assessment protocols, and these were currently too health orientated. Also, the tacit knowledge of the inspectors should be made acceptable in the inspection process.

The certification bodies did not consider it feasible, under current regulations and certification guidelines, that inspectors should give advice as part of the inspection process. It was suggested that this issue could be solved by introducing an improved and comprehensive reporting system into the inspection protocol, so that reporting and guidance could be followed up.

CONCLUSIONS AND RECOMMENDATIONS

The following conclusions and recommendations were made:

1. Whilst there is indication that the existing organic standards do not require changes, certification bodies need to ensure that misinterpretation of standards does not lead to welfare problems. They need to actively encourage and ensure better implementation of standards that require good husbandry, choice of suitable breeds, good feeding practices and access to natural behaviour.
2. Certification bodies need to embark on a joint mission to formalize welfare assessment in connection with organic inspection; to clarify the current confusion on the requirements of formal health plans; and to clarify the difference between advice and guidance in connection with inspection visits.
3. Veterinary support on organic farms needs to be strengthened and veterinary involvement at certification body level should be obligatory.
4. There should be careful consideration given to formal welfare assessment at farm level. Encouragement of farmer "ownership" of the process and ensuring that the results are useful for the farmer as a management tool are paramount for the success of this process.

ACKNOWLEDGEMENTS

The authors would like to acknowledge the contribution of all vets, inspectors, advisers and producers who were involved in the project. The project was funded by the Scottish Executive Environment and Rural Affairs Department (SEERAD).

REFERENCES

HARPER G.C. and HENSON S.J. (1998) *Consumer Concern about Animal Welfare and the Impact on Food Choice: comparative literature review.* EU FAIR CT98-3678.

HOLMBERG H. (2000) *Rapport konsumentundersokning om ekologiska produkter/KRAV.* LUI Marknadsinformation AB.

MAIN D.C.J., WHAY H.R., GREEN L.E. and WEBSTER A.J.F (2003) Effect of the RSPCA Freedom Food scheme on the welfare of dairy cattle. *The Veterinary Record*, 153, 227-231.

THAMSBORG S.M., HOVI M. and BAARS T. (2000) What to do about animal welfare in organic farming? A report on the animal welfare discussion at the 2nd NAHWOA workshop. In: *Diversity of livestock systems and definition of animal welfare, Proceedings of the 2nd NAHWOA Workshop, Cordoba, 8-11, January 2000.* 161-165.

Productivity and Nutrient Composition of Multi-Species Swards

A. HOPKINS
IGER, North Wyke, Okehampton, Devon, EX20 2SB, UK

ABSTRACT

Replicated cattle-grazed plots of perennial ryegrass/white clover (PRG/WC) and multi-species pasture (MSP) swards were compared on an organic farm over two harvest years. Herbage accumulation from the MSP swards was similar to that from PRG/WC swards in both years, and each had similar proportions of grass and dicot species, of which white clover contributed *c.* 28% of total DM yield in the PRG/WC and *c.* 5% in the MSP sward. Herbage from MSP had lower mean total N content than PRG/WC, but higher concentrations of Na, K and Mg. Compared with the mean for all dicots (including clover), grass had lower concentrations of N, Ca, Mg and Na. Among individual species, N concentrations were highest in white clover, Ca was highest in chicory and plantain, Na was highest in yarrow and ribwort plantain, K was highest in yarrow but was low in plantain, and Mg was higher in white clover and chicory than in grass.

INTRODUCTION

Forage herbs and drought tolerant grasses have long been considered beneficial components of swards (Elliot, 1943; Foster, 1988). Within the context of low-input systems generally, and organic farming in particular, multi-species swards containing various sown grass and dicot species (legumes and forage herbs) may offer potential advantages over botanically simple perennial ryegrass-clover leys in terms conservation of macro-nutrients, supply of trace elements, improved drought tolerance and possible animal health benefits. They may also improve total or seasonal DM yield (Daly *et al.*, 1996). While previous experiments have compared the herbage yield and nutrient composition of sown ryegrass or ryegrass-white clover with botanically diverse permanent swards under both low and high inputs (e.g. Hopkins *et al.*, 1990; 1995) few studies have investigated multi-species sown swards. The aim of the experiment reported here was to compare the productivity and nutrient content of a sown multi-species sward with a perennial ryegrass/white clover ley under organic grazing conditions.

METHODS AND MATERIALS

The study was carried out in a 3.9 ha paddock on organic farmland in Gloucestershire, UK (51°39'N/ 2°9'W, 130 m elevation). The soil was a stony clay loam; pH was 6.9, P index 1, and K index 2, and at the start of experimentation the sward was a 2-year-old perennial ryegrass-white clover (PRG/WC) ley, used mainly for dairy cattle grazing. In late June 1996 four replicated plots (12m x 6m) were allocated at random within the paddock and a 6m x 6m sub-plot of each was ploughed and cultivated, then broadcast-sown with a multi-species pasture (MSP) seeds mixture (Table 1), the remaining area of each plot representing the existing sward. The MSP mixture reflected attributes appropriate for the conditions of the site and the overall management objectives of the farm. These included potential agronomic advantages (herbage nutrient composition, N fixation, drought tolerance, presence of condensed tannins) and potential for enhanced biological diversity. The sown areas were protected by temporary fencing for 15 weeks, during which some hand weeding was carried out, the sward then topped with a reciprocating mower, then grazed as part of the common management of the paddock. Exceptionally dry weather resulted in a slow establishment of the sown plots.

Table 1. Species and seed rates used for the multi-species sward

Cocksfoot *(Dactylis glomerata)*	10 kg/ha
Red fescue *(Festuca rubra)* §	7 kg/ha
Tall fescue (*Festuca arundinacea)*	7 kg/ha
White clover *(Trifolium repens)* Kent wild white	2 kg/ha
White clover *(Trifolium repens)* cv. Menna	2 kg/ha
Chicory *(Cichorium intybus)* cv. Puna	4 kg/ha
Common birdsfoot trefoil *(Lotus corniculatus)* §	3 kg/ha
Greater birdsfoot trefoil *(Lotus pedunculatus)* cv. Maku	3 kg/ha
Yarrow *(Achillea millefolium)* §	1 kg/ha
Common knapweed *(Centaurea nigra)* §	0.5 kg/ha
Ribwort plantain *(Plantago lanceolata)* §	0.5 kg/ha
Salad burnet *(Sanguisorba minor)*	1.0 kg/ha

Species marked § were of UK native origin

In each of the two subsequent years, herbage accumulation was assessed by measuring growth under exclosure cages. Cages, measuring 2.75 x 1.25 x 0.6 m high, were placed at randomly determined coordinates on each of the eight sub-plots at the start of the growing season, the area under each having first been trimmed to a stubble height of *c.* 2.0 cm. At each cut herbage was again harvested to *c.* 2.0 cm, using hand shears and a 1.0 m x 0.5 m sampling frame, commencing on 27 April. This was repeated at intervals (of 3-5 weeks, according to seasonal growth) to mid-October, and each cage was moved on to a newly pre-trimmed area of the plot after each cut. Herbage accumulation as total dry matter (DM) yield was determined for each sub-plot at each cut by oven-drying of sub-samples. In addition, sub-samples of herbage at each harvest were hand sorted to determine botanical composition, and mineral analysis (concentrations of N, Ca, Mg, K, Na and P) was carried out on grass and non-grass fractions for each sub-plot, as well as for individual species.

RESULTS

There were no statistically significant differences in annual herbage yield between the two sward types, or in the respective yields of the grass and dicot components (Table 2). In the PRG/WC swards white clover contributed most of the dicot component (*c.* 28% of total annual DM). Legume contribution in the MSP sward was low (*c.* 5%) in both harvest years; dicots which showed the best establishment and contributed most to the herbage were yarrow and chicory, followed by ribwort plantain.

The two sward types differed in overall nutrient composition. Total herbage N concentration was higher in the PRG/WC than the MSP swards, attributed to N fixation from the greater amounts of white clover. However, concentrations of Na, K and Mg in the total herbage harvested from MSP swards were higher than from PRG/WC, particularly in year 1 (Table 3).

Table 2. Annual dry matter yield (t/ha) from perennial ryegrass/ white clover (PRG/WC) and multi-species (MSP) swards, as total and grass / dicot components.

Herbage	Sward type PRG/WC	MSP	s.e.d.	F. prob.
Year 1				
Grass only	6.65	6.43	1.441	0.880
Dicots only	3.40	4.73	1.292	0.380
Total yield	10.05	11.16	0.761	0.245
Year 2				
Grass only	7.51	7.27	0.700	0.541
Dicots only	3.47	2.90	1.055	0.627
Total yield	11.22	10.17	1.506	0.536

Table 3. Concentrations of N and mineral elements (g/kg DM) in herbage harvested from perennial ryegrass/white clover (PRG/WC) and multi-species (MSP) swards (derived from bulked samples of 6 cuts each year, 4 replicates per treatment). Values having different superscripts within the same year were significantly different, *P*<0.05 or greater.

	Sward type			
	Year 1		Year 2	
Element	PRG/WC	MSP	PRG/WC	MSP
N	34.0 ^{a}	28.6 ^{b}	34.6 ^{a}	26.5 ^{b}
Ca	8.10 ^{a}	8.29 ^{a}	5.78 ^{a}	5.04 ^{a}
Mg	1.90 ^{a}	2.18 ^{b}	1.55 ^{a}	1.39 ^{a}
K	24.5 ^{a}	27.9 ^{b}	25.4 ^{a}	27.4 ^{a}
Na	4.61 ^{a}	6.90 ^{b}	5.89 ^{a}	6.26 ^{a}
P	4.09 ^{a}	3.99 ^{a}	3.20 ^{a}	2.87 ^{a}

There were also significant differences in herbage mineral composition between the total-grass and total-dicot components, as well as between the different dicot species. Based on data from the two years, grass herbage had lower N concentrations than dicots (means of 27.7 and 36.7 g/kg DM, respectively, *P*<0.001); lower Ca (4.3 and 5.2 g/kg DM, *P*<0.001); lower Mg (1.48 and 2.16 g/kg DM, *P*<0.001) and lower Na (5.4 and 6.6 g/kg DM, *P*<0.05).

Analysis of herbage samples for grass (not sorted by species) and individual dicot species (white clover, chicory, yarrow and ribwort plantain) harvested in summer (cuts 3 + 4) of year 2, showed that N concentrations were highest in white clover: *c.* 45 g/kg DM, cf. *c.* 27-32 g/kg for grasses and forage herbs. Ca concentrations were markedly higher in chicory and plantain than grass (12.9 g/kg DM for both dicots, cf. grass at 4.0 g/kg and white clover at 10.5 g/kg). Na concentrations were high generally, and were highest in yarrow and plantain (*c.* 9.0 g/kg DM, cf. *c.* 6.0 g/kg for grass and white clover, and 6.6 g/kg for chicory). K concentrations were highest in yarrow (37.0 g/kg DM) and lowest in plantain (17.4 g/kg), with chicory, white clover and grass intermediate (23.0 and 30.0 g/kg DM, respectively). Mg concentrations were generally higher in white clover and chicory (1.9 g/kg) than grass (1.4 g/kg). There were no differences in P concentrations between any of the species examined.

DISCUSSION AND CONCLUSIONS

Despite the relatively poor contribution of white clover in the MSP swards during the period of assessment, herbage accumulation from the two sward types was remarkably similar, annually and seasonally. The experiment evaluated only one type of MSP sward and other combinations of species might yield different levels of herbage production, depending on the site conditions; e.g. Daly *et al.* (1996) found the inclusion of lucerne in an MSP mixture gave an additional herbage yield advantage. Further work, with different types of multi-species mixtures on different types of sites, is therefore suggested.

The concentrations of N and herbage minerals recorded here were within the general range reported from other grassland sites in the UK. Site conditions have been shown to be more important than sward type in affecting herbage Ca, Mg and Na, but within a site the sward composition and its age affect herbage nutrient composition (Hopkins *et al.*, 1994). There is increased interest in the use of multi-species swards to improve herbage nutrition, and chicory is one species which, compared with perennial ryegrass, has previously been reported to have higher concentrations of Ca, K, Na and total N, as well of other trace elements important for livestock health (Barry, 1998). The presence of chicory, ribwort plantain and yarrow in the present study was shown to enhance Ca, Na and Mg. Other dicot pasture species may merit investigation, but choice of species for farmers will depend on commercial availability.

ACKNOWLEDGEMENTS

This work formed part of a joint BBSRC, NERC, ESRC-funded project. Assistance from the management and staff of Duchy Home Farm is also gratefully acknowledged.

REFERENCES

BARRY T.N. (1998) The feeding value of chicory *(Cichorium intybus)* for ruminant livestock. *Journal of Agricultural Science, Cambridge*, 131, 251-257.

DALY M.J., HUNTER R.M., GREEN G.N. and HUNT L. (1996) A comparison of multi-species pasture with ryegrass-white clover pasture under dryland conditions. *Proceedings of the New Zealand Grassland Association*, 58, 53-58.

ELLIOT R.H. (1943) *The Clifton Park System of Farming*, 5th edn. London: Faber and Faber.

FOSTER L. (1988) Herbs in pastures. Development and research in Britain, 1850-1984. *Biological Agriculture and Horticulture*, 5, 97-133.

HOPKINS A., GILBEY J., DIBB C., BOWLING P.J. and MURRAY P.J. (1990) Response of permanent and re-seeded grassland to fertilizer nitrogen. 1. Herbage production and herbage quality. *Grass and Forage Science*, 45, 43-55.

HOPKINS A., ADAMSON A.H. and BOWLING P.J. (1994) Response of permanent and reseeded grassland to fertilizer nitrogen. 2. Effects on concentrations of Ca, Mg, K, Na, S, P, Mn, Zn, Cu, Co and Mo in herbage at a range of sites. *Grass and Forage Science*, 49, 9-20.

DIY Faecal Egg Counts in Sheep

GILLIAN BUTLER
Tesco Centre for Organic Agriculture, University of Newcastle upon Tyne,
Nafferton Farm, Stocksfield, Northumberland, NE43 7XD, UK

ABSTRACT
This paper gives preliminary results of a study monitoring the use of *Fecpak* kits for assessing faecal egg counts (FEC) in sheep on nine organic or in-conversion farms between June and October in 2003. The dry weather conditions resulted in relatively few recordings being made, although four of the farms did identify and prevent what could have been serious *Nematodiris* outbreaks in early June. Generally there was good agreement between farm- and lab-based assessments, albeit on a limited number of samples. The main reasons given for tests being carried out were: routine (37%), interest (30%) and dirty lambs (27%), with only 3% of tests prompted by lambs not thriving. Over half of the tests (57%) resulted in no management action being taken, 13% led to selected lambs being wormed, 23% to all lambs being wormed with only 7% followed by worming of ewes. Results and comments are encouraging, and the plan is to continue this exercise until the spring of 2005.

INTRODUCTION
Internal parasites can compromise sheep welfare and performance under organic management. Organic standards encourage a range of approaches to minimize this impact; sound nutrition, selective breeding and clean grazing, thus minimizing reliance on anthelmintics. These approaches will only succeed, and drug use reduced, if progress can be measured and evidence provided that parasitic burdens are at tolerable levels.

Generally, faecal egg count (FEC) is taken as an indication of parasitic burden and some producers routinely send faecal samples to veterinary laboratories for screening. On-farm counting offers a number of advantages over central laboratories, largely due to the relatively rapid availability of results. Sheep can be held during assessment, allowing worming, if necessary, before returning to grazing. On-farm monitoring of FEC using *Fecpak* kits has been used in New Zealand since the early 1990s, driven by concerns over anthelmintic failure. Some organic farmers in the UK have adapted this technique, although uptake has been relatively slow.

The ability to make rapid, multiple tests either on groups or individual sheep can be used as a tool for strategic worming as well as for assessing progress due to breeding, grazing management and/or nutrition. Tests on individual sheep could also be used to improve genetic resistance to parasites within closed flocks. This study looks at the adaptation of on-farm assessment of FEC on nine organic or in-conversion sheep farms using *Fecpac* kits, presenting a preliminary report of the farmers' experiences.

MATERIALS AND METHODS
Of the nine farms involved, six had the dedicated use of a *Fecpak* kit. The remaining three, within five miles of each other, shared a single kit and the suggestion was to leave this at one site and other farmers could bring samples to this base. A training day was

organized initially to give instructions in the use of the technique and discuss other aspects of management important in parasite control on organic farms.
Test procedures can be summarized in the following steps. Fresh faeces need to be collected from the animals under test. With individual sheep the procedure is straight forward, but for a group or *mob* test, equal volumes of faeces need to be collected from representatives within this group. The faeces are weighed and diluted with water before a homogenized aliquot is removed and mixed with concentrated saline solution. After filtering, this is pipetted onto the double layered microscope slide for inspection. The saline solution causes the eggs to float, making them relatively easy to focus on, identify and count. The raw data on numbers of eggs within the marked grid of the slide are converted to 'eggs per gram' of faeces, making it comparable with other techniques. Although individual species cannot be identified, farmers are able to differentiate between eggs from *Strongyle* worms and the larger eggs of *Nematodiris*. Farmers completed a record sheet each time a test was carried out, collecting information on:

1. Weather (tick the most appropriate box to describe the weather in the week leading up to the test).
2. Reason for carrying out the test (tick the most appropriate box).
3. Was this a group test or individual sample?
4. Number of *Strongyle* and *Nematodiris* eggs counted in each chamber of the microscope slide.
5. Management action (tick the most appropriate box or boxes).
6. Product used.
7. Did all run smoothly ?
8. Time taken for test (from collecting sheep until clearing away after finished).

One test in six was sub-sampled and sent to CBS Technologies for assessment of parasite egg counts. These were allocated at random and predetermined in the recording handbook and involved a duplicate aliquot of the diluted homogenized faeces being decanted and dispatched to the labs by first class post. The laboratory used is responsible for carrying out all faecal egg counting for the MLC ram performance monitoring and the Northern Upland Sheep Strategy.
The training for this project was carried out at the end of May 2003 and individual record sheets were collected from farms 20 weeks later, in the middle of October.

RESULTS

A total of thirty record sheets were collected from six farms. No records were collected from three farms, although it is known that one of these has been using the kit and submitting samples to the laboratory. The two farmers sharing the kit and travelling to carry out tests only did so on three occasions in total, fewer than those with easy access to equipment.
Table 1 presents the reasons which farmers gave for carrying out the tests. These tend to fall into three main categories with only one test for 'lambs not thriving' and one test ('other') carried out after worming, in order to test efficacy of drenching.

Table 1. Reasons given by farmers for carrying out the faecal egg counts

Category	Personal interest	Dirty lambs	Lambs not thriving	Individual record	Routine	Other
% of recordings	30	27	3	0	37	3

Just over half of the tests resulted in no further action by the farmer, with only 43% of tests leading to worming of lambs or ewes, as shown in Table 2.

Table 2. Management actions taken following tests

Action	Nothing	Selected lambs wormed	All lambs wormed	Ewes wormed	Other
% of recordings	57	13	23	7	0

Figure 1. Faecal egg counts recorded on farms

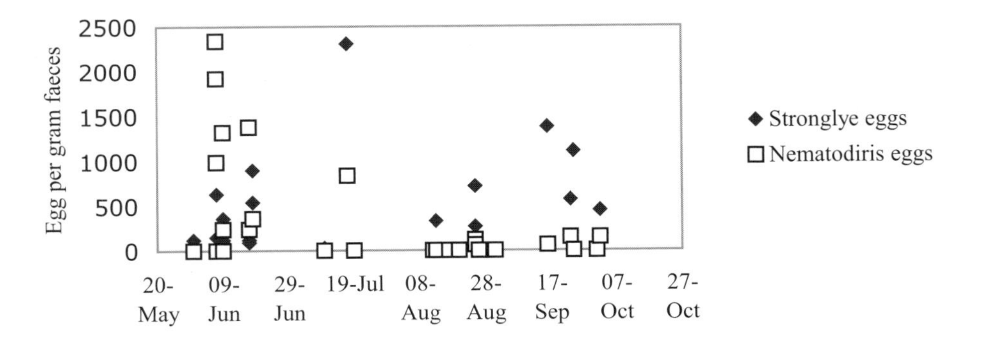

The numbers of eggs recorded on farms are presented in Figure 1. Four of the farms recorded very high levels of *Nematodiris* in early June and were able to treat lambs (and ewes) before welfare and performance suffered. Also, eight tests recorded *Strongyle* eggs over 500 per gram of faeces throughout the season, which could compromise performance.

Unfortunately, the question on time taken to carry out the tests was poorly recorded, with only nine out of the thirty completed. These ranged from 20 – 90 minutes, with an average of 38 minutes. There were only eight samples sent to the laboratory and five of these (62.5%) showed close agreement between the farm and laboratory results. In one case the laboratory recorded higher results and the farm records were higher in the other two cases.

DISCUSSION

The relatively small number of tests carried out makes it unwise to read too much into these preliminary results, although findings are encouraging. Generally, comments were very positive from six of the farmers who were keen to continue, although no contact has been made with those failing to submit results. The number of completed records is disappointing, but on the whole, the weather was dry throughout the summer of 2003, reducing the chances of lambs picking up infective larva from the pasture and exhibiting symptoms of parasitic gastro-enteritis. Farmers became more confident with the technique as the season progressed and it became more familiar. Relatively few tests were carried out by farmers sharing equipment, raising doubts about whether this will allow the potential for the technique to be exploited fully.

Identifying the high output of both *Nematodiris* and *Strongyle* eggs allowed early treatment of lambs and/or ewes, reducing any negative impact on performance and welfare. In addition, expedient treatment will limit pasture contamination and the threat to lambs subsequently grazing these fields. In the case of *Nematodiris*, this advantage will be carried over to the following year when the population of infective larva emerging in the spring will be reduced. With relatively few samples submitted to the laboratory for quality control, it is not possible to comment on the reliability of the on-farm test as an alternative. There was good agreement in five of the eight samples although the discrepancies with the other three may need closer investigation. Over half of the tests resulted in not worming sheep given the results of the FEC test, but it is unclear how this might compare if no testing had been carried out. This will be investigated in future.

The plan is to continue this study, taking recording through the winter months and into the spring and early summer of 2005. Two producers in the group breed Blackface rams and are planning to measure variation in FEC in individuals under common management in order to assess the scope for improving flock resistance to parasites through selection.

ACKNOWLEDGEMENT

This work forms part of a project aimed at improving organic red meat production, funded by the Red Meat Industry Forum (RMIF) and I am indebted to the farmers for their help; Steve Ramshaw, Jake Elder, Frank and Julie Charlton, Colin and Michelle Anderson, Bill Brown, Les and Jenny Pole, Kenneth and Lara Porter and John Rowell.

The Potential Value of Different Nurse Crops for Organic Systems and their Influence on the Undersown Swards

R.F. WELLER, P.J. BOWLING and J. VALENTINE
IGER, Trawsgoed Research Farm, Aberystwyth, Ceredigion, SY23 4LL, UK

ABSTRACT
The value of a range of spring-sown cereals and forage peas as nurse crops in organic systems was assessed in relation to forage production and their influence on the establishment and yield of undersown ryegrass/red clover swards. Silage DM yields of over 9 t/ha were obtained from two husked oats and the naked oat cv. Bullion. Yield of the UK spring barleys cvs. Riviera and Dandy were 0.7-0.8 t/ha lower, and yields of 8.6 and 8.8 t/ha were obtained from the triticale cv. Purdy and the forage pea cv. Canis. Grain, which is the most nutritious fraction, constituted a greater proportion of yield in barley than in the other cereal species. There was a clear inverse relationship between silage yield and aftermath growth. The highest yields of the undersown swards were recorded when barley was the nurse crop, and the lowest when oats and triticale were used. This was attributed to the latter crops being later maturing and taller growing, thus reducing light penetration to the undersown sward. However, there was little difference between nurse cereal crops in their clover content in the following spring, with a tendency for the swards undersown to oats and triticale to have fewer broad-leaved weeds than those undersown to barley. The swards undersown to peas had low clover content and higher weed levels.

INTRODUCTION
Although winter diets for organically managed dairy cows are based primarily on grass/clover silage, the inclusion of whole-crop cereal silage and forage peas in the diet has the potential to improve total feed intake and diet quality (Kristensen, 1992). Whole-crop cereals can be widely grown in the UK, providing extra forage when either grown alone or undersown to provide an effective cover crop for the establishment of grass/clover leys. Forage peas produce silage with a high protein content, leading to lower feed costs as the requirement for high-priced protein concentrate feeds is reduced. In addition, the peas provide a significant input of nitrogen into the system via N-fixation. Studies to date on whole-crop cereals have focused on winter wheat and, to a lesser extent, barley. Little work has been carried out to determine the potential yields from other cereal species, including husked oats, naked oats and triticale. Cereal varieties are bred primarily as grain crops and there may be significant differences between them when they are harvested at an earlier stage of growth, in terms of both forage yield and quality.

In the current study different species and varieties of cereal crops and forage peas were evaluated to determine their yield potential as whole-crop forages and also their influence as cover crops on the yield of the undersown grass/clover ley.

MATERIALS AND METHODS
The study was carried out in the 2000/01 period at Trawsgoed Research Farm near Aberystwyth, with the spring-sown crops established in an experimental design with 18 treatments and 3 replicates. The different forage species and varieties that were evaluated

during the study are shown in Table 1. The forage crops were drilled on 21 March 2000, and a hybrid ryegrass/perennial ryegrass/red clover mixture drilled on all plots at a rate of 30 kg/ha on 4 May 2000. As the seed weight varied between both the species and varieties of the forage crops, different seed rates were sown. The different cereal species and varieties were harvested at a similar growth stage rather than on the same calendar date. Forage pea plots were harvested when the pods were well formed and before the crop lodged.

Table 1. Forage species and varieties compared and the seed rates used.

Species	Variety	Treatment code	Seed rate (kg/ha)
Husked oat	Triple Crown	1	107
Husked oat	Stork	2	124
Husked oat	SW923687	3	111
Husked oat	Banquo	4	114
Husked oat	Amigo	5	114
Naked oat	Bullion	6	99
Spring barley	SWA95096	7	143
Spring barley	SWA99254	8	143
Spring barley	Verner	9	143
Spring barley	SWN971157	10	121
Spring barley	Riviera	11	140
Spring barley	Dandy	12	144
Triticale	Purdy	13	173
Triticale	Taurus	14	140
Forage peas	Canis	15	196
Forage peas	Grande	16	196
Forage peas	Carneval	17	196
Forage peas	Karita	18	196

RESULTS AND DISCUSSION

The growth stages, dry matter (DM) content, total DM yield of whole crop and grain yield from the different treatments are shown in Table 2. The yield of the forage whole-crop was determined at the first harvest, with the yields of the undersown Italian ryegrass/red clover ley measured 40 days later, and also in May of the following year. The yields shown are the means from three replicates per treatment. The cereal plots were harvested at the 83-88 growth stage (Zadoks *et al.*, 1974) when the grain was in the early dough to hard dough stage of maturity. The highest whole-crop yields of over 9.0 t DM/ha were recorded from the husked oats (cvs. Stork and Banquo) and naked oats (cv. Bullion) plots. The lowest yields were recorded from the barley varieties SWA99254, SWA95096 and cv. Verner. However, the proportion of grain, which is the fraction with the highest energy content, was greater in the spring barley varieties (36-50%) compared with the husked oats (23-34%), naked oats (15%) and triticale (27-31%).

It is important to note that as the plots were undersown with a grass/clover ley the seed rates used for the cereal and pea treatments were lower than those normally used for cereal crops not undersown. Therefore, the potential whole-crop yields from both the cereal and pea treatments would have been higher if the crops had not been undersown. The pea yields varied from 7.03 to 8.79 tonnes.

Table 2. Total whole-crop forage and grain yields of nurse crops.

Treatment	Growth stage	Harvest date	Whole crop DM %	Whole crop yield	Relative yield*	Grain yield	Per cent grain in whole crop
			(%)	(tDM/ha)	(%)	(t/ha @ 15%mc)	(%)
1	84	25 July	29	7.75	132	1.78	23
2	84	25 July	33	9.15	155	3.12	34
3	83	25 July	31	7.98	135	2.65	33
4	84	25 July	31	9.04	153	2.68	30
5	83	25 July	34	8.26	140	2.72	33
6	83	25 July	32	9.01	153	1.33	15
7	87	19 July	44	6.46	110	3.24	50
8	88	19 July	38	5.89	100	2.13	36
9	88	19 July	41	6.89	117	3.10	45
10	87	19 July	43	8.08	137	3.91	48
11	87	19 July	39	8.35	142	3.49	42
12	83	19 July	41	8.33	141	3.39	41
13	85	11 Aug	36	8.65	147	2.66	31
14	85	11 Aug	34	8.29	141	2.24	27
15		19 July	24	8.79	149	1.79	20
16		19 July	25	8.36	142	1.25	16
17		19 July	27	7.06	120	1.22	17
18		19 July	26	7.03	119	0.98	14

Undersowing the cereal and pea treatments allowed the effective establishment of a grass/clover ley and also increased the annual production of forage to over 10.0 t DM/ha in the cvs. Stork, Banquo (husked oats), Bullion (naked oats) and Canis (peas) treatments. There was an inverse relationship between the yields of the whole-crop forages and the first-cut yields from the hybrid ryegrass/perennial ryegrass/red clover ley, with the highest yields from the ley recorded when spring barley varieties were sown as the cover crop (average 1.60 t DM/ha) compared with an average of 0.98 and 0.97 t DM/ha from the husked oats and triticale varieties, respectively. The higher yields are attributed to the lower plant height and earlier maturity of the barley plants, compared with the oats and triticale treatments, leading to the benefits for the undersown ley of a less dense canopy and reduced competition from the cover crop. The swards undersown to peas had both higher weed populations and low clover contents. The yields from the ley in May of the following year showed no significant differences between the treatments. However, there was a tendency for the swards undersown to oats and triticale to have fewer broad-leaved weeds than those undersown to barley.

REFERENCES

KRISTENSEN V.F. (1992) The production and feeding of whole-crop cereals and legumes in Denmark. In: B.A. Stark and J.M. Wilkinson (Eds) *Whole-Crop Cereals*. pp.21-38. Marlow, UK:. Chalcombe Publications.

ZADOK J.C., CHANG T.T. and KONZAK C.F. (1974) A decimal code for the growth stage of cereals. *Weed Research*, 14, 415-421.

Alternative Control of Parasites of Organic Livestock: Effect of Pasture Environment on Gastrointestinal Parasite Development and Survival

C.L. MARLEY[1], R. COOK[1], J. BARRETT[2], N.H. LAMPKIN[3] and R. KEATINGE[4].
[1]IGER, Plas Gogerddan, Aberystwyth, SY23 3EB, UK
[2]Institute of Biology, University of Wales, Aberystwyth SY23 3DA, UK
[3]Institute of Rural Sciences, University of Wales, Aberystwyth SY23 3AL, UK
[4]ADAS Redesdale, Otterburn, Newcastle-upon-Tyne NE18 1SB, UK

ABSTRACT

Studies in New Zealand and the UK have shown that certain alternative forages, such as *Cichorium intybus* (chicory) and *Lotus corniculatus* (birdsfoot trefoil), reduce parasitic infestation in sheep. The mechanisms for these effects may include the provision of a pasture environment which is unsuitable for parasite development or survival. Pots of chicory, birdsfoot trefoil and perennial ryegrass were cut to a height of 50 mm before samples of sheep faeces, containing 10,385 *Cooperia curticei* eggs were added. Replicate pots of each forage type were destructively sampled on day 8, 16, 20, 28 and 37 to extract the helminth larvae. There were differences among the forages on day 16, with the highest and the lowest number of larvae being extracted from the ryegrass and chicory, respectively ($P < 0.05$). The findings indicate that the structure of the sward canopy of chicory and birdsfoot trefoil may affect development and survival of *C. curticei* eggs and reduces the numbers of infective stage larvae on pasture, and is therefore one of the mechanisms by which these forages may affect parasitic infections in grazing livestock.

INTRODUCTION

Conventionally, farmers rely upon the routine use of anthelmintics to control helminth parasites in livestock and, historically, their use has proved highly cost-effective. However, their routine use in organic livestock systems is restricted by the organic standards. Instead, organic systems rely on using management practices, including mixed grazing and clean grazing systems to control internal parasites in their livestock.

Chicory and some forage legumes have been found to reduce internal parasites in growing lambs (Marley *et al.*, 2003a; b). The mechanisms for these effects are not fully understood but may involve a combination of a) the direct or indirect effect of secondary plant compounds (e.g., condensed tannins); b) improved mineral and/or protein status of the livestock; and c) the provision of a pasture environment which is unsuitable for parasite development or survival (Moss and Vlassoff, 1993; Scales *et al.*, 1994; Niezen *et al.*, 1998).

Alternative forages may produce a micro-climate in the sward environment that is unsuitable for parasite development or survival, thus breaking the lifecycle of these parasites and reducing their ability to infect grazing livestock. An understanding of these mechanisms is needed before the use of alternative forages to control internal parasites in livestock can be fully implemented on organic farms.

The aim of this experiment was to investigate the effect of plant structure, and thus pasture environment, on the development and survival of helminth parasites on different forages.

MATERIALS AND METHODS

Fifteen pots of chicory (cv. Grasslands Puna), birdsfoot trefoil (cv. Leo) and perennial ryegrass (cv. Napoleon) were established using sowing rates that were equivalent to field rates of 18, 20 and 32 kg/ha, respectively. The pots were 25 cm in diameter and contained sterilized compost (Humex™, Humex, Carlisle, UK).

At the start of the experiment (Day 0), the forage in each pot was cut to a height of 50 mm from soil level and placed in an environmentally-controlled glasshouse in a randomized block design. A further screen of 'guard' pots, sown with the same forages as used in the experiment, were placed around the perimeter of the block of pots, in a randomized manner, to ensure a uniform environment for all of the experimental pots. A 3.1g sample of sheep faeces, containing a single helminth parasite species (*Cooperia curticei*) (kindly provided by The Moredun Research Institute, Edinburgh, UK) was well mixed and then sprinkled evenly onto the soil in each experimental pot. The faecal sample had an egg count of 3350 per g (wet faeces), thus providing a total of 10,385 helminth eggs per pot.

The pots were then maintained in the glasshouse for a total of 37 days. An optimum temperature (23 - 28°C) and humidity (57 – 68%) for helminth egg development and survival were maintained using an automatic ventilation system, together with automatic overhead (which watered the forages with fine spray for 10 s every 2 h), and bench (operating for 20 s every 3 h) watering systems. The temperature and humidity were electronically recorded every 15 min throughout the experimental period.

Replicate pots of each forage type were destructively sampled on day 8, 16, 20, 28 and 37 of the experiment, between 07:30 - 08:45 on each occasion. If a delay before larval extraction was anticipated the samples were stored at 10°C, before being processed within 48 h. Nematodes (parasitic and free-living) were extracted from the forages after soaking in water with a few drops of Decon 90 (Decon Laboratories Limited, Sussex, UK), by a modification of the method given in the MAFF Technical Booklet (1997). The total dry weight of forage used was determined by placing each forage sample in an oven at 80 °C for 48 h following larval extraction. Twenty percent of the nematodes extracted were counted in a Doncaster counting dish (W.J. Cox, Tring, UK) under a dissection microscope at x 40 magnification (Olympus Optical Company, Southall, UK). This was repeated and, after checking that the coefficient of variation between the two sample counts was within 10%, the mean value was recorded. The total number of parasitic larvae was then counted by further reducing this sample to 1.5 ml and centrifuging at 7000 g for 5 min. using a bench centrifuge (Denver Instrument Company, USA). The supernatant was removed and the remaining 'pellet' was examined on a microscope slide. The larvae were killed by passing the slide over an alcohol flame for a few seconds and covered with 22 x 22mm cover-slip. Either 10% or all (if the total count was less than 100) of the larvae were examined under x100 using a compound microscope (Olympus Optical Company, Southall, UK) to determine the ratio of free-living to parasitic nematodes in the sample.

The total number of parasitic larvae per kg DM forage was calculated and, as the data sets were not normally distributed, the data were analysed using a Kruskal-Wallis one-way analysis of variance.

RESULTS

The results for the total number of infective stage *C. curticei* parasite larvae per kg forage DM found at different time points during the experimental period are shown in Table 1. Analysis of the data for total number of larvae only showed a significant effect of forage on day 16, with the highest and the lowest number of larvae being extracted from the ryegrass and chicory, respectively ($P < 0.05$).

The data from the automatically recorded environmental conditions inside the glasshouse showed that the temperature did not exceed 36.6°C or fall below 12.4°C, and that the relative humidity was between 28 - 98 % during the experimental period.

Table 1. Total number of infective stage *Cooperia curticei* larvae extracted at various time points from three different forages. Median values are given, with associated range in parenthesis. All values g/kg DM.

	Chicory		Birdsfoot trefoil		Perennial ryegrass	
Day 8	7232	(6338-13521)	0	(0-1028)	11048	(0-13636)
Day 16	21947	(20368-51656)	55016	(35421-102139)	111976	(80731-147532)
Day 22	48767	(19286-56251)	18051	(6638-57279)	26845	(10644-39335)
Day 28	12542	(2291-13715)	5143	(2019-9169)	14962	(11210-42938)
Day 37	15168	(9628-20556)	13181	(4108-15483)	19572	(13186-48378)

DISCUSSION

There were never as many *C. curticei* larvae extracted from chicory or birdsfoot trefoil as from ryegrass at day 16. This indicates that both these forages have reduced the development and/or survival of these larvae compared to ryegrass. The data also showed that that the maximum number of larvae from chicory was on day 22, whereas the maxima for birdsfoot trefoil and ryegrass were on day 16. This suggests that chicory may be reducing the rate of larval development compared to the other forages. The results of this study are in agreement with the findings of Moss and Vlassoff (1993) and Scales *et al.* (1994), who showed that chicory had fewer parasitic larvae on its leaves compared with perennial ryegrass/white clover. The effect of birdsfoot trefoil on parasitic helminth development and/or survival has not been shown previously.

There are many possible reasons for these findings. Levine and Andersen (1973) found that changing environmental conditions, temperature and humidity affects larval development and/or survival. The primary reason given by Moss and Vlassoff (1993) was that chicory foliage structure was less suitable for larval development and migration. Niezen *et al.* (1998) made similar proposals that differences in the number of larvae on different herbage was the result of trichomes (tiny outgrowths from the epidermis of the plant that do not contain vascular tissue) present on the leaves and stems of certain herbage species. These trichomes may maintain a surface of moisture on the leaves and stem, which may aid the movement of larvae vertically (Croll, 1970). The results of Niezen *et al.* (1998) showed that Yorkshire fog had the least vertical migration, suggesting that trichomes may actually interfere with larval migration or that the position of moisture films on trichomes may reduce larval migration. However, Niezen *et al.* (1998) did not refer to the potential effect of the trichomes on chicory in this context.

It should be noted that the results of this study are based only on one helminth species and it has been shown that forages may affect different helminth species in different ways. The results of an outdoor study showed that chicory had fewer *Teladorsagia* species but not fewer *Trichostrongylus* species compared to various grasses, *Trifolium repens* (white clover) and *Medicago sativa* (lucerne) (Niezen *et al.*, 1998). Furthermore, the experiment reported here was conducted under controlled glasshouse conditions to determine the effects of the structure and sward canopy of chicory and birdsfoot trefoil on the development and survival of parasitic larvae without the influence of variables that can alter larval development in the field. For example, forage species could influence the number and type of nematode antagonists in the ecosystem of the pasture sward (Sayre and Walter, 1991). These factors must be considered when relating the results of this indoor study to the field situation.

CONCLUSIONS

This experiment has shown that the structure and sward canopy of chicory and birdsfoot trefoil alters the development and survival of *Cooperia curticei* helminth parasites from eggs into infective stage larvae. Thus, this is one mechanism by which these forages may affect parasitic infections in grazing livestock. Further studies examining the effects of the trichomes on chicory on the vertical migration of parasitic larvae are now needed to clarify whether the effect of chicory on larval numbers on pasture is due to the trichomes on this forage reducing larval movement and migration.

REFERENCES

CROLL N.A. (1970) *The Behaviour of Nematodes*. Edward Arnold Ltd, London.

LEVINE N.D. and ANDERSEN F.L. (1973) Development and survival of *Trichostrongylus colubriformis* on pasture. *Journal of Parasitology*, 59, 147-165.

MAFF (1997) *Manual of Veterinary Parasitological Laboratory Techniques*, 18. MAFF Reference Book No. 18. MAFF, HMSO Publications, London, UK.

MARLEY C.L., COOK R., KEATINGE R., BARRETT J. and LAMPKIN N.H. (2003a) The effect of birdsfoot trefoil (*Lotus corniculatus*) and chicory (*Cichorium intybus*) on parasite intensities and performance of lambs naturally-infected with helminth parasites. *Veterinary Parasitology,* 112, 147-155.

MARLEY C.L., FRASER M.D. FYCHAN R. THEOBALD V.J. and JONES R. (2003b) The effect of legume forages on helminth parasites in grazing lambs. *Proceedings of the 7th British Grassland Society Research Conference*, Aberystwyth, 2003, 43-44.

MOSS R.A. and VLASSOFF A. (1993) Effect of herbage species on gastro-intestinal roundworm populations and their distribution. *New Zealand Journal of Agricultural Research,* 36, 371-375.

NIEZEN J.H., CHARLESTON W.A.G. HODGSON J., MILLER C.M., WAGHORN T.S. and ROBERTSON H.A. (1998) Effect of plant species on the larvae of gastrointestinal nematodes which parasitise sheep. *International Journal for Parasitology,* 28, 791-803.

SAYRE R.M. and WALTER, D.E. (1991) Factors affecting the efficacy of natural enemies of nematodes. *Annual Review of Phytopathology,* 29, 149 - 166.

SCALES G.H., KNIGHT T.L. and SAVILLE D.J. (1994) Effect of herbage species and feeding level on internal parasites and production performance of grazing lambs. *New Zealand Journal of Agricultural Research,* 38, 237-247.

Organic Forage Seed Production Systems: from Research to Farm Scale Demonstration

A.H. MARSHALL and H. MCCALMAN
IGER, Plas Gogerddan, Aberystwyth, Ceredigion, SY23 3EB, UK

ABSTRACT

Forage production is the key to predominantly grassland-based organic systems and rotations, and the removal of current derogations to EU organic standards presents significant challenges to organic livestock farmers. Of these, meeting the demand for organically produced forage seed will be one of the most difficult. The National Assembly for Wales now has an organic target of 10% by 2005 and the production of sufficient quantities of organic seed of appropriate varieties of relevant species will be hard to achieve. Conventional herbage seed production in Wales has declined but it has been grown successfully in the past. Following farmer discussion group meetings on grass and clover varieties, and the sourcing and cost of organic seed, a feasibility project to tackle some of the practical challenges of organic seed production was set up with local farmers. Building on small-plot studies at IGER, where organic seed crop management techniques are being developed, the initial challenges and progress of this project are discussed. Results will become available over the next two years as the project develops.

INTRODUCTION

Organic systems of forage production for feeding ruminants are based on a grass plus legume-based sward with regular reseeding (Lampkin, 1990), placing a high demand on seed of appropriate varieties. Organic standards require farmers to use organically grown seed, although a derogation exists at present to permit the use of non-organic seed when suitable organically grown seed is not available. In practice, farmers are required to sow grass seed mixtures which contain a minimum percentage of organically grown seed. For the 2004 season, the minimum percentage of organically grown seed required in mixtures will be 50%, but this is being increased annually. The National Assembly for Wales has an organic target of 10% by 2005 and the production of sufficient quantities of organically produced seed of appropriate varieties of relevant species to meet this target will be difficult (Marshall and Humphreys, 2002). Organic farmer discussion group meetings held at IGER and commercial farms identified their concerns about appropriate grass and clover varieties and highlighted the difficulty of sourcing and cost of organic seed. Conventional grass seed crops require mineral nitrogen at precise stages of crop development to stimulate flowering, ensure good seed filling and produce high seed yields. In organic systems, nitrogen can be supplied by application of animal manure or by using the N fixed by forage legumes, either by relying on the residual N in the soil following a legume, or by using forage legumes sown as companion crops with the grass seed crop (Aamlid, 1999).

Small-plot trials have been established at IGER to develop strategies to overcome some of the problems in organic grass seed production, including the potential of white clover as a source of nitrogen during grass seed production. A feasibility project to tackle some of the practical challenges of organic seed production was set up with local farmers to build on the small-plot studies at IGER, give an economic appraisal of organic forage seed production and assess the potential for organic grass seed enterprises in Wales. This paper reports results of some of the small-plot trials and the challenges and progress in the farmer led feasibility project.

SMALL PLOT TRIALS

Providing sufficient nitrogen to the developing seed crop is a significant challenge for the organic grass seed producer. We have investigated the potential of white clover as a companion crop for different grass species, and its capacity to supply the amount of nitrogen necessary to produce reasonable seed yields.

On 23 July 2001, 3 m x 1.4 m plots of the perennial ryegrass (*Lolium perenne* L.) cv. AberDart, hybrid ryegrass (*L. boucheanum* Kunth.) cv. AberLinnet and Italian ryegrass (*L. multiflorum* L.) cv. AberComo were sown at IGER, Aberystwyth on soil of the Rheidol series at a seed rate of 12 kg/ha, 18 kg/ha and 18 kg/ha, respectively, in rows 20 cm apart. Plots were sown with white clover cv. AberAce (small leaved) or AberHerald (medium leaved) at 3 kg/ha in the same row as the grass (T1), sown between the grass rows (T2) or between alternate grass rows (T3). Although not sown on certified organic land these plots were treated as if they were organic and received no fertilizer nitrogen or chemical weed control. Control plots, sown in an adjacent block within the same experimental field, were treated as a conventional seed crop and received fertilizer N and herbicide at a rate comparable with conventional treatments. Seed production of the treatments was measured during the 2002 growing season. Plots of AberComo and AberLinnet received a silage cut in May. All plots were then allowed to flower and set seed prior to seed harvest. Two weeks before the seed harvest a 450cm^2 quadrat was removed from each plot and the number of reproductive tillers, spikelets per tiller and seeds per spikelet counted. All plots were harvested with a Hege small-plot combine, seed was dried in linen bags over cold air, threshed and seed weight per plot and mean individual seed weight was determined.

No differences in heading date or in general crop development were observed between the plots sown with white clover and the controls. Generally, however, the seed yields were low compared with the plots receiving conventional seed crop management and there were signficant differences in seed yield components. The number of reproductive tillers of the grasses when grown with white clover was comparable with the control in most treatments (Table 1). Only AberDart in T1 and AberLinnet in T2 and T3 had significantly fewer reproductive tillers than the control plots. In contrast, the number of seeds per floret was significantly less than in the control plots in all of the treatments, ranging from 55% to 80% of the controls. This was reflected in the harvested seed yields which were significantly lower in the treatments with white clover than in the control plots. In T1, T2 and T3, AberDart had a significantly higher seed yield than AberLinnett and AberComo, with the harvested seed yield in T2 nearly 80% of the control plots.

Seed yields were greatest in T2, where white clover was sown between the drills of the grass seed crop. Weed content was relatively low in all of the treatments and there were few significant differences between the treatments or between the treatments and the control. Nitrogen content of the grass foliage in early spring has been used as an indicator of the seed yield potential of perennial ryegrass (Rowarth and Archie, 1975). The nitrogen content of the grass foliage was lower in the plots sown with white clover than in the plots receiving conventional levels of fertilizer N. However, the number of reproductive tillers and spikelets per tiller (data not shown) of the grasses sown with white clover and those receiving fertilizer N was comparable, but there were fewer seeds per floret seed yield suggesting that there was insufficient N for seed filling.

Table 1 Number of reproductive tillers, seeds per floret and harvested seed yield of AberComo, AberDart and AberLinnet sown with white clover in the same drill (T1), between each grass drill (T2) or between alternate grass drills (T3).

Treatment	Variety	Reproductive tillers	Seeds per floret	Seed yield
		% of conventional control		
T1	AberComo	98	64	32
	AberDart	86	59	58
	AberLinnet	101	66	43
T2	AberComo	102	65	43
	AberDart	98	55	77
	AberLinnet	85	71	42
T3	AberComo	106	66	40
	AberDart	99	60	63
	AberLinnet	92	80	44

In the UK, conventional grass seed growers defoliate Italian and hybrid ryegrass in the spring prior to the seed harvest (Marshall and Hides, 1999) to remove excessive leaf growth that can impair harvestability. On farms with livestock, the forage removed at defoliation is also used to make high quality silage. The low seed yield of AberComo and AberLinnet in comparison with the perennial ryegrass variety AberDart suggests that nitrogen levels within the system may be insufficient for regrowth after the silage cut. Where a silage cut is taken, an application of animal manure may be required to supplement the nitrogen removed in the silage. There was no significant difference between the treatments or clover variety in the weed content of the seed sample suggesting that the weed problem was no worse under this system than a conventional system. Weed content in the second harvest year will be monitored.

FARM SCALE DEMONSTRATION

Integrating the results from small-plot trials into organic farming systems has been the focus of a feasibility project conducted on commercial farms. Four farmers were recruited from within discussion groups with a range of farm types and systems. Field plots were designed in discussion with the farmers to address some of the important challenges facing organic forage seed production. Each farm has focused on different aspects to provide a range of demonstration points and assess the feasibility of different approaches to organic forage seed production. The topics explored included:

- the use of white clover sown between or within the grass rows as a nitrogen source for a hybrid ryegrass seed crop (with one half cut for silage and farmyard manure applied during the season, the other with no manure applied). The seed yield of the area without silage cut was 640 kg/ha with a weed content no greater than a conventional seed crop.

- the potential of different fertility building legumes (white clover/red clover/ vetch/lupins) to provide nutrients to the following grass seed crop;
- the response of different grass species (perennial ryegrass/hybrid ryegrass/timothy) to a red or white clover fertility building phase;
- integration of herbage seed crops into a whole organic farm system.

To optimize input of stakeholders, the participating farmers hosted a meeting with Organic Seed Certification and NIAB seed certification personnel to explore the issues in organic forage seed production and develop a better understanding of the challenges involved for all. The farmers also visited a recently converted conventional herbage seed producer in England and exchanged information and ideas on organic seed production and approaches to growing and harvesting organic arable and grass crops.

The decision-making process for the management of plots has been guided to a large extent by the participating farmers and other group members. Input from other interested parties (NIAB and organic certification bodies) is invaluable so that the problems and challenges are being tackled together and a greater understanding is achieved. On one site weed control is an important issue and on another capitalizing on fertility build up stimulated a good discussion, particularly when taking into account the practicalities of fitting herbage seed production into a crop rotation that fits the farm. The work is on-going and the interest and enthusiasm of the farmers has continued to increase following on-farm meetings and discussions.

ACKNOWLEDGEMENTS

This work was part funded by the National Assembly for Wales Farming Connect programme.

REFERENCES

AAMLID T. S. (1999) Organic seed production of timothy (*Phleum pratense*) in mixed crops with clovers (*Trifolium* spp.). In: *Proceedings from the 4th International Herbage Seed Conference,* Perugia, Italy, pp. 28-32.

BOELT B. (1997) Undersowing *Poa pratensis* L., *Festuca rubra* L., *Festuca pratensis* Huds., *Dactylis glomerata* L. and *Lolium perenne* L. for seed production in five cover-crops I. The yield of cover-crops and the seed yield of the undersown grasses. *Journal of Applied Seed Production,* 15, 41-47.

LAMPKIN N. (1990) *Organic Farming.* Ipswich: Farming Press Books.

MARSHALL A.H. and HIDES D.H. (1999) Maximizing seed yields of tetraploid hybrid ryegrasses (*Lolium x boucheanum* Kunth.). *Journal of Applied Seed production* 17, 35-42.

MARSHALL A.H. and HUMPHREYS M.O. (2002) Challenges in organic forage seed production. In: J. Powell (ed.) Proceedings of the UK *Organic 2002 Conference,* 26-28 March 2002, Aberystwyth, pp. 95-96. Organic Centre Wales, University of Wales Aberystwyth.

ROWARTH J.S. AND ARCHIE W.J. (1995) A diagnostic method for prediction of seed yield in perennial ryegrass. *Proceedings of the International Herbage Seed Conference,* 3, 64-67.

What do Organic Farmers Look for when Choosing Herbage Seed?

H. MCCALMAN
IGER, Plas Gogerddan, Aberystwyth, Ceredigion, SY23 3EB, UK

ABSTRACT
Forage production is key to organic farming and selection of herbage seed mixtures is important to the system and for economic viability. The removal of the derogation for use of non-organic seed and feed will mean that reseeding decisions will be an increasingly important part of the organic farming system. As part of a study to review the current testing regimes for new herbage varieties, organic farmers were amongst those asked to complete a short questionnaire to give their views on what was important on their farm when choosing seed for reseeding or renovating grassland. Results from this are outlined and the implications for future seed and breeding programmes discussed.

INTRODUCTION
Organic systems of forage production for feeding ruminants are based on a grass-legume based sward with regular reseeding (Lampkin, 1990), placing a high demand on seed of appropriate varieties. Value for Cultivation and Use (VCU) trials play an important role in directing the forage grass and clover breeders by determining the type of varieties that are added to the UK National List. As grass and clover seed accounts for only 1% of the seed purchased by farmers, funds available to seed companies for testing and marketing new candidate varieties are limited. The last 30–40 years of the twentieth century saw intensification of farming with increasing polarization of livestock and arable production associated with greater use of fertilizers and agrochemicals. During this period the area of land subject to rotational grassland and arable cropping decreased, and the area in long term leys or permanent grass increased. These factors all tended to diminish the market volume for herbage seed. Patterns in the change of organic land use suggest that although crop rotations are key to many systems, 92% of all organic land in 2002 was under permanent grassland. However, the area of grass leys on many farms has increased and reflects the importance of fertility building and quality forage within the system (Soil Association, 2002; Wilkins *et al.*, 2003). At the same time there was increasing dominance of perennial ryegrass, especially in medium- and short-term leys. More recently the use of fertilizers has decreased slightly, and there has been growth in organic production. Wales and England both have organic action plans to promote the growth of organic farming and for setting targets (Welsh Agri-food Partnership, 1999; Defra 2002). Other agricultural policy directions promote more diversity in land use and, whilst it is not yet clear what impact CAP reform may have on farming systems and crop distribution, it seems likely that herbage seed selection decisions will place new priorities on the plant breeder and seed producer. A greater importance of legumes in an industry more dependent on the market and less driven to high stocking rates is one clear impact. During organic discussion group meetings held in Wales, farmers have been voicing concerns over the lack of appropriate varieties and limited availability of organic seed. As part of a study to review the current testing regimes for new herbage varieties, organic farmers were amongst those asked to give their views on what was important on their farm when choosing seed for reseeding or renovating grassland. This paper summarizes

results from this survey and discusses implications for future seed and breeding programmes.

METHODS

A survey was carried out which was targeted at farmers who had participated in IGER's *Grassland Technology Transfer Programme* or who attended discussion group meetings at agricultural shows and events in Wales during 2003. There was a range of questions relating to reseeding and seed selection choices and the results presented relate to the responses from organic farmers. A total of 275 farmers, including 109 organic farmers based in Wales and participating in Farming Connect activities, responded to a questionnaire. The questionnaire requested information on farm area, enterprises, areas with different types and ages of grass swards, fertilizer use (on conventional farms), use of grass and clover seed for different reseeding and renovation situations, use and perceived importance of white and red clovers, and or present use of alternative grasses or forage herbs. Respondents were also asked to select the three characteristics most important for seeds mixtures for grazing, silage and cutting-plus-grazing. A copy of the questionnaire can be obtained from the author.

RESULTS

1. The three most important characteristics when choosing a seeds mixture

Persistency was the most highly valued trait (22%), followed by total annual yield (20%), early spring growth (13%), digestibility (12%), ground cover (9%), mid-season growth (9%), heading date (5%) and late-season growth (3%). Generally, these accord with the current weightings for VCU testing decisions. However, early spring growth was considered to be an important characteristic and this currently is given no weighting in grasses and only a low weighting in white clover.

2. How important are clover-rich swards to you?

As might be expected, clover-rich swards were rated as 'very important' by organic farmers, particularly for swards for grazing (Table 1).

Table 1. Responses to perceived importance of clover-rich swards?

Score	Grazing	Cutting	Cutting+Grazing
1	88	85	71
2	5	5	6
3	4	6	15
4	0	1	5
5	2	2	3

(where the scores: 1 = 'very important' and 5 = 'not important at all')

3. Do you use red clover?

More than half the organic farmers used red clover with a further one-fifth considering its use. This highlights its growing importance in organic systems. For comparison, Table 2 gives information on the response of the conventional farmers that were also surveyed. Although 24% of the organic farmers never used red clover, the conventional farmers were slightly less positive about its use than organic farmers, and one-quarter of all the farmers may do in the future.

Table 2. Respondents' use of red clover (as % within each farm category)

	Organic	Conventional	All farms
Never	24	28	26
Sometimes	39	33	35
Always	15	12	13
May do	21	28	25

4. Which of the following do you use, or have you used in the past?

Timothy was used by 58% of the farmers responding, cocksfoot by 33% and meadow fescue by 16%. Of the more novel species, chicory was used by 6%, which was a greater proportion than herb mixes (3%), yarrow (2%) or lotus (1%). Twenty-two percent did not use any of these other species.

DISCUSSION

Current testing procedures attach no weighting for early spring growth in grasses, and only a low spring-growth weighting in white clover, though these were considered to be highly valued characteristics by the farmers surveyed. It is important that the needs of farmers and end users are taken into account when the testing system is revised. These characters are an integral part of enabling organic farmers to be sustainable both ecologically and economically. Where breeding effort addresses these needs it is imperative that the testing system that ensures their place on the National Recommended List gives recognition for valuable characteristics. When revising weightings for future testing this should be incorporated appropriately to ensure that new varieties meet the needs of twenty-first century UK farming, whether organic or low input, and address these needs. However, considering that the clearly most important issues were persistence and yield it is perhaps not remarkable as both these factors have great importance predominantly grassland farms and livestock holdings.

The survey did not contain a sufficiently high proportion of arable-livestock mixed farms, where short- and medium-term leys are important, and which may result in a different rating to features of sward longevity. As the organic farming sector increases, and also as the number of farmers entering agri-environmental schemes (e.g. Countryside Stewardship/Tir Gofal) also increases, this may result in an increase in the mixed arable-livestock farm type. This could then become a feature deserving some further investigation.

The importance of clover in organic systems and value attached to the use of red clover means the views of organic farmers should be taken into account by breeders when planning programmes of genetic improvement.

It might have been expected that the importance of a high clover content would be similar in cutting and grazing, as it supports the fertility building phase of a crop rotation. The greater importance attached to white clover for grazed swards may be a reflection of limited use of rotational cropping on organic farms in Wales; many of the respondents would have been from all-grass farms with reseeding limited to renovation, rather than full reseed as part of a crop rotation. In addition, when current derogations to EU livestock standards are removed in 2005, and all feed will be 100% organic, the need for home-grown protein-rich forage is likely increase the importance attached to clover-rich swards for cutting.

There was considerable interest in the use of other species, particularly timothy. This supports a commonly held belief that organic farmers value diversity in the swards for both animal health and diversity of fauna.

It is important that future testing takes account not only of the characteristics valued by organic farmers, but that the varieties are tested under organic conditions. Current testing regimes use high levels of fertilizer nitrogen.

Where breeding effort addresses these needs it is imperative that the testing system that ensures their place on the National Recommended List gives recognition for valuable characteristics. From other parts of this study it seems that both persistence and early spring growth are also considered highly desirable by conventional farmers (Wilkins *et al.*, 2003). When revising weightings for future testing this should be incorporated appropriately to ensure that new varieties meet the needs of UK farmers in the twenty-first century, including organic and low input.

ACKNOWLEDGEMENTS

This work was funded by DEFRA.

REFERENCES

DEFRA (2002) *Action Plan to Develop Organic Food and Farming in England.*

LAMPKIN N. (1990) *Organic Farming*. Ipswich: Farming Press Books.

SOIL ASSOCIATION (2002) *Annual Report*. Bristol: Soil Association.

WELSH AGRIFOOD PARTNERSHIP (1999) *Welsh Organic Food Sector - a strategic action plan.*

WILKINS P.W., MCCALMAN H.M., MARSHALL A.M., BENTLEY S. and MITCHELL R. (2003) *The implications of recent trends in grassland farming for Value of Cultivation and Use (VCU) trials of forage grass and clover varieties.* Report for Defra. IGER, UK.

Current Sheep Dipping Practices to Control Ectoparasites on Welsh Organic Sheep Farms

D. FROST and B.M.L. McLEAN
ADAS Pwllpeiran, Cwmystwyth, Aberystwyth, Ceredigion, SY23 4AB, UK

ABSTRACT

Under current organic livestock standards organic farmers are prohibited from using organophosphates (OPs) but are permitted to use synthetic pyrethroids (SPs) to control ectoparasites such as scab mites. However, SPs are known to have high aquatic toxicity, and concerns have been raised as to the impact of organic farmers' SP use. At present there is little information on how many organic farmers use these permitted chemicals as part of their flock management or how they dispose of their dip. A survey of both organic and conventional sheep farmers in Wales was carried out between December 2002 and April 2003. The survey was designed to investigate farmers' current ectoparasite control and treatment practices. As part of the survey current practices for dip disposal were also investigated. Of the organic sheep farmers surveyed, 58% stated that they treated their flocks for ectoparasite infestations, 29% used pour-on formulations to treat ectoparasites, 16% used plunge dipping, 11% used sheep showers and 5% used injectable products. All organic farmers who carried out plunge dipping used a SP-based product. When asked how they disposed of spent dip, 83% of organic farmers using dip claimed to dilute spent dip before spreading to land, and 16% claimed to add a bacteriostat to the spent dip. However, none carried out any other treatment of spent dip such as the addition of slaked lime.

INTRODUCTION

Organophosphate (OP) based dips offer a broad spectrum control against all major ectoparasites in the UK. However, there is continuing controversy over the safety to dip operators and the environment. The use of OPs is prohibited by United Kingdom Register of Organic Food Standards (UKROFS – superseded in 2003 by ACOS, the Advisory Committee on Organic Standards), primarily because of concerns about mammalian toxicity. Organic farmers are permitted to use synthetic pyrethroids (SPs) (dip products and pour-on products) and/or macrocyclic lactones (injectable products) to treat/control ectoparasites providing a derogation has been obtained from the certifying body. SPs have been shown to be considerably more toxic to aquatic organisms than OPs. The environmental impact of SPs is not limited to levels in dip but also to residues in sheep fleeces. SPs may be removed from the wool by rainfall and then deposited in the local environment. Recent results from the Environment Agency show that the majority of sheep dip pollution incidents involve SP dips. Concerns have been raised that as organic farmers can only use SP based dips they may be contributing to the high levels of SP based pollution incidents. However, there is little information as to current practices amongst organic sheep farmers to control/treat ectoparasites or on how these farmers dispose of the spent dip. In order to address this lack of information a survey was carried out by ADAS Pwllpeiran to investigate current practices amongst sheep farmers in Wales.

METHODOLOGY

A survey form was designed and trialed at an Organic Open Day at ADAS Pwllpeiran. The form was then revised and trialed at the Royal Welsh Agriculture Society Winter Fair 2002. Further modifications were made and the final survey form was used in both a postal and telephone survey. For comparison purposes both conventional and organic farmers were included in the survey.

RESULTS

In total 134 questionnaires were completed with 96 completed by conventional sheep farmers and the remaining 38 being completed by either in-conversion organic farmers or fully registered organic farmers. Fifty-eight per cent of organic farmers surveyed treated their flocks for ectoparasites compared to 67% of conventional farmers.

In the survey, farmers were asked to list the ectoparasites for which they treated their flocks. Results are shown in Table 1.

Table 1. Percentage of organic and conventional farmers who treated their flocks for different ectoparasites.

Method	Organic	Conventional
Flystrike	52.6	58.3
Scab	21.1	61.5
Other		

The percentage of organic farmers who listed flystrike as a parasite for which they treated their flocks was comparable with that of conventional farmers. However, a much lower percentage of organic farmers than conventional farmers listed scab as a reason for treatment. Less than 10% of conventional and organic farmers surveyed treated for other ectoparasites and, of those that did, lice and ticks were listed as the main ectoparasites.

Farmers were also asked to indicate if they treated their flocks for a combination of ectoparasites or for single parasite control (scab-only' or 'flystrike only' etc) (Table 2).

Table 2. Percentage of organic and conventional farmers using different types of treatment against flystrike and scab.

		Organic	Conventional
Flystrike treatment	Flystrike only	36.8	7.3
	Combined treatment	15.8	51.0
	Total	52.6	58.3
Scab treatment	Scab only	2.6	11.5
	Combined treatment	18.4	51.0
	Total	21.0	62.5

Despite a large percentage of conventional farmers listing scab as an ectoparasite to be treated, relatively few of these treated for 'scab only'. The majority treated for scab in combination with other ectoparasite control. A similar pattern for scab was seen amongst organic farmers surveyed. Conventional farmers also tended to treat for flystrike in combination with other ectoparasites. Similarly, despite other ectoparasites such as lice and ticks being listed by both conventional and organic farmers, neither of these two parasites were treated singly but were always treated in combination with other ectoparasites. However, an exception to this pattern can be seen in the relatively high proportion of organic farmers who treated for 'flystrike only'.

Where farmers answered yes to the treatment of ectoparasites they were then asked to list treatment methods and frequency of treatment (Tables 3 and 4).

Table 3. Preferred dipping method of farmers who treated their sheep annually.

Method	Organic	Conventional
Plunge dipping	22.7	76.1
Pour-on	54.5	16.5
Combination of treatments	5	19

Table 4. Frequency of treatment of conventional farmers who treated their sheep annually.

Method	Organic	Conventional
Once per year	54	42
Twice per year	27	52
Three or more times per year	14	7

Of farmers who treated for ectoparasites, a much higher percentage of the conventional farmers listed plunge dipping as the preferred dipping method, compared with organic farmers. In contrast, the use of pour-on products was higher amongst organic farmers. A lower proportion of organic than conventional farmers used a combination of treatments (Table 3).

All organic farmers who carried out plunge dipping operations used a SP-based product, whereas of the conventional farmers who carried out plunge dipping, only 39% used an SP-based product. Overall, 16% of the organic farmers surveyed used an SP-based dip product, compared with 24% of conventional farmers surveyed. Thirty-eight percent of all conventional farmers surveyed used an OP-based dip product.

Where farmers carried out plunge dipping, 53% of conventional farmers diluted spent dip before spreading to land, compared with 83% of organic farmers who carried out plunge dipping operations. Of all the farmers surveyed who carried out plunge dipping, only one treated dip with slaked lime before spreading.

Farmers were asked if they used a mobile dipping contractor to carry out dipping operations. Five per cent of all organic farmers surveyed used a mobile dipping contractor, compared with 23% of conventional farmers surveyed. On all organic holdings where contractors were used they were also responsible for the disposal of spent dip, but this was the case on only 74% of the conventional holdings using contractors. On 42% of the holdings using contractors the spent dip was removed from the farm, whereas on 32% of holdings the spent dip was spread on the farm land. It is interesting to note that

26% of farmers using contractors in this survey did not know how the contractor disposed of spent dip.

DISCUSSION

The percentage of sheep farmers treating their flocks for ectoparasite infestations was remarkably similar for organic (58%) and conventional (69%) farmers, but there was a marked difference in the species of ectoparasites being treated. On conventional farms scab was seen as the most important threat, whereas on organic farms flystrike was seen as the greater threat.

From the evidence of several of the completed survey forms however, some farmers appeared to be using incorrect treatment methods. Some were using injectable products to treat blowfly and/or lice infestations but injectables are licensed only for scab treatment. Some were using pour-on products to treat both blowfly and scab but pour-on products are licensed only for blowfly treatment and prevention. In some cases farmers were using multiple treatment methods to treat ectoparasite infestations. The use of multiple treatments or incorrect treatments is largely a reflection of the plethora of products available, and suggests a lack of understanding as to which products are appropriate and licensed for the treatment of specific ectoparasites.

The survey found that SP dips were used by a lower percentage of organic sheep farmers than conventional sheep farmers. As there are fewer organic sheep farmers overall, there is little evidence that organic sheep farmers contribute disproportionately to the level of SP-based pollution incidents.

ACKNOWLEDGEMENTS

The authors would like to acknowledge Organic Centre Wales and Farming Connect for funding this work.

Modelling Organic Dairy Production Systems

P. NICHOLAS[1], S. PADEL[1], N. LAMPKIN[1], S. FOWLER[1], K. TOPP[2] & R. WELLER[3]
[1]Institute of Rural Sciences, University of Wales Aberystwyth, SY23 3AL, UK
[2]Scottish Agricultural College, Edinburgh, EH9 3JG, UK
[3]IGER, Trawsgoed Farm, Aberystwyth, SY23 4LL, UK

ABSTRACT
In this study, a large number of organic dairy production strategies were compared in terms of physical and financial performance through the integrated use of computer simulation models and organic case study farm data. Production and financial data from three organic case study farms were used as a basis for the modelling process to ensure that the modelled systems were based on real sets of resources that might be available to a farmer. The case study farms were selected to represent a range of farming systems in terms of farm size, concentrate use and location. This paper describes the process used to model the farm systems: the integration of the three models used and the use of indicators to assess the modelled farm systems in terms of physical sustainability and financial performance

INTRODUCTION
The justification for model building and, in this instance, the use of existing models in systems research, is that experimentation on real systems can be very site specific, time consuming and expensive. However, modelling cannot exist without data from observation and experimentation, which is necessary to develop and validate models. An integrated approach is therefore required.

A number of studies developing and using simulation models for dairy systems have been undertaken (Topp and Doyle, 1996a; Topp and Doyle, 1996b; Topp and Hameleers, 1998). Only one (Häring, 2003) looked specifically at different organic dairy systems. In Häring's research, however, the modelling simulated the effects of potential policy developments on the profitability and adaptation strategies of organic dairy farms, rather than looking at the physical farm, which is the subject of this work.

METHODOLOGY
Three differing milk production strategies were selected to model: arable and forage combined (Arable); forage only (Forage) and forage with purchased concentrates (Purchased). They represented the extreme variants of strategies currently found on organic farms throughout the UK and Europe, ranging from predominantly self-sufficient (high resource use efficiency relying on either home produced arable feeds or forage) through to relying heavily on imported feeds (generally more intensive, profitable systems).

The Arable and Purchased strategies are demonstrated on two experimental organic dairy systems at the IGER farm, Ty Gwyn, located in mid Wales. Data from the 1999 season from these two systems were used for validation of the models in an earlier study (Nicholas *et al.*, 2002). Elements of the three contrasting strategies (Arable, Forage and Purchased) are also represented on three case study farms A, B and C (Nicholas *et al.*, 2002). Sufficient physical and financial data were collected from these farms to allow them to be used for the modelling exercise. These case study farms formed the basis for the main body of modelling.

The complete methodology is described in Nicholas *et al.* (2002). In summary, three management strategies (Arable, Forage and Purchased) were applied to three different sets of farm resources: A, B and C (base resources); which comprised areas in forage/cereal crops, cow numbers, calving date, milk yield per cow and milk quota. Each of these combinations was modelled in three different climatic environments (Devon, Pembrokeshire and Shropshire) and with two different concentrate feeding strategies (flat rate and production based concentrate feeding).

In summary, there were a total of 54 model runs:
3 base resources x 3 scenarios x 3 climates x 2 concentrate feeding strategies.

Three existing models were used, which together could model an entire organic dairy system. Two models simulated the physical processes on the farm (Dairy Systems Model (DSM) (Topp and Hameleers, 1998) and Feedbyte (Schofield *et al.*, 1998), and the other, a whole-farm model, combined data outputs from the first two models to provide nutrient budgets and gross margins for the systems as a whole (OrgPlan (Padel, 2002)). The following steps were used for the modelling (Nicholas *et al.*, 2002).

Step 1: Simulate grazing season
Milk and silage production during the grazing season were modelled with the DSM. The total amount of silage that was harvested and available to feed during the winter period was calculated, as was milk production from pasture and any concentrates that were fed during the grazing season. The difference between milk production during the grazing season and total quota was taken as the target milk production for the housing period. This target milk production was required as an input for FeedByte.

Step 2: Determine winter feed requirements
FeedByte was used to determine feed requirements during housing and to develop a winter feed ration to meet the target milk production for the housing period defined in Step 1. There was a maximum of eight feeds (forages and concentrates) available depending on the scenario being modelled (Nicholas *et al.*, 2002).

Step 3: Feasibility of the combined system
From these two modelling procedures, both summer and winter-feed requirements to produce milk to quota were identified. It was then necessary to calculate whether there was sufficient land/feed resources available to provide the feed required (based on standard organic yields for the various crops), and hence determine the viability of each system. Farm systems were also rejected at this stage if the quantity of concentrate in the diet on a daily basis was higher than 40% of daily dry matter intake, the maximum allowed under organic regulations. A total of 12 of the 54 model systems were discarded based on this criterion.

Step 4: Calculate whole farm gross margin and nutrient budgets
The outputs from the DSM and Feedbyte from the viable systems were entered into OrgPlan to model whole farm gross margins and farm-gate nutrient budgets.

Step 5: Calculate financial and sustainability indicators
From the outputs of OrgPlan, a number of sustainability and financial indicators were calculated and analysed for each model run to attempt to identify the best systems in

terms of profitability and resource use sustainability (Nicholas *et al.*, 2002). The financial indicators were whole farm gross margin per hectare (GM/ha), per cow (GM/cow) and per litre of milk (GM/l) and kg of purchased concentrate per £ gross margin (kg/GM). Resource use indicators were P and K farm gate balances, kg of purchased concentrate per litre of milk produced (kg/l) and the N surplus per litre of milk produced (g N/l). For both the financial and resource use outputs the modelling provided a score of between 0 and 20 for each system – the higher the score, the better.

RESULTS

Climate

All Pembrokeshire and Shropshire systems were compared and whilst they both had very similar financial scores, resource use scores were higher in the Pembrokeshire climate. Devon Forage systems (the only viable Devon systems as Devon Arable and Devon Purchased were discarded in Step 3 above) were compared with Pembrokeshire and Shropshire Forage systems and again, financial scores were very similar (12 or 13 out of a possible 20), but resource use score was lower for Shropshire Forage (11) than for Devon Forage (15) and Pembrokeshire Forage (14).

Farming Strategy

Financial score in particular, and resource use score to a lesser extent, were, on average, lower for the Purchased systems than for the Arable or Forage systems. The key indicator that caused the total resource use score to be low for Purchased strategy systems was kg purchased concentrate per £ gross margin (kg/£). Both the Arable and Forage systems scored 5 out of 5 for this indicator because no purchased concentrate was used in either system. The Purchased strategy, however, relied entirely on purchased concentrates and therefore scored poorly. This indicator was included specifically for the purpose of incorporating a measure of self-sufficiency into the resource use score.

The other indicator that contributed to the low resource use score of the Purchased strategy systems was GM/l of milk. Whilst, on average, the Purchased systems have exactly the same milk yield as the Arable systems, gross margin is lower for Purchased systems because of the high cost of purchasing concentrate feeds as opposed to growing them at home. The organic compound feed used only in the Purchased systems had a particular influence on gross margin, as it was the most expensive concentrate.

When the Devon Forage systems were compared with the Pembrokeshire and Shropshire Forage systems, the key difference between the three averages was in the potassium score. The Devon Forage systems had the best potassium score (4), followed by Pembrokeshire (3) and Shropshire (1) Forage systems. All other individual indicators were very similar.

Base Resources

The financial score for base resource set C was significantly greater than for resource sets A and B. The key factors resulting in a higher financial score for base resource set C were GM/cow and GM/ha. This was primarily due to the fact that quota was used to determine annual milk yield. For resource set C the annual quota per cow was much higher than for sets A and B, and the average stocking rate for all the resource sets was 2.0 cows/ha. On a per cow basis, milk yield for resource set C ranged from 6291 to 8204 l, compared with 4839 to 5149 l for base resource sets A and B. Concentrate price did not influence this result because the average for each base resource set was taken across the three farming strategies (Arable, Forage and Purchased) to which all inputs are attached.

CONCLUSIONS

The aim of the dairy modelling project, which was to compare a broad range of organic dairy systems in terms of financial and resource use performance, has been fulfilled. By no means have all the combinations of variables and systems been modelled, but the results provide useful information to organic dairy system managers as to what effect systems changes might have on financial and resource use performance of their farms.

In terms of overall performance, the results of the modelling suggest that as a general rule the management strategies were ranked Arable first, followed by Forage and lastly Purchased. Also, systems based on C resources, which have high yielding, autumn calving cows, tended to outperform systems based on A and B resources (moderately yielding, spring/split calving cows). Neither climate, nor concentrate feeding strategy greatly influenced the overall ranking of the systems.

On an individual farm level, a farmer's objectives need to be identified before selecting the "best" system for their specific set of resources. What the modelling provides is basic financial and resource use performance data for each set of physical resources under three management strategies. Based on these data, farmers can identify which system might best meet their goals and then undertake further analysis looking at how that system might fit into the wider farm context and economic environment, considering such aspects as fixed costs, farmer expertise in crop/forage husbandry and concentrate and milk price.

REFERENCES

HÄRING A.M. (2003) Organic dairy farms in the EU: Production systems, economics and future development. *Livestock Production Science,* 80, 89-97.

NICHOLAS P.K., PADEL S., FOWLER S., LAMPKIN N., TOPP C.F.E. and WELLER R. (2002) The validation of a computer simulation model for use in organic dairy farms in the United Kingdom. In: Powell, J. (Ed), *UK Organic Research 2002: Proceedings of the COR Conference*, 26-28th March 2002, Aberystwyth, UK, 199-200.

NICHOLAS P.K., PADEL S., FOWLER S. and LAMPKIN N. (2002) *Organic Dairy Farm Modelling 2001/2002.* Final Report to DEFRA Project Code OF0146. Institute of Rural Studies, University of Wales, Aberystwyth, UK.

PADEL S. (2002) *Conversion to Organic Production Software.* Final Report to DEFRA Project Code OF0159. Institute of Rural Studies, University of Wales, Aberystwyth, UK.

SCHOFIELD J.L., CALDER J.M., FRASER I.R., LEWIS M., OLDHAM J.D., OFFER N.W. and ROOKE J.A. (1998) FeedByte – Ration Formulation and Evaluation. In: *Proceedings 7th International Conference on Computers in Agriculture*, Orlando, Florida, USA, 26-30 October 1998. American Society of Agricultural Engineers (ASAE), St Joseph, USA, 903-909.

TOPP C.F.E. and DOYLE C.D. (1996a) Simulating the impact of global warming on milk and forage production in Scotland: 1.The effects on dry matter yield of grass and grass/white clover swards. *Agricultural Systems,* 52, 213-242.

TOPP C.F.E. and DOYLE C.D. (1996b) Simulating the impact of global warming on milk and forage production in Scotland: 2.The effects on milk yields and grazing management of dairy herds. *Agricultural Systems,* 52, 243-270.

TOPP K. and HAMELEERS A. (1998) Mechanistic Modelling of Dairy Production Systems. In: Tijskens, L.M.M., Hertog, M.L.A.T.M. (Eds), Proceedings of the International Symposium on Applications of Modelling as an Innovative Technology in the Agri-Food-Chain Model-It. *Acta Horticulturae,* 476, 157-163.

Partner Farms: Understanding the Importance of Grass-Clover in both Livestock and Arable Production

U. PRINS, J. DE WIT and T. BAARS
Louis Bolk Instituut, Hoofdstraat 24, 3972 LA Driebergen, The Netherlands

ABSTRACT

Partner farming is a concept of co-operation between organic farms specialized in different enterprise types (e.g. livestock and arable). As a partner farm, each highly specialized farm can benefit from the advantages of a mixed farming system without losing the advantages of its specialization. With partner farms many issues are addressed that challenge the organic sector: reduction of conventional inputs (e.g. manure, concentrate feed), regionalized production, and traceability of inputs. Grass-clover plays an important role in meeting these challenges, by building sustainable crop rotations for both livestock and arable farms, not only providing an essential N-input on arable farms, but also reducing weeds and serving as a product which can be traded for animal manure. Viable solutions to the challenges of the organic sector can only be found by involving farmers in the projects and using a variety of farming system research techniques.

INTRODUCTION

Dutch agriculture is characterized by highly specialized, intensive farms pressured by high land prices. Production is maintained by using a lot of manure and feeding high inputs of concentrates to animals. In the organic sector the situation is similar, although it is slightly more extensive than in conventional agriculture. A large part of the manure and concentrate feed used is, even now, still of conventional origin. In the future, however, this will have to change. In 2005 all concentrate feed must be of full organic status, and in 2002 a process was initiated in the Netherlands gradually to abandon the use of conventional manure. Dutch organic agriculture will face an immense challenge to maintain the necessary high net return on the land, while reducing the use of manure and readjusting to higher prices and lower availability of concentrate feed. Using data from Dutch organic agriculture in 2001, it is estimated that on average ±65 kg N/ha is available in organic animal manure for the whole organic sector. This represents a substantial shortfall compared with current rates of manuring (140-170 kg N/ha) (Prins *et al.*, 2002). Cooperation between livestock farmers and arable farmers (partner farms), with grass-clover as a key element, offers a potential solution.

RESEARCH METHODS

In the Partner Farm project, conducted by the Louis Bolk Institute between 1996 and 2003, farmers were actively involved in the action research. Different research techniques were used to enable the farmers to participate and contribute in the development. As a result not all findings are thoroughly statistically based, but are embedded in organic farming practice; technically, economically, practically and ethically. The emphasis of the research techniques used lies with on-farm activities and experimentation (replicated experiments, non-replicated trials and observation of, and reflection on, innovative actions of farmers, including pattern recognition) and discussions between farmers and researchers (interactive learning and transferring hypotheses derived from trials or desk studies to farming practice) (Baars, 2002a).

GRASS-CLOVER AND LIVESTOCK FARMING

Before the use of conventional manure was restricted by law, dairy farmers used most of the on-farm produced manure for their own land. Many of them thought 30-40 m^3/ha/year of slurry was necessary to maintain grassland production. In the Partner Farm projects livestock farmers started to experiment with lower fertilization rates so that they could provide their arable partner farm with organic manure. To their surprise, DM-yield was hardly affected by reducing manuring rates to 10-15 m^3/ha/year provided the clover content in the sward was sufficiently high (on average 35-50% of the DM yield). Experiments even showed that when fertilization with animal manure was abandoned completely, grass-clover production decreased only slightly (10 instead of 12 t DM/ha), provided that soil potassium and phosphate levels were maintained (Baars, 2002a and 2002b). The reduced N-input by manure is compensated by an increase in clover content of the sward and an increase in N-fixation. For livestock farmers herbage production at low fertilization rates (10-15 m^3/ha) is now considered the optimum, also because it seems that a small input of manure is better for soil life. Maintaining a good clover content in the sward is essential to maintain production at these fertilization rates, but it enables a considerable amount of animal manure to be made available for the arable partner farm.

GRASS-CLOVER IN AN ARABLE ROTATION: Not just simple economics

Although manure can be made available by livestock farmers, manure will become less available with the banning of conventional manure. Instead of 130-140 kg N/ha used at present, only 65-70 kg N/ha will have to suffice. The incorporation of leguminous crops in the rotation will become more and more important (Baars, 1998). On arable farms, grass-clover may seem to have little attraction since net financial returns are very low (close to zero) compared with other legumes such as beans and peas. However, in practice, grass-clover seems to be growing increasingly popular amongst arable farmers. One of the reasons for this is that grass-clover serves as a product to trade for manure. Livestock farmers are much more willing to sell a substantial part of their manure at reasonable prices to their partner farms than to an anonymous market. Obtaining grass-clover herbage in return strengthens the bond and triggers the livestock farmer to experiment on his own farm in reducing the rate of manuring. A second advantage of grass-clover over other leguminous crops is the reduction of weeds. Controlling perennial and annual weeds is given high priority, mainly by arable farmers that converted to organic farming some years ago. As grass-clover is a vigorously growing crop that is cut four times a year, it is one of the few crops that effectively reduces the weed seedbank. For this reason cereals, which are known to increase weeds in the seedbank, are increasingly substituted by grass-clover in the rotation, even though this means a reduction in net return. The farmers accept the loss of income from this crop as it will be compensated by higher yields in the following crop and lower weeding costs. In 2002 the experimental farm OBS in Nagele calculated that the substitution of oats by grass-clover resulted in an increase in production of the following crop (onions) which was estimated at 7.1 t/ha (19%), and a reduction in hand-weeding of ±60 hours/ha. This represents an increase in net return of ±€2500/ha, which more than compensates for the loss in net return (±€700/ha) due to the replacement of oats by grass-clover. This is true for crop rotations where most of the revenues come from high yielding, intensive crops like onions, cabbages, lettuce, carrots and spinach. Grass-clover improves the soil, leaving a good soil structure, clean of weeds and with increased soil fertility, making good yields possible for the following high-value crop. In these rotations grass-clover is also appreciated as it needs little attention from the farmer. Contractors do most of the work, so all the farmers' attention can be focused on the cash crops. Thus, despite the net return of grass-clover being close to zero, it is increasingly seen as a vital building block of a healthy profitable arable rotation.

MANAGING GRASS-CLOVER IN AN ARABLE ROTATION

Management of grass-clover in an arable rotation differs from that on livestock farms since it plays a different role. While grass-clover is the main crop for livestock farmers, it is a secondary crop for the arable farmer. On arable farms grass-clover needs to suppress weeds and improve soil fertility. Therefore, specific mixtures of grass-clover are required and fertilizer strategies may also differ. On arable farms grass-clover swards are rarely retained for more than one season. Thus, it has to be an easily established, highly productive mixture. Red clover is the preferred legume for these mixtures since DM production is higher than from mixtures with just white clover (see Table 1). The biggest difference in production occurs in the first two cuts, as red clover starts to grow earlier than white clover.

Table 1. Effect of sowing date and seed mixture on DM yield and weed suppression

Sowing date	Mixture	1st & 2nd cut yield (tDM/ha)	Total yield (tDM/ha)	Weed biomass in 1st cut (tDM/ha)
Late summer	Grass/WC/RC	9.4a	13.4a	0.14
	Grass/WC	7.0b	10.9b	0.24
Spring	Grass/WC/RC	6.9a	8.9a	0.95
	Grass/WC	4.9b	6.8b	1.34

Means with different letters, within columns and within same sowing date, are significantly different at $P<0.05$ or above. Field and conservation losses not taken into account.

Quick establishment in spring is also essential for suppressing weeds. In general, only the first cut contains a considerable amount of weeds and early spring development is essential to suppress the weeds. This means that early developing grass varieties should be chosen. The need for weed suppression in spring also gives high priority to sowing of the grass-clover in late summer of the previous season, instead of spring sowing. Not only is the production of a one-year grass-clover ley much lower with spring sowing (7-9 t DM/ha instead of 11-14 t DM/ha) but weed development in spring is also much higher (see Table 1). We have observed spring-sown leys which have had to be mulched in order to control the weeds. However, this can result in the delayed development of the young grass-clover ley and the rapid reappearance of the weeds. If the mulching is repeated several times this results in a severely delayed first cut at the end of July, or even August, and subsequently very poor DM yields over the year. Thus, in the Dutch situation, late-summer sowing in the previous year is very important for this type of ley although this should be done as early as possible (Baars and Veltman, 2000).

Table 2. DM and Net-N yield of fertilized and non-fertilized grass clover

Fertilization	Nutrients applied (kg/ha)	Total yield (tDM/ha)	Clover yield (tDM/ha)	Net N yield (kg/ha)
With slurry	196N, 26P, 267K	10.79	3.78	143
No slurry	35P, 195K*	9.96	5.17	321

* potassium salt as 'K-60' and P as superphosphate. Both are not allowed in organic agriculture.

Next to weed suppression, grass-clover serves as an important input of soil fertility. This is of growing importance as manure is becoming increasingly scarce and has an impact on the fertilization strategy for the grass-clover crop. Because manure is becoming a scarce resource, it may be better employed on high value cash crops than on low value crops like grass-clover. Furthermore, the absence of fertilization on the grass-clover only slightly reduces the annual DM yield, but highly increases the net N-yield as can been seen in Table 2. This is another illustration of the difference in management resulting from the different roles of grass-clover in arable and livestock farming.

CONCLUSIONS

The management of grass-clover is linked to the socio-economic and legislative environment of the farming system. Understanding the complexity of the role which grass-clover plays in different parts of this farming system is therefore essential. Farming system research, using a wide range of research techniques and involving farmers as equal partners in the research and development process, makes it possible to understand better the management choices and leads to more viable solutions.

REFERENCES

BAARS T. (1998) Modern solutions for mixed systems in organic farming. In: Keulen H. van, Lantinga E.A. and van Laar H.H. (eds.) *Proceedings of an International Workshop on Mixed Farming Systems in Europe*, Dronten, Wageningen, The Netherlands, 25-28 May 1998. Ir. A.P. Minderhoudhoeve-series no. 2. pp. 23 - 29.

BAARS T. (2002a). *Reconciling scientific approaches for organic farming research.* PhD dissertation. Wageningen University and Research Centre, the Netherlands.

BAARS T. (2002b) N-fixation of white clover affected by clover yield and type of manure. In: Durand J.L., Emile J.C., Huyghe C. and Lemaire G. (eds) *Multi-functional grassland*, 19th General Meeting of the European Grassland Federation, La Rochelle, France, pp.654-655.

BAARS T. and VELTMAN L. (2000) Adapted grass/clover mixtures for ley farming - a participatory approach to develop organic farming systems. In: Søegaard K. *et al.*(eds) *Grassland farming – balancing environmental ands economic demands*. Proceedings 18th General Meeting of the European Grassland Federation, pp.542-544.

PRINS U. *et al.* (2002) *Koppelbedrijven: het gemengde bedrijf op afstand (Partner Farms: mixed farming at a distance).* Louis Bolk Instituut, Driebergen, 18 p.

Health and Welfare Benchmarking as a Tool for the Development of Dairy Herd Health Plans on Organic Farms

J. BURKE[1], S. RODERICK[1], J.N. HUXLEY[2], H.R. WHAY[2] and D.C.J. MAIN[2]
[1] Organic Studies Centre, Duchy College, Rosewarne, Camborne, Cornwall, TR14 0AB
[2] Department of Clinical Veterinary Science, University of Bristol, Bristol, BS40 5DU

ABSTRACT

Legislation requires organic farms to establish an animal health plan to assist in delivering positive health management. An established method of dairy herd welfare benchmarking using animal-based measures was applied to 15 organic dairy farms in south-west England during the winter of 2002-03. The objective was to aid farmers and their veterinarians in the iterative process of health and welfare planning through the identification of weaknesses in management systems. During an initial farm visit parameters reflecting behaviour and injury as well as current herd health plans were assessed. Participants and their veterinarians were invited to attend a workshop to discuss the benefits, and utilization, of the benchmarking results. The response of farmers to the benchmarking process, results and outcomes was evaluated by qualitative research interviews. All 15 farms had a herd health plan in place. Although the results of the herd assessments showed a considerable range across the 15 farms, no single farm performed badly for all criteria measured. Ten individuals implemented changes to improve cow welfare, six discussed the results with their veterinarian and of these, three reviewed the herd health plan in the light of the results. Re-assessment of all herds was scheduled for 2003-04.

INTRODUCTION

Legislation requires all organic farms to establish a plan for maintenance of health and control of diseases appropriate to individual farm circumstances that can assist in delivering positive health management on livestock farms. The health plan should bring together information about the farm and its animals and outline management strategies to prevent and control herd health problems. However, there is evidence that farmers regard health plans as additional paperwork needed for inspection purposes alone and that health planning is already a feature of day to day farming (Huxley *et al.*, 2003).

Welfare may be viewed as a measure of the animal's state as regards its attempts to cope with its environment and may be improved through human intervention (Broom and Johnson, 1993). Since the health of an animal is affected by its environment and the potential for this environment to create circumstances resulting in injury or disease, health is an important component of welfare. Whay *et al.* (2003) describe a method of animal welfare assessment using animal-based measures that focuses on how the animal is coping with its environment, and thus enables comparisons of production systems with different resource provisions. However, farmer acceptance of this type of assessment and benchmarking is essential for its successful introduction as a management tool.

Qualitative research interviews are a recognized method to explore the 'lived world' of the research subjects and to obtain descriptions of specific situations and action sequences from the subject's world (Kvale, 1996). This technique has been used in a range of subject areas including an agricultural context (Vaarst *et al.*, 2002).

The aim of this study was to investigate the potential benefits of animal welfare assessment benchmarking as a tool for the development of health and welfare plans. The objectives were to aid farmers and their veterinarians in the iterative process of health and welfare planning through the identification of strengths and weaknesses in management systems and, through comparison with other farms, demonstrate what might be achieved.

MATERIALS AND METHODS

Herd assessment

Fifteen organic dairy farms in south-west England were visited during the winter housing period of 2002-03. All farms approached agreed to participate in the study. The farm assessment included a review of the current herd health plan. The herd welfare assessment involved a sequence of animal based measures that could be visually assessed. These included the number of cows in the herd considered to be idling, assessments of rising restrictions and the flight distance of ten randomly selected animals. Within each herd, 20% of individuals were then assessed in detail, including body condition score, cleanliness, coat condition, rumen fill, claw overgrowth and conformation and signs of injury or trauma. All herd members were scored for lameness as they exited the milking parlour. A four-point scale described by Whay *et al.* (2002) was used and only animals classified as lame or severely lame were included in the analysis. The sequence of observations is described in detail by Whay *et al.,* (2003). A single trained observer carried out the observations, all of which were made on unrestrained animals. Results for each parameter were analysed using median and range values for each herd.

Workshop

All participants and their veterinarians were invited to attend a workshop where they were presented with the results for each assessed parameter on their farm, clearly benchmarked against the mean of all farms. The interpretation of results and the relevance of individual parameters were discussed in the context of re-evaluating current herd health plans.

Qualitative interviews

Qualitative research interviews were conducted with each participating farmer between July and September 2003 in order to assess individual responses to the herd assessment process and the benchmarking results. The interviews were structured around an interview guide consisting of four thematic questions: initial thoughts and expectations; experiences of the project; the influence of the project; and what should happen next. All interviews were performed by the same person and recorded on mini-discs. They were fully transcribed and analysed using the grounded theory approach (Kvale, 1996; Strauss and Corbin, 1997; Vaarst *et al.,* 2002).

RESULTS

Herd assessment

All 15 farms had a herd health plan in place. Ten had completed the plan with the help of their veterinarian, one with an organic adviser and four without any external advice. The plans had been in place on the farm for between 2 and 48 months (median 17 months). Farmers' opinions about herd health plans were mixed and varied across the 15 farms.

Some suggested that it made them think about animal health and welfare, whilst others considered it to be a waste of time and just more paperwork.

Although the results of the herd assessments showed a considerable range across the 15 farms, no single farm performed badly for all criteria measured. The full results are too numerous to be included here and are described in detail by Huxley *et al.*(2003).

In order to illustrate the range across herds, examples of scores representing measures of nutrition, lameness, integument, environmental injury and behaviour are given here:

- thin cows (body condition score < 2): median 2.8%; range 0.0% to 30.0%
- lame cows: median 24.2%; range 6.8% to 55.6%
- dirty hind limbs (at any level of severity): median 100%; range 80.0% to 100.0%
- hock hair loss (at any level of severity): median 78.6%; range 0.0% to 90.0%
- maximum flight distance: median 2.0m; range 1.5m to 4.0m.

These parameters should not be taken in isolation as they are presented as examples from a total of twenty seven used in the herd welfare assessment.

Qualitative interviews

Farmers were asked to describe their initial thoughts and expectations when asked to participate in the study. Four farmers thought it sounded interesting, useful or beneficial, whilst two hoped to learn more about health planning and health plans. Three farmers were less enthusiastic about 'yet another survey' but even so, agreed to participate. Six farmers expressed interest in comparison with other organic farms. Although comparison between organic and non-organic herds was not the objective of this study, there was an expectation amongst farmers that it would provide evidence of superior health and welfare in organic dairy herds.

When asked about their experience of involvement in the project a number of farmers expressed disappointment at their results, with lameness, hock injury and dirty hind limb figures causing the greatest concern. They considered that the scoring was 'hard', 'very strict', even 'harsh', that 'it was done with a very critical eye' and that 'every fine detail was taken into account'. However, some commented that this might be a good thing. Those who had attended the workshop considered that they had benefited from the feedback and discussion. Generally, farmers felt that they had got on well with the assessor and were appreciative of the fact that he was a qualified veterinary surgeon with a specialist interest in dairy production.

Question three explored the way in which farmers had been influenced by the project. Ten farmers had made changes in the light of the results, ranging from changes to housing to reduce injuries, increasing bedding or yard scraping to improve cow cleanliness, paying more attention to cows feet and watching cows 'differently' and 'more critically'. Two other farmers were considering making changes whilst three were not intending to make any changes. A total of six farmers had discussed the results with their veterinary adviser and, of these, three had reviewed their herd health plan, in the light of the results, with the assistance of their veterinarian.

When asked what should happen next, nine farmers volunteered that they were keen to participate in a repeat assessment to identify any effect of changes implemented on cow welfare on their farms. When asked directly, the remaining six farmers said they would be happy to take part in a second assessment.

DISCUSSION

Systematic welfare assessment across farms allows welfare benchmarking of the results. This process also enables farmers to examine the weak points in their production systems and try to find their own solutions. Qualitative interviews, recording feedback from the assessment process were essential for understanding the impact and outcomes of the intervention, as well as highlighting ways in which the process might be improved for future use. It was apparent that farmers in this study took the results of the herd assessment very seriously and were clearly motivated by the process. As a result of participation in the study ten individuals implemented changes to their system or to the way they inspected their cows with the aim of improving cow welfare. It was also apparent that the interest expressed in repeating the assessments and benchmarking, in order to identify where welfare improvements had been made, was also related to a desire to improve both individual herd performance and benchmarking results. The potential for using animal welfare assessment benchmarking as a tool for the development of health and welfare plans was illustrated by the action taken on three of the participating farms. Interestingly, these farms already had full health plan support from their veterinarians.

Some of the results (e.g dirty hind limbs) describe the prevalence of those measures present on the farms at any level of severity (i.e. mild, moderate and severe). Consequently, for some measures the prevalence appears high. Thus, this method of benchmarking allows for improvement even in potentially acceptable situations. However, it is as yet unclear as to the relevance and relative impact of each of the measured parameters on animal welfare, and hence some care is required when interpreting results, particularly those that are subjective in nature. Further to this, the general suitability and application of these parameters to organic systems has not yet been explored. This issue was also raised by some of the organic farmers interviewed. The rumen fill of an animal on a largely roughage-based diet is an example, as is coat condition under a system that promotes outdoor access.

ACKNOWLEDGEMENTS

Funded by Objective One (EAGGF and DEFRA), Mole Valley Farmers Ltd. and the Organic Milk Suppliers Co-operative Ltd. The authors thank the participating farmers.

REFERENCES

BROOM D.M. and JOHNSON K.G. (1993) *Stress and animal welfare.* London: Chapman & Hall.

HUXLEY J., BURKE J., RODERICK S., MAIN D.C.J. and WHAY H.R. (2003) Animal-based welfare assessments: benchmarking as a tool for health and welfare planning in organic dairy herds. *Veterinary Record.* (Submitted)

KVALE S. (1996) *Interviews. An introduction to qualitative research interviewing.* Sage, USA.

STRAUSS A. and CORBIN J. (1997) *Grounded Theory in Practice.* USA: Sage Publications.

VAARST M., PAARUP-LAURSEN B., HOUE H., FOSSING C. and ANDERSEN H.J. (2002) Farmers' choice of medical treatment of mastitis in Danish dairy herds based on qualitative research interviews. *Journal of Dairy Science,* 85(4), 992-1001.

WHAY H.R., MAIN D.C.J., GREEN L.E. and WEBSTER A.J.F. (2002) Farmer perception of lameness prevalence. *12th International Symposium on Lameness in Ruminants,* Orlando, 355-358.

WHAY H.R., MAIN D.C.J., GREEN L.E. and WEBSTER A.J.F. (2003) Assessment of the welfare of dairy cattle using animal-based measurements: direct observations and investigation of farm records. *Veterinary Record,* 153(7), 197-202.

Sustaining Animal Health and Food Safety in European Organic Livestock Farming

M. VAARST[1], S. PADEL[2], M. HOVI[3], D. YOUNIE[4] and A. SUNDRUM[5]
[1] Danish Institute of Agricultural Sciences, PO Box 50, DK – 8830, Tjele, Denmark
[2]Institute of Rural Sciences, University of Wales Aberystwyth, SY23 3AL, UK
[3]VEERU, Dept. of Agriculture, University of Reading, Reading, RG6 7AR, UK
[4]Scottish Agricultural College, Craibstone Estate, Bucksburn, Aberdeen, AB21 9YA, UK
[5] Dept of Animal Nutrition & Health, University of Kassel, Nordbahnhofstr. 1a, Witzenhausen, D-37213, Germany

ABSTRACT
In Europe, organic livestock production has experienced rapid growth in the past decade. Whilst emphasizing the importance of a systems approach to animal health and welfare protection, organic livestock production standards place considerable restrictions on the use of many animal health inputs that are routinely used in conventional production systems. Recommended practices in the European organic livestock standards such as closed herds and flocks and improved health security on farms, also include extensive production systems (e.g. free range production) that expose livestock to increased disease challenge. Organic livestock production faces major challenges with regard to harmonization and successful integration of organic animal husbandry into the whole organic production system. Major questions about food quality and safety exist. Significant diversity between farming systems and between different countries in Europe, including pre-accession countries, should be taken into account in developing farming systems that all comply with common EU standards, but are in harmony with their geographic and cultural localities. A newly initiated EU network project (SAFO) with 26 partners in Europe, focuses on the integration of animal health and welfare issues with food safety aspects. It is the purpose of this paper to outline the aims of the project, and to give examples of problem areas which are in focus in this network project.

INTRODUCTION
The IFOAM Basic Standards state that one of the basic principles of organic agriculture is "to give all livestock conditions of life with due consideration for the basic aspects of their innate behaviour". Organic systems are designed to achieve a balanced relationship between the components of soil, plant and animal, and the animal herd forms an important part of the organic farm. Animals are living individuals with senses and behavioural needs, which place a responsibility on the organic farmer to take care of their well-being. The production system is not sustainable if animal welfare is not high and there is evidence of pain, disease or distress as a result of an inadequate system. Animals living in harmony with their surroundings and experiencing a good quality of life seem to be obvious elements of a system which favours these values. Apart from these philosophical considerations, there are practical reasons why animal health and welfare must have a high priority in organic livestock production. When marketing organic products, the perception that organic livestock have been able to perform more of their natural behaviours and have benefited from higher welfare standards than animals on conventional farms is an important selling point. These perceptions have been reported

among consumers in various European countries. An EU-funded Concerted Action project, Network for Animal Health and Welfare in Organic Agriculture (Hovi, 2001) identified that organic livestock production faces major challenges with regard to harmonization and successful integration of organic animal husbandry into the whole organic production system.

Certain values and ideas form the framework for organic production, such as naturalness, harmony at all levels of production, local circulation of resources and the precautionary principle. Positive animal health in organic production is to be achieved without the use of conventional veterinary medicines on a routine basis. This provides a safeguard for the health of human consumers of organic livestock products. Such products are less likely to contain drug residues, such as antibiotics and hormones, as a result. Reduced levels of antibiotic resistance in indicator bacteria have been reported in organic broiler flocks compared with conventional ones. On the other hand, food safety must be safeguarded in relation to the presence of zoonotic organisms, such as *Salmonella*, *Campylobacter* and *Escherichia coli*. There have recently been accusations that organic products are more likely to be contaminated by such bacteria than conventional ones because of the reliance on animal wastes as fertilizers. There is little documented evidence to support this contention, although researchers found a higher prevalence of *Campylobacter* in organic broiler flocks than in conventional and extensive indoor flocks in Denmark.

The rapid growth of organic livestock production in Europe in the last decade has not been without problems. Common EU standards for organic animal production were implemented in 2000 (EU Regulation 1804/1999) (CEC, 1999). The implementation process is an ongoing effort to harmonize the standards under widely differing circumstances and farming conditions in the member states. Significant diversity between climate and farming systems, e.g. between southern and northern European countries, needs to be taken into account in developing farming systems that all comply with common EU standards, but are in harmony with their geographic and cultural localities. Furthermore, the introduction of pre-accession member states from Central and Eastern Europe into the EU will increase the need to harmonize standards and make them practically applicable in these countries. As the development of organic livestock production standards should be driven and informed by research that reflects the practice and experience of organic farmers in each locality, it is important to create and maintain opportunities for the research and wider stakeholder community to exchange views and information.

Recommended practices in EU Regulation 1804/1999, such as restrictions on the use of veterinary drugs, closed herds and flocks and improved health security on farms, also include extensive production systems (e.g. free range production), which may expose livestock to increased disease challenge. Major questions about food quality and safety exist, in conventional, as well as in organic livestock production. These issues have not been a focus of any previous EU research initiatives.

OBJECTIVES

The objective of the SAFO network project is to improve food safety and animal health in organic livestock production systems in existing and pre-accession member countries of the European Union. This will happen through exchange and active communication of research results and conclusions between researchers, policy makers, farmers and the

wider stakeholder community, including consumers.
The main intermediate objectives of the project are:

- To identify important food quality characteristics linked to organic livestock products, and improve food quality, including food safety with regard to zoonoses, drug residues and the development of anti-microbial resistance in the food chain. This objective also includes aspects of food processing quality with regard to animal health and welfare in organic livestock production systems,
- To develop strategies for implementing and harmonizing organic livestock production standards in existing and candidate member countries, and
- To improve the interaction between researchers, farmers, certification bodies and policy makers in order to guarantee the development of organic livestock standards that are driven by inputs from all stakeholders in the EU.

ELEMENTS OF THE NETWORK ACTIVITIES

The SAFO Network works on different levels in order to form a true platform for exchange and building up of knowledge. The Website (www.safonetwork.org) provides a platform for global dissemination of the shared knowledge and experiences, which are to be published under the formal communication structure of SAFO. As one of the project tasks is to work actively in standard development, a specific effort will be made with regard to communication with policy makers, administrators of organic standards and organic certification bodies and farmer organizations.

The formal and active sharing and learning experience of the project is a series of five workshops. The first workshop was held in Florence in September 2003. The second workshop was held in March 2004 with the title 'Development of organic livestock farming: potentials and limitations of husbandry practice to secure animal health, welfare and quality'. These are to be followed by a third workshop in Poland (in September 2004) with the title: 'Enhancing animal health security and food safety in organic livestock production' and the fourth workshop (April-May 2005) will be about quality and safety of organic livestock food and other livestock products. The final workshop will be held in Brussels and focus on conclusions of the SAFO-project and technology transfer.

The Proceedings with papers from the workshops are produced in hard copy and will be freely available electronically on the web-site. The workshop proceedings will be summarized and translated into partner country languages other than English and published as user-friendly farmers leaflets/articles for dissemination to advisory services, farmers and policy makers in each of the twenty-one partner countries. In addition to these formally published summary articles, there will be substantial informal dissemination of SAFO-generated information in partner countries, since many of the partners have direct contact with farmers, advisory services, policy makers and agricultural students, through individual contact and in formal meetings.

At the end of the project, workshops will be held in five of the pre-accession countries.

DISCUSSION

Many challenges exist in the organic sector to ensure good animal welfare and a high health status, and safe animal food products from the farm. These challenges exist in the areas of:

1) Giving animals a good life – this is an important part of the process quality of the animal food products.
2) Ensuring a high health status in herds – which will also help to minimize the use of all drugs, and reduce the risk of any disease including zoonoses.
3) Making animal production a part of the whole farm, in order to ensure harmony between different production areas of the farm, and to develop the idea of a whole organic farming system, comprising both animals and crop production.
4) Processing food of high product quality and with no artificial or synthetic substances.

The consequences of the 'organic way of living', such as outdoor life, local circulation of feed stuffs and manure, and attempts to later weaning, all form challenges, for which solutions and compromises must be found in the highly diverse farming sector of Europe today. Further challenges with respect to food production, food quality and food safety exist in relation to organic production, where 'good raw material', high quality, natural and traceable products are wanted.

The diversity across Europe with regard to markets and conditions for animal husbandry, form a major challenge for the development of organic food production. One big challenge is to extend this discussion not only to the Western European countries of the EU, but also to emerging member states in Central and Eastern Europe. Many of these countries have a large livestock production sector that, in many cases, is suitable for conversion to organic production. These countries need to be involved in the on-going development of Regulation 1804/99 as early as possible, in order to harmonize production standards in all member states and to take into consideration the specific local needs in some of these countries.

Further development of organic livestock production towards a system that offers safe products and meets the consumers perceptions on animal health and welfare standards requires further work in terms of research and development.

REFERENCES

CEC (1999). Council Regulation No. 1804/1999 supplementing Regulation No 2092/91 on organic production. *Official Journal of the European Communities* 42, L222, 1-28.

HOVI M. (2001) Final Report, Network for Animal Health and Welfare in Organic Agriculture (NAHWOA), EC FAIR CT 4405.

MANMOD: A Decision Support System for Managing Manure Nutrients from Organic Livestock Production

C. FAWCETT[1], M. SHEPHERD[1], J. WEBB[1] and L. PHILLIPS[2]
[1]ADAS, Wergs Road, Wolverhampton, WV6 8TQ, UK
[2] IOR-Elm Farm Research Centre, Hamstead Marshall, Newbury, RG20 OHR, UK

ABSTRACT

MANMOD is a Windows®-based Decision Support System (DSS) for organic farmers and consultants that provides a straightforward means to calculate NPK nutrient flows in an organic livestock system. It also quantifies how changes to management practise can influence these flow pathways. The MANMOD DSS allows an iconographic-based model representation of individual farm manure management systems to be readily constructed from a library of system components using a 'drag and drop' operation. Simulation of the model provides a NPK nutrient balance at each stage of the system, including details of any nutrient losses to the wider environment as well as a whole farm summary. The model structure and model parameters can be easily modified within the DSS to allow rapid scenario testing. MANMOD is a tool that both farmers and their advisors will find useful when planning their organic farming systems.

INTRODUCTION

Manure is an important source of organic matter and nutrients in organic farming systems, principally nitrogen (N), phosphorus (P) and potassium (K). Careful management is required during storage, handling and land-spreading to ensure the most efficient use of the nutrients in the farming system and to limit emissions of nitrate (NO_3), ammonia (NH_3), nitrous oxide (N_2O), methane (CH_4) and P to the wider environment. With a likely increase in the organically farmed area, information is needed on best practices for manure management in organic systems to minimize the environmental impacts of these systems.

OBJECTIVES

The objective of this project was to develop a robust model to calculate the integrated effects of management practices during housing, storage and spreading on NPK flows in manure production. Previous work had developed a simple N spreadsheet model for a manure management system (Shepherd *et al.*, 1999). The aim was to develop and improve this into a new model incorporating the most recent emission factors and linking to existing models for assessing field losses. The new model was to be incorporated into a prototype Decision Support System (DSS) that would calculate NPK fluxes associated with each aspect of the livestock system, and provide options to explore the impact of management change at key stages in the manure management process.

MODELLING APPROACH

Algorithms for each stage of the manure management system had to be developed. The challenge was providing an appropriate level of complexity to accurately calculate the nutrient fluxes that are responsive to management change, whilst minimizing the need for environmental data. It was important that the model could take into account the interactions between the different management stages. For example, if a change in housing management increased the ammonium-N component of the manure entering storage, the risk of ammonia volatilization during storage would be higher. At the housing, hard-standing and storage stages, algorithms used emission factors that were obtained from NARSES and MARACAS (Webb and Misselbrook, in press). One further refinement was an estimate of ammonia volatilization losses during composting, adapted from the early work undertaken by Shepherd *et al.*(1999). At the field stage, the MANNER model (Chambers *et al.*, 1999) was used to identify the gaseous emissions and nitrate leaching following the application of manure either directly from the animal or via spreading. The flow of P and K was simply based on mass balance calculations using standard excretion factors. It was felt that this approach was justified because P and K do not undergo the same range of losses as N.

MANMOD DSS

The MANMOD DSS was developed to allow an iconographic-based model representation of individual farm manure management systems to be readily constructed from a library of system components using a 'drag and drop' operation. This allows the user to construct a diagram of connecting components or 'nodes' (e.g. manure source, housing system, storage system) which direct and limit the flow pathway of nutrients through the farming system. Each component or node represents a key stage of the system. Such an approach is ideal for rapidly constructing models of complex systems such as this. The components used in MANMOD are: manure source, housing system, hard-standing area, storage system, applicator, field system, environment node, and manure import and export.

The properties of each node can be changed to reflect the management status of each stage (e.g. slurry store with cover). Simulation of the model provides a NPK nutrient balance at each node in the system, the details of any nutrient losses to the wider environment (e.g. ammonia volatilization or nitrate leaching), and nutrient output (i.e. the amounts of N, P and K that will be transferred from that component of the system to the next). A whole farm summary is also presented.

To maximize technology transfer, the MANMOD model-engine was developed as an ActiveX control, which allows it to be embedded into other development code. By adopting this development approach, the scientific algorithms can be readily reused. It also facilitates scenario testing, where the model can be simulated multiple times with changes to parameters and the structure.

MODEL VALIDATION

Validation of the DSS was undertaken by: calculating nutrient flows through typical manure management systems; comparing calculated cattle manure composition with standard data; and scenario testing, where individual management factors were altered to study the effects on the system. The output from the model on test simulations are sensible and represents interactions between management processes in a logical manner. In more detailed analysis, a simple dairy system was constructed and simulated, firstly based on a slurry system and secondly as a straw based system. The resulting calculated nutrient contents of slurry and FYM are presented in Table 1 and are compared with standard figures taken from the industry standard RB209 (Anon, 2000) and typical values for cattle manure from organic holdings (Shepherd *et al.*, 2002).

Table 1. Comparison of calculated cattle manure nutrient composition with standard values as published in RB209 and typical values for cattle manure from organic holdings. Nutrient content (kg/m^3 or kg/t):

Manure	Source	N	P_2O_5	K_2O
FYM	MANMOD	4.8	2.1	4.9
	RB209	6.0	3.5	7.2
	Organic	5.2	2.9	6.3
Slurry	MANMOD	4.1	2.3	5.2
	RB209(10% DM)	4.0	2.0	5.0
	Organic (8%DM)	2.5	1.2	2.5

Despite the complex transformations the model has to represent, the comparison with standard values in RB209 was reasonable. The comparison was better for 'conventional' slurry than for FYM. This was not surprising given that a slurry system is more straightforward, and does not have the complicating factors of interactions with straw during housing and manure storage. However, the FYM calculation seemed to consistently underestimate nutrient content for N, P and K. One possible reason could be the mass loss calculation, and this needs further investigation. In comparison to the standard values for manure from an organic holding, the simulated values tend to have a higher nutrient concentration. For FYM, because MANMOD calculated smaller nutrient concentrations than those reported in RB209, and because concentrations are smaller for the organically produced manures (Shepherd *et al.*, 2002), there appeared better agreement between MANMOD and organically produced manures. However, the better agreement between calculated nutrient contents and those from organic holdings is a coincidence, given that nutrient excretion rates in the model are based on 'conventional' livestock. This is further supported by the fact that the organically produced slurry had nutrient concentrations of about a half of those calculated by MANMOD. Future work will integrate data for organic livestock into the DSS when it becomes available.

To further test if MANMOD was sufficiently sophisticated to be able to determine differences between changes in management, adjustments to individual elements were made to observe the changes in N fluxes. None of the practices had effects on P and K fluxes. Converting to a slurry system increased ammonia losses during the housing and storage (open slurry store) phase. Increasing the amount of straw only had a small effect on the N losses before land-spreading. The calculation method assumes only a relatively small amount of N immobilization during the housing and storage phases. A 10-fold increase in straw addition decreased ammonia losses in the model by 4% when the manure was stockpiled. Further experimental data are required to confirm the assumptions in our calculation. The largest effect came from composting the manure, compared with stockpiled (undisturbed) manure, as active composting (stirring/heating) encourages ammonia loss. This is clearly critical to the N balance of the system. The basis of the calculation described above was supported by a previous literature review (Shepherd *et al.*, 1999). However, this is so critical that the relationship should be further confirmed. The slurry store management had a significant effect on ammonia losses during this phase. Compared with an open store, a 'covered store' decreased losses to 80% and a 'crusted' slurry halved N losses.

CONCLUSION

The software has been developed such that manure systems can quickly and simply be constructed. This has proven to be an effective method of representing complex systems. The output from the model on test runs looks sensible and represents interactions between management processes in an logical manner. Agreement of predicted manure nutrient concentrations for cattle FYM and slurry was reasonable, but was less satisfactory for organically produced manure. Considering the complexity of the systems the model can represent, significantly more model validation is required. This will be undertaken when more data is available and will be the focus of future studies.

REFERENCES

ANON. (2000) *Fertilizer Recommendations for Agricultural and Horticultural Crops* (RB209). Norwich, UK: The Stationery Office.

MARACCAS - Model for the Assessment of Regional Ammonia Cost Curves for Abatement Strategies. http://www.huxley.ic.ac.uk/research/AIRPOLL/MARACCAS/

CHAMBERS B.J., LORD E.I., NICHOLSON F.A. and SMITH K.A. (1999) Predicting nitrogen availability and losses following application of manures to arable land: MANNER. *Soil Use and Management,* 15, 137-143.

SHEPHERD M.A., BHOGAL A., LENNARTTSON M., RAYNS F., JACKSON L., PHILIPPS L. and PAIN B. (1999) The environmental impact of manure use in organic farming. MAFF Commissioned Review.

SHEPHERD M.A., PHILIPPS L., JACKSON L. and BHOGAL A. (2002) The nutrient content of cattle manures from organic holdings in England. *Biological Agriculture and Horticulture,* 20, 229-242.

WEBB J. and MISSELBROOK T.H. (In press) A mass-flow model of ammonia emissions from UK livestock production. *Atmospheric Environment.*

SESSION 3

CROPPING SYSTEMS

Ecological Cropping Systems – An Organic Target

PROFESSOR MARTIN WOLFE
Elm Farm Research Centre at Wakelyns Agroforestry, Fressingfield,
Suffolk IP21 5SD, UK

ABSTRACT

Organic agriculture has expanded rapidly but is now in danger of reaching a plateau of productivity and of contributions to ecosystems. There is an urgent need for major development of ecologically sound cropping systems using knowledge from ecological sciences and from appropriate breeding and selection. One of the most important aspects of successful natural ecosystems is their fine-grained plant diversity. Parallel examples in agricultural systems are considered including variety and species mixtures together with other intercropping approaches. These are integrated into potential organic cropping systems that eliminate the need for separate rotation phases for production and fertility building. At the highest organizational level, such approaches to diversity can be integrated into organic agroforestry systems. Stress is laid not only on the value of such diversified systems for sustaining their own productivity, but also on ways in which such systems can contribute to integration with other ecosystems and hence to increases in ecosystem services.

INTRODUCTION

Organic agriculture has expanded rapidly over the last thirty years. The area certified as organic or in conversion has increased greatly (though much of the converted area is permanent pasture) and sales have exploded (though an uncomfortably large proportion is from imported produce). Much of this change has been achieved through relatively simple modifications of currently conventional agriculture systems that help to reduce negative ecological impacts, although gross production per hectare over a whole rotation is significantly less. Organic agriculture does offer other positive ecological gains (biodiversity, energy use, decentralization, etc.) relative to conventional, but there is much still to be learnt from the positive ecological impacts and sustainability of natural (unmanaged) ecosystems. Indeed, application of advances in ecological sciences, both within the cropped area and by integrating agriculture with its surrounding ecosystems, is essential. Organic agriculture must become more ecologically oriented, not only as a way of farming, but to blend more with other ecosystems while simultaneously influencing outputs (marketing, food, energy, recycling). A fundamental starting-point in this direction is in the development of ecological cropping systems.

Current organic practice in terms of cropping systems has settled into a limited range of relatively short rotations based on monoculture cropping. For example, a recent survey (A. Lamy, S. Tehard, personal communication) recorded that the commonest cropping system is a monoculture rotation:

Winter cereals – pulses – spring cereals – two year ley

In which, most commonly, the cereals are wheat and the pulses are beans.

Variants on this theme include either more winter cereals and more ley, or, more cereals with a winter and two spring cereal crops, each cereal crop being separated by other crops including pulses. But if we continue only in this way, there is a danger of reaching a plateau of productivity with little scope either for increases in outputs or decreases in inputs.

This would also imply a plateau of contributions to and from ecosystem services (those processes and functions that occur naturally in healthy ecosystems and that benefit human society; Costanza *et al.*, 1997). Examples of ecosystem services include carbon sequestration, nutrient cycling, pest and disease buffering, pollination and water cycling and purification. Integrating more of the principles and materials from natural into agricultural ecosystems should help to increase and sustain those services. Examples would be by increasing the activity of soil micro-organisms or providing safe habitat for a wide range of bee species.

MODELS FOR NEW DIRECTIONS: USING DIVERSITY

To develop an ecologically richer agriculture, we need a model to work towards which would be close to a successful natural ecosystem. Here, we are helped considerably by the work of David Tilman and his colleagues (Tilman, 1988). They have shown how grassland communities can be highly productive, season after season, with inputs limited to sun, soil, air and water. Their studies have revealed the increased productivity of the plant community compared with that of its components grown as sole stands. Some of the main mechanisms responsible for such improvements are:

a) increased range of functions as the number of community components increases
b) niche differentiation among the components improves exploitation of the local environment
c) complementation of functions among different components.

Such mechanisms need to be dynamic because of the changing nature of the environmental challenge. This is provided partly by frequency changes within the community to ensure the best available population structure for each set of circumstances. However, to ensure long-term adaptability, there is also the potential for genetic change within the community; new assemblages of characters are frequently available.

Other areas of ecological research are revealing many more mechanisms operating in the interactions among organisms than were previously known. Some examples are: emerging knowledge of mycorrhizal functions in soil, the positions of beneficial invertebrates in food webs, the structure and function of semiochemicals in plants and animals (Chamberlain *et al.*, 2000), self-medication in animals (Engel, 2002) and even the use of noise by some insects to protect themselves against others. All of these developments, and others, represent potential new opportunities in ecological agriculture, particularly if they are considered not as individual 'silver bullets', but rather to provide and encourage synergistic interactions in a wide range of directions improving the overall robustness of the system.

MODELS FOR NEW DIRECTIONS: BREEDING FOR DIVERSITY

There are two genetical requirements that are fundamental to any major shift towards the use of diversity in cropping systems. The first is for crop genotypes that are able to thrive together (ecological combining ability). This may be achieved to some extent by selection among existing genotypes but is more likely to need appropriate selection within breeding programmes. The second is to maintain a level of genetic variability in the varieties used to provide in-built buffering against environmental variation, which is greater in organic /ecological agriculture relative to conventional.

One approach to both of these needs is DEFRA project AR0914. This project has developed a series of composite cross populations in wheat, the F1 generation of which has been planted at four different field sites to produce a series of F2 populations for harvest in 2004. The source of the composite cross bulk populations is a series of crosses made in all combinations among 20 wheat varieties selected on the basis of their large-scale, long-term success either as high-yielding or as high grain quality varieties. The bulks will be allowed largely to self-select under conventional, integrated or organic management for several years. This should provide unique insights into wheat evolution under different selection regimes and, simultaneously, valuable genetic resources adapted to those different regimes. From the organic selection regime, it is hoped to produce segregating population samples that are both adapted to organic production and that retain adaptability to the environmental variability that is inevitable under organic farming.

One important comparison will be between the population based on high quality varieties and the parents of that bulk grown as pure stands, to determine whether or not the bulk provides useful quality, for example, through complementation of characters from different parents. A second relates to the stability of quality: is the quality population more stable than the parents in terms of grain quality when grown across different environments?

If the populations provide useful material for cropping directly, there will also be questions concerning the larger scale production of the material. Can it be traded and under what basis? Can new rules be introduced or will it depend on the development, for example, of appropriate farmer clubs? Indeed, if we are to embrace fully the approaches that will improve both productivity and sustainability, we will need to be prepared for appropriate changes in the framework of administrative, legal and marketing structures.

BUILDING DIVERSITY IN PRACTICE

The underlying principle of the model approach described above is fine-grained diversity. In agricultural terms this means diversity of crop and non-crop plants to provide many different functions and outputs. This can be built up from the level of the gene in a simple scheme:

A. Crop diversity

Crop diversity can be considered at the level of the gene (e.g. lines or genotypes of a particular variety with different resistance genes), variety (within each species, different varieties cover a range of characteristics), or species.

B. Spatial arrangement

This refers to the ways in which the different forms of crop diversity, noted above, can be arranged in space. These are listed in order of decreasing variety/species interaction:

Mixed – (lines, varieties or species) planted in an ordered arrangement (Weiner *et al.*, 2001) or at random
Row – alternating forms of diversity in row planting
Strip – alternating forms of diversity in strip planting

Plot or Crop area (from small plots to larger fields)

Monoculture in large fields is the arrangement most likely to encourage spread of pests and diseases. However, arranging even a small number of varieties in small rather than large blocks can have a large effect in delaying epidemic development (S. Phillips, personal communication). Reducing plot size to that of a single plant will maximize such effects.

C. Temporal arrangement
This refers to the ways in which the different forms of crop diversity, noted above, can be arranged in time. The principal forms are rotation, which ensures diversity between seasons, and sequential, involving diversity within a season. Rotation may be regarded as 'unnatural' but is particularly important for restriction of weeds, soil-borne pathogens and animal parasites. It helps to deal with the unwanted features of natural ecosystems such as major shifts in plant succession and loss of species.

It is important to stress that all forms of arrangement can be used simultaneously, for example, in our own organic agroforestry systems, where rotation of different crop species and varieties grown as mixtures is carried out in strips alternating with tree strips (mixed species), each of which has a mixed understorey of non-crop plants.

Examples of crop diversity arranged in random mixtures
A large body of experimental evidence (Finckh *et al.*, 2000) confirms that random mixing of varieties can have a large effect in delaying development of foliar diseases in particular, though positive effects with soil-borne diseases were also demonstrated. The largest effects with foliar diseases are observed with plants that are individually small (such as cereals) and pathogens with relatively flat spore dispersal gradients (for example, mildews and rusts) (Garrett and Mundt, 1999).

Table 1. Organically grown mixture of three wheat varieties showing high yield stability over four years.

Variety	2000	2001	2002	2003	Means
Mixture	108	106	111	113	110
Hereward	106	111	84	120	104
Shamrock	91	111	106	98	101
Malacca	103	77	110	82	95
Pure variety means t/ha	4.07	2.53	3.99	3.47	3.51

A recent example of organically grown wheat mixtures is shown in Table 1. The commonest diseases over the trial period were powdery mildew, leaf rust and *Septoria tritici* although disease levels were generally not high. The mixture reduced the leaf cover of all diseases by about half. From Fig. 1, the yield of the mixture varied less from year to year than did that of the other components and, on average, was higher. In two years (2001, 2003), one of the components yielded more than the mixture but this was not predictable.

Precise reasons for the improved yield of the mixture are not known, though from previous work with fungicide-treated controls we can predict that an important part of the effect was due to reduced disease. Other data suggest that such variety mixtures might also restrict weed development. This could be due, at least in part, to reduced disease which would allow the crop plants to be more vigorous and thus better able to compete with weeds.

Although much of the earlier work on mixtures was completed on small plot trials, the positive benefits increase as the scale of mixture production increases. For example, large-scale (350,000 ha) use of mixtures of barley varieties in the former German Democratic Republic was highly successful (Finckh *et al.*, 2000). Also, in China, large-scale trials with rice mixtures to reduce the effects of blast were taken up with great enthusiasm by local farmers (Zhu *et al.*, 2000; Wolfe 2000).

Cereals have been the major target for mixture research, but similar effects have been observed for an increasingly wide range of annual and perennial crops. Recently, under organic conditions, we have found reductions in late blight in potato variety mixtures (S. Phillips, personal communication). This work highlights the importance of ecological combining ability among varieties in that some varieties performed better when mixed with some others. For example, there is positive synergism in mixtures of Cara and Sante because Cara performs better when not competing with itself and it also helps to protect Sante from blight, so improving the performance of the latter. Thus recommendations on growing mixtures need to be based on appropriate field trials, and breeding programmes could be designed to select varieties that are particularly suited to production in mixtures.

The question of selection applies also to mixtures of species which have shown considerable potential in terms, again, of yield stability and restriction of diseases, pests and weeds (e.g. Bulson *et al.*, 1990). Many different combinations are possible, driven often by the potential use of the combined crop product, such as wheat and beans or barley and peas for feed, either as silage or grain. Interestingly, many successful examples involve species of cereal and legume which are complementary in terms of both agronomy and human diet.

Examples of other forms of spatial arrangement

An important ecological objective for intercropping is the notion of simultaneous cultivation of production crops with fertility-building crops, largely to avoid periods without a cash return from part of the cropping area (e.g. www.intercrop.dk). Our own preliminary work in intercropping is reported in two MAFF projects (OF 0173, OF 0181). The first was concerned with intercropping cereals and legumes and the second with vegetables and legumes. Some of the lessons from the two projects were:

a) White clover is not only an aggressive competitor but it allows little release of accumulated nitrogen while it remains alive.

b) Some crops are better-suited than others to legume intercropping. For example, among cereals, oats is well able to cope with clover competition whereas triticale is not. Among vegetables, root and leaf beet crops thrived whereas onions, leeks etc., did less well.

c) Any yield reductions in intercropping can be offset against later gains in the rotation due to the previous presence of clover and other crops (see below). The important point is that the clover crop occupies the inter-row spaces (about 60% of the field area) that would otherwise be bare or occupied by weeds, giving more efficient land use.

d) Pests and diseases occurred in the systems but only to minor levels. Slugs also occurred but were of little consequence. Weeds were severely restricted by the clover bands.

e) Specialized machinery is needed (see reports) for example for mowing the bands of clover grown between the crop bands. This is essential to delay ingress of grass weeds into the clover. The clover bands made the system easily workable.

Diversity in time: building a cropping system

For most field vegetables, a rotational break of 4-6 years is needed between crops of the same species. This often translates to a six year rotation overall in which two of the years are used for fertility building based on legumes or legume mixtures. Cash income is thus limited to two-thirds of the cropping cycle and fertility building to only one third.

With the vegetable-legume intercrop, fertility was not made directly available to the current vegetable crops. However, crop performance following three years of vegetable rotation intercropping was outstanding. The intercrop area was ploughed in during the early spring of 2003 and planted with potato trials. From these trials, mixtures of Cara and Sante averaged 32.1 t/ha in a dry season with no irrigation (Cara alone averaged 27.3 and Sante 31.1 t/ha). In other words, the fertility building role of the vegetable-clover intercrop was highly effective which means that any deficiency in vegetable production has to be discounted against more efficient land use during the rotation and high yields from the subsequent crop(s) (see point c. above). In a conventional organic production system, production of potatoes after three years of vegetables would not be an economic option.

From the experience with potatoes following vegetable-legumes, it seems likely that the potato crop benefited from the nitrogen and phosphorus made available from the legume crop and mycorrhizae but possibly also from potassium which may have been dissolved through citrate release from the beet crops (basic soil analysis Index values for P and K were around 1).

Similar results are being obtained from cereal – clover intercropping. Yields from winter wheat, oats, barley and triticale grown in a clover ley were less than those grown without clover, but the overwintering aftermath from the intercropping is a dense clover ley which will provide high fertility for subsequent spring cereals. An obvious advantage during the intercrop phase was strong suppression of weeds relative to plots with no intercrop despite mechanical weeding in the latter plots. Less obvious were other probable advantages of the cover crop including provision of habitat, reduction in leaching etc.

These results indicate the potential for development of improved cropping systems which could follow rotations such as:

Vegetable – legume intercrop (3 years, possibly more), potatoes (1 year), cereal – legume intercrop (2 years), then return to the vegetable - legume intercrop.

The vegetable part of the system would be based on a rotation among Alliums, Betae, Brassicas, Crucifers, Legumes and Umbellifers grown in the smallest convenient areas of each. The overall six year rotation would not require a separate fertility building phase, though a ley could be introduced as required if there was a need for livestock production. The cereal proportion in the rotation could be increased by reducing the vegetable cropping frequency. Alternatively, different parts of the farm could have a high vegetable or a high cereal frequency using the same basic system. Other options should evolve, varying, for example, the legume species (or mixture) and, perhaps more importantly, increasing the range of useful species incorporated at each stage of the rotation.

QUALITY AND MARKETING

A crucial question that follows any increase in the diversity and complexity of cropping systems is the impact of such systems on marketing. However, the ideal for any organic system is local and decentralized systems of marketing whether for individual consumers or other processors and users. Such systems depend on a wide range of produce. This can be achieved only through localized diversity of production which should help to ensure continuity because of maintenance of soil quality and protection from depredation by pests and diseases. These aspects can be improved further through development of local marketing co-operatives: one example is the recently formed Eostre Organics in East Anglia whose twelve organic growers supply directly markets, box schemes and school and hospital interests.

At the finer levels, mixing represents the greatest potential complication in terms of crop quality, particularly for cereal variety mixtures (large-scale mixed potato bulks can be separated into their components by Optical Character Recognition systems). However, it can be argued that variety mixtures may be beneficial for quality because a range of samples of a particular cereal mixture is likely to be more stable than a similar range of samples of any of the components. The major stipulation would be that the mixture components should be of similar quality or have desired compensatory qualities. The outstanding example of spring malting barley mixtures in the former GDR shows what can be achieved in this way; high quality malt was produced and a large proportion exported successfully to western Europe.

ORGANIC AGROFORESTRY: THE ULTIMATE SYSTEM?

The discussion so far has centred on incorporation of diversity into systems of cropping annual species. Agroforestry opens up the potential for integrating perennial species into cropping systems. There are many possible advantages including more effective nutrient cycling, carbon sequestration, shelter for crops, animals and humans, greater opportunities for pest and disease control, a wider range of products on the farm, better spread of labour use through the year, a large increase in biodiversity and extended aesthetic interest.

Although many different forms of integration of crops and trees are possible, the most common is some form of alley cropping in which a suitably sized crop alley is sited between lines or bands of trees running in a north-south direction. Such systems of 'production hedges' update the older northern European agroforestry system of hedgerows (Gordon and Newman, 1997). Organic (ecological) agroforestry systems should be highly diversified to maximise the generation of ecosystem services from farmland.

Organic agroforestry systems being tried at Wakelyns Agroforestry, include mixed hardwood standards (Ash, Hornbeam, Italian Alder, Oak, Small-leaved Lime, Sycamore, Wild Cherry) planted in randomized sets. One variant includes dispersed apple trees to reduce the rate of spread of apple pests and pathogens. Other examples are based on mixed fruit species with fruit and nut shrub understorey, or walnut and plum with clover understorey. There are also two hardwood coppice systems, one based on a hazel population and a second on a willow variety mixture. Alternate willow hedges are coppiced each year. Cut willow is air-dried in the field for one year, delivering the estimated equivalent of more than 12 t/ha dry wood per annum. Some stems are sold for craft use, but most are now used for heating.

From observation, it is difficult to say precisely which gains are being achieved and to what extent. Carbon sequestration is clear with little obvious impact on crop production. Reductions are limited to about one metre from the tree line, and are species dependent, for example, wheat appears more sensitive than oats to tree competition. Also, coppiced willow appears more sensitive to crop competition than do non-coppiced species. In terms of nutrient cycling, as the trees develop, there is an increasing spread of leaf litter across the cropping alleys. Complete food webs have established in the tree strips evidenced by vole and barn owl activity. The range of bird species is considerable and probably increasing (R. Fuller, personal communication). This means that the tree strips are acting as 'beetle banks', providing habitat for beneficial invertebrates. The agroforestry areas have become aesthetically interesting, sufficiently so as to attract the eye of some local artists.

In collaboration with Sheepdrove Organic Farm, a tree alley system on that farm has been integrated with poultry production (silvo-poultry system). The tree strips in this case include a range of standards, hedge plants and shrubs selected partly for shelter and partly for diversity of food. In addition, there is a complex herbage understorey to provide a range of plants with potential medicinal value including anti-bacterial, anthelminthic, expectorant, calming and other qualities. This herbage system was devised by Cindy Engel (see Engel, 2002).

CONCLUSIONS

Recently, there have been numerous and diverse developments in ecological sciences that are improving our understanding of both natural and agro-ecosystems. There is now a need for greater application of established and emerging ecological principles into agricultural systems, most obviously through improvements in cropping systems.

Examples given include increases in spatial and temporal diversity at the levels of genes, varieties and species, up to the development of fully integrated agroforestry systems. Future developments will include more precise use of such diversity based on, for example, further discoveries in chemical, physical and biological signalling systems and in other interactions among plants and animals. Integrating such examples into cropping systems can provide new approaches to organic farming that improve and stabilise productivity while increasing the ecological benefits of the farming system. Such applications of ecology are essential, partly because of their importance in replacing synthetic inputs in agriculture, partly to improve the contribution of agriculture to ecosystem services, and partly to provide better integration of natural and agro-ecosystems because of their inter-dependence.

These ecological and agricultural elements need to be brought together in a modern form of certified organic agriculture. This should produce a multi-faceted output well beyond improvements in biodiversity and yield, including a balanced and diverse output for ideal dietary needs, energy savings, energy generation, wildlife, support and development of rural society, more aesthetically appealing countryside and greater opportunities for education and leisure in the natural and farmed environment (see also Tudge, 2003).

Such changes will pay for themselves in the long-term, but in the short term there is a need for greater investment into appropriate ecological and cropping research. This is consistent with current Government policy on sustainability, biodiversity and renewable energy. But policy development needs to go much further in this direction not only in the UK but also in relation to reform of the Common Agricultural Policy.

REFERENCES

BULSON H.A.J., SNAYDON R.W. and STOPES C.E. (1990) Effects of plant density on intercropped wheat and field beans in an organic farming system. *Journal of Agricultural Science,Cambridge,* 128, 59-71.

CHAMBERLAIN K., PICKETT J.A. and WOODCOCK C.M. (2000) Plant signalling and induced defence in insect attack. *Molecular Plant Pathology*, 1, 67-72.

COSTANZA R., D'ARGE R., de GROOT R., FARBER S., GRASSO M., HANNON B., LIMBURG K., NAEEM S., O'NEILL R.V., PARUELO J., RASKIN R.G., SUTTON P. and VAN DEN BELT M. (1997) The value of the world's ecosystem services and natural capital. *Nature*, 387, 253-260.

ENGEL C. (2002) *Wild Health*. London: Weidenfeld and Nicholson.

FINCKH M.R., GACEK E.S., GOYEAU H., LANNOU C., MERZ U., MUNDT C.C., MUNK L., NADZIAK J., NEWTON A.C., de VALLAVIEILLE-POPE C. and WOLFE M.S. (2000) Cereal variety and species mixtures in practice with emphasis on disease resistance. *Agronomie*, 20, 813-837.

GARRETT, K.A., MUNDT, C.C. (1999) Epidemiology in mixed host populations. *Phytopathology*, 89, 984-990.

GORDON A.M. and NEWMAN S.M. (Eds.) (1997) *Temperate Agroforestry Systems*. Wallingford: CABI.

TILMAN D. (1988) *Plant strategies and the Dynamics and Structure of Plant Communities*. Princeton NJ: Princeton University Press.

TUDGE C. (2003) *So Shall We Reap*. London: Allen Lane.

WEINER J., GRIEPENTROG H.-W. and KRISTENSEN L. (2001) Suppression of weeds by spring wheat (*Triticum aestivum*) increases with crop density and spatial uniformity. *Journal of Applied Ecology,* 38, 784-790.

WOLFE M.S. (2000) Crop strength through diversity. *Nature,* 406, 681-682.

ZHU Y., CHEN H., FAN J., WANG Y., LI Y., CHEN J., FAN J., YANG S., HU L., LEUNG H., MEW T. W., TENG P. S., WANG Z. and MUNDT C.C. (2000) Genetic diversity and disease control in rice. *Nature,* 406, 718-722.

Observations on Agronomic Challenges during Conversion to Organic Field Vegetable Production

P.D. SUMPTION, C. FIRTH and G. DAVIES
IOR-HDRA, Ryton Organic Gardens, Coventry, CV8 3LG, UK

ABSTRACT

Prior to conversion, growers perceived agronomic challenges such as controlling weeds, pests and diseases as their biggest concerns. Through monitoring of commercial farms since 1997, the DEFRA-funded 'Conversion to Organic Field Vegetable Production' project aimed to provide information to growers on the economic and agronomic performance of farms during conversion. Many of those technical challenges of production have been successfully overcome. While there have been some problems with nitrogen availability, especially where insufficient attention has been paid to fertility building and also on lighter land, there have been few major changes in soil nutrient levels. Investment in new weeding equipment and new techniques has enabled growers to reduce costs of controlling weeds. Perennial weeds do not appear to be increasing though there have been localized problems. Seasonal effects, such as wet weather, have had a strong influence on levels of weeds, pests and diseases, and the cost of their control. All growers have had to adapt to new techniques and new crops, but those with vegetable growing experience and those growing crops they had grown before have adapted quickest to the technical challenges of production. While there are still agronomic challenges to overcome, the concerns of growers have shifted towards the marketing issues.

INTRODUCTION

Since the mid-1990s many new producers have converted to organic field vegetable production in response to the rapid increase in the market and the introduction of conversion grants from the government (Firth *et al.*, 2003). The aim of this paper is to make some preliminary observations on the agronomic challenges these farms faced during the conversion process and to make some initial judgements as to whether these have been overcome.

HDRA has been leading the DEFRA-funded 'Conversion to Organic Field Vegetable Production' project which has monitored ten farms converting from conventional production to an organic system which includes organic field vegetables, since 1997. The farms range in size from 20 to 1900 ha with target conversions of between 4 and 700 ha, and are on varied soil types and geographical locations. They have converted from predominantly arable, livestock and intensive vegetable systems. In 2002 each farm grew between 2 and 115 ha of organic vegetables. The principal concerns of many growers at the start of conversion were weed control, pest and disease pressure and plant nutrition.

METHODS

HDRA staff and OAS advisers have visited the farms on average six times per year during and after conversion. Information has been gathered through a combination of farmers' records, interviews with farmers and detailed field assessments. Weeds, pest and disease incidence have been monitored through regular crop walks. Quadrat samples were used to assess individual weed occurrence and percentage total weed cover. Presence and levels of damage due to pests and diseases were recorded at the same time. Records of farm operations are taken for farm strategies on weed management and pest and disease control. Soil samples were taken in the spring of each year to monitor soil nutrient status, pH, soil organic matter and trace elements. Yields from crops were assessed from farm records. All this information has been used to build up case studies of each farm system.

OBSERVATIONS

Good soil management and a healthy soil are at the heart of successful organic production. Building fertility through the use of leguminous crops is a central component of organic farming systems. In the rush to convert with new government conversion grants and a seemingly booming organic market some of the growers neglected, initially, to pay sufficient attention to the fertility-building period. Mistakes or inappropriate practices such as failing to use legumes can lead to conversion-specific yield reductions (Lampkin *et al.*, 2002). Four of the farms had completely stockless systems in which the success of the ley phase is crucial. Establishment and management of leys was a new technique to the farmers converting from conventional arable or horticultural production. Poor establishment, allowing leys to grow too tall (the mulch preventing regrowth and permitting weeds and crop to seed), the removal of silage, and therefore of plant nutrients, off the farm, and poor incorporation of the leys have been issues on some of the farms. Two farms had nitrogen-stressed crops in the latter stages of the rotation due to shortage of time in, and quality of, fertility building. Generally in those phased conversions, later blocks of land have been given more extended fertility-building, of up to two years, after following advice. Farmers appear increasingly to be recognizing the importance of fertility building and the use of cover crops.

Problems relating to plant nutrition have been mainly on the light sandy soils (two farms) both of which had problems of nitrogen availability for crops towards the end of the first season of organic cropping (even after a full two-year fertility building). On one of these farms composting of the manure has helped, while the other received derogations for a restricted input of pelleted poultry manure.

Monitoring of soil nutrient changes has shown that there has been very little change in soil organic matter levels or major nutrient levels. Soil phosphorus (P) is an issue on six of the farms with low or marginally low levels (according to the Elm Farm Research Centre (EFRC) interpretation) and though there are some declines in readily available P, there is very little evidence of any long-term decline of potentially available reserves. Monitoring of the stockless system (with no inputs) at HRI Kirton has shown an initial decline in soil available P with levels stabilising and later increasing. There were also declines in soil potassium (K), which may be levelling out.

Soil compaction, while also a problem in conventional production and not due to conversion itself, has been a problem. In the first two years of organic cropping, problems due to soil compaction were observed on several farms, following two very wet autumns when crops had been harvested in wet conditions. In 2002 it was not observed as a problem. However, the percentage of fields recording high levels of manganese, which may indicate compaction, according to the EFRC soil analysis interpretation, rose from 24% in 1999 to 54% in 2003. Soil analysis also indicated that the biological activity of the soil was out of balance, with associated problems of micro-nutrient availability for between 61% and 74% of fields sampled between 1999 and 2003.

Weeds were perceived by growers as being their biggest technical challenge prior to conversion. Weeds have generally been managed effectively, with a few exceptions. The challenge has been to keep costs down by reducing the hand labour necessary. Adoptions of new technologies such as the finger-weeder have made a significant impact, with many crops such as brassicas being managed exclusively with mechanical weeding. Although weed control equipment is the biggest proportion of conversion-related investment, ranging from 3% to 69% on six of the farms (average 37%), these investments have led to savings being made. Farmers have had to adapt to new techniques and new equipment. Instances of problems included: 1) wet seasons when weeding windows were missed, or weeds have grown fast and/or re-grown after mechanical weeding operations; 2) on farms with large weed-seed banks, on fertile black fen soils and where there have been high weed seed returns in the past; 3) coping with expansion - under-estimation of management time needed in organic systems and being under-staffed during rapid expansion of organic cropping.; 4) poor management of grass leys with seed returns and poor incorporation of leys; and 5) early crops, especially those under fleece have been difficult to manage.

Weeding costs can be significantly reduced in an efficient organic vegetable system but timing is critical and adverse weather conditions can have a huge impact on costs. On one farm the average weeding costs were £1625/ha in 2000, which was a wet season, and £949/ha in 2002, which was drier, though experience and new equipment would also have contributed to these lower costs.

Perennial weeds are a big concern for growers. On two farms converting from arable systems couch has been a problem with bastard fallowing necessary to reduce weed levels and also a change in the rotation. This has had a cost. Creeping thistle has only been a problem on two farms and does not appear to be increasing. Docks have only been a problem on one farm converting from a pasture-based system and also do not appear to be increasing.

A wide diversity of pests and diseases has been observed on the reference farms, with some of the most significant losses being due to vertebrate pests such as rabbits, pigeons and geese. Of all pest and disease damage considered severe 21% was due to these large vertebrate pests. Losses due to disease have also been significant, accounting for 42% of severe damage. In some cases this has had an impact on the rotation, e.g. not being able to grow onions after *Sclerotium cepivorum* (white rot). Brassicas were the most problematic crops to grow having a wide range of pest and disease problems, but also accounting for 42% of the severe rated problems. It is not surprising therefore that the greatest efforts taken to combat pest and diseases have been in brassicas, with a higher cost for these crops than others. The strategies on the farms have polarised, with farms either having intensive spraying regimes with high use of potassium soap, *Bacillus thurungiensis* (Bt), sulphur and copper, or no control measures at all. The farms with the most use of sprays tend to be the larger farms with bigger field sizes supplying into the packers and supermarkets. These are also the farms with the biggest risks (if crop is rejected there may be no alternative market). As specifications for organic produce have got tighter (Firth *et al.*, 2003), appearing to approach zero tolerance of pest or disease damage or presence, so have the number of sprays used increased on these farms. There also appears to be a correlation between plot size and severity of pest or disease attack, with more on larger plots (Davies *et al.*, 2002). There is considerable seasonal variation with wet summers such as 2000 bringing problems of slugs, *Phytopthora infestans* (Potato late blight) and *Septoria apiicola* (Celery leaf blight). Drier summers such as in 2002 have seen fewer problems.

DISCUSSION

Seasonal weather effects play a huge role in the extent that weeds, pests and diseases are problems. Wet summers can be particularly problematic and costly and dry seasons, though not without challenges can provide a grower with more management options and opportunities. This seasonal variation makes it difficult to evaluate any long-term trends. Although growers appear to be mastering many of the technical challenges of crop production, it remains to be seen in the long term whether issues such as perennial weeds, soil structure and compaction and retention of nutrients will become more important. These will continue to be monitored on five of the farms in the new HDRA-led Sustainable Organic Vegetable Network Project, funded by DEFRA, together with five new farms. The market plays a crucial role and a drive for growers to specialise in certain crops and to meet very high specifications could force growers into standardised and intensive management practices that sacrifice diversity. Growers on a smaller scale supplying direct markets and those with a diversity of marketing are better placed to withstand these pressures. Many factors beyond the growers' control, such as soil type, past land use (weed seed burden and soil-borne disease) and intensity of production in the surrounding area will all have influence on the success of conversion.

CONCLUSIONS

Conversion is often described as a steep learning curve. There are many new techniques to master. Arable growers are not used to growing fertility-building crops, and the demands of rotation mean that many growers have a more complex cropping plan post-conversion with new crops that they have not grown before. Some growers had little or no experience of growing vegetables. Not surprisingly, those with experience of field vegetables and those that grew crops that they had previously grown conventionally, had most success.

Growers have tackled many of the technical challenges of conversion as they have adapted to or learnt new skills. Weed control has generally been managed effectively through conversion, though with some problems as they have expanded their cropping areas and learnt new techniques. Many of the problems of pests and diseases can and have been controlled effectively, though often with direct rather than cultural methods. The market will continue to pressure growers to reduce costs and tight specifications will also mean that the risks for large-scale growers tied into one market will be high.

REFERENCES

FIRTH C., GEEN N. and HITCHINGS R. (2003) The UK Organic Vegetable Market. Coventry: HDRA.

LAMPKIN N., MEASURES M. and PADEL S. (2002) 2002/03 Organic Farm Management Handbook. Aberystwyth: Institute of Rural Studies, University of Wales.

DAVIES G., SUMPTION P., CROCKETT M., GLADDERS P., WOLFE M. and HAWARD R. (2002) Developing improved strategies for pest and disease management in organic vegetable production systems in the UK. In: *The BCPC Conference Pests and Diseases 2002, Volume 2,* 547-552.

Systems Thinking in Organic Research: Weeds are the Symptom

G. DAVIES and R.J. TURNER
IOR-HDRA, Ryton Organic Gardens, Coventry, CV8 3LG, UK

ABSTRACT
Taking the idea that 'weeds are the symptom' when formulating projects for organic weed research projects will naturally lead to taking a 'soft systems' approach to diagnosing and solving problems of direct relevance to growers. Such an approach allows a logical grasp and analysis of complex and uncertain problems where there are no simple answers, as exemplified by weed management options in organic farming systems. It should stimulate a co-learning approach between stakeholders (including farmers and researchers), incorporating formal scientific knowledge as well as more informal farmer knowledge and experience. Challenges for researchers include understanding other stakeholders' perspectives on weed management decisions by looking at linkages within a system, adapting their research methodology to take account of the emergent properties that become apparent at different levels in a system and understanding the relevance of research in context. Some of these challenges are illustrated in the framework of a participatory investigation of the management of weeds in organic production systems.

INTRODUCTION
Weeds, loosely defined as 'naturally occurring non-crop vegetation', are an important component of agroecosystems. They are traditionally seen as major constraints to crop production, especially in organic farming systems, and a somewhat intractable problem. However, several authors have recently asked if we are addressing the right issues (e.g. Barberi, 2001) or have pointed out some the positive benefits weeds can have in the farming system (e.g. Marshall *et al.*, 2003). Many authors writing on the subject are agreed that researchers need to take at least an interdisciplinary approach (e.g. Liebman and Davis, 1999) and most practical advice hinges on the need to work 'with nature', using experience, strategic use of appropriate control methods, and long term planning, largely underpinned by scientific knowledge, to manage weeds in organic systems (e.g. Kristiansen, 2000). In this paper we ask if recent ideas about analysing complex systems can aid in understanding weeds as components of organic farming systems, and whether this necessitates a different approach to weed management research, and ultimately weed management practice. Many of the ideas have arisen in the context of a DEFRA participatory project (OF0315) on weed management in organic production systems.

FARMING SYSTEMS
Generally, complex systems are defined in terms of the relationships or linkages between entities within a defined boundary, and we attribute meanings to the whole system that we do not attribute to the parts (the so called 'emergent properties' of a system). Where we draw the boundaries to a system depends on our perspective. Integrated management programmes in agriculture have often resulted in a 'hard systems' approach to defining farming systems, that is using tools and models to analyse inputs, outputs and control points in a system defined by the researcher, e.g. models of nutrient flows. However, this is increasingly being replaced by a 'soft systems' approach that sees farming as a human

activity system defined by the participants (Gibbon, 2002). A more detailed discussion of such soft systems thinking is provided elsewhere in these proceedings (Davies and Gibbon, 2004) but from such a description, it follows that systems are in effect models that we use to give meanings or understanding to a situation.

As soft systems, farming systems are characterized by non-linear complexities such that it is often not possible to form definitive statements of problems. There are usually many people who have a stake in the problem to be solved, and they will have many different perspectives on the problem, which makes the process social rather than purely technical. It follows that there are no objectively 'right' answers to problems, so that stakeholders will often accept the most promising sounding solutions for any given situation. They also accept that the constraints on the solution are likely to change over time and that the problem solving process is often constrained by resources (e.g. in time, effort or money). In such circumstances problems can often only be understood by developing and testing (interim) solutions, thus creating a cycle of experimentation, experience and critical reflection that advance knowledge in any particular area (Kolb, 1984). Much recent farming system research effort has gone to developing a toolkit of methods that can be used to gain perspectives on systems from many differing points of view (Packham, 2003) and to allow the creation of conditions for such learning environments, in effect to encourage systems thinking.

SYSTEMS THINKING AND WEEDS

Weeds are an integral part of agroecosystems. There are many complex interactions and perspectives on weeds, which are increasingly been seen as 'ecological goods' (Gerowitt *et al.*, 2003) as a wider range of stakeholders voice their opinions on farming systems. Ecologically, the role of weeds can be considered to be well understood, especially aspects like competition with crop plants, the behaviour of weed species (germination, reproduction, dispersal etc.) and seed bank performance. It is also becoming increasingly recognized that weeds can play a vital role in nutrient cycling, mediation of pest and disease interactions with crop plants, protection against soil erosion, and modification of soil ecology and structure amongst many other ecological effects. Economically the effect of weeds has traditionally been understood through their effects on yield and monetary returns, i.e. their negative impact on farm economic performance. However, it has not generally been easy to define their more social aspects; the perceived need to have clean fields on the part of growers, the aesthetic value of weeds to ramblers, and imparting a feeling of countryside to rural dwellers being just some of the myriad of factors that might be taken into consideration.

When viewed in this light it begins to be difficult to pin down weed problems and to calculate the pay-offs between these different aspects of weeds within farming systems. A soft systems or systems-thinking type approach might help to reconcile different perspectives. We consider this to be especially important in organic farming systems where consideration should be given to all facets of systems performance.

WEEDS AS SYMPTOMS

Symptoms are the emergent properties that we use to diagnose systems functioning, e.g. as traditionally used to identify the cause of disease. Weeds seen in this light then become the emergent properties of a farming system at some level. It is therefore necessary to understand the underlying structural causes of weed problems in order to promote effective actions and to avoid treating the symptoms of the problem rather than the cause. From the previous discussion of soft systems approaches, this will imply a need to examine linkages and flows within a farm to understand causes, but crucially will also need to define the perspectives of various stakeholders in order to decide on the appropriate level to treat weed problems. This could for instance, mean the difference between looking for a technical solution to a particular weed, as compared to changing a perspective and regarding a weed as a resource.

RESEARCH ON WEED MANAGEMENT

Research in weed control has been very effective at devising technical solutions to weed problems in tightly defined conventional systems, and has recently begun to model complex hard system approaches (Grundy and Turner, 2002). However, we would argue that organic researchers should take a soft systems approach to organic weed management research, which will allow a flexible and diverse approach to weed management in organic systems. It will also stimulate a co-learning approach that should draw in knowledge from both farmers and researchers, but also wider groups of stakeholders. Crucially, such a process should look at novel ways of engaging stakeholders with an interest in organic weed management

Some of the challenges for organic researchers with this approach lie in understanding other stakeholders' perspectives on organic farming systems and defining these boundaries. Another challenge for researchers is to adapt their research methodology to take account of the emergent properties that become apparent at different levels in a system and understand the relevance of their research in this context.

A PARTICIPATORY WEED RESEARCH PROJECT

A recent DEFRA-funded project (OF0315) has begun to explore ways of engaging stakeholders in such a research process through a participatory investigation of weed management in organic farming systems. It has done this by first working to engage farmers in the research process (HDRA, 2003). The project began with an open remit to consult organic stakeholders, and especially organic farmers, on organic weed management issues. An initial mailshot to all organic farmers was met with a huge response and a database of some 200 interested farmers was formed, of which fifty or so attended a stakeholder day that also included researchers and advisers. During this open workshop weed management issues were discussed and prioritized so that by the end of the meeting it was decided to take forward three priority themes using a focus group approach. The themes were: 1) docks (and perennial) weeds; 2) system studies; and 3) knowledge collation and dissemination.

Subsequently, each theme has been developed by a focus group, which in each case has decided on group role and activities within project budgets. Subsequent to this, the project has taken three main approaches that run through each theme: 1) collection and collation of existing knowledge from both a practical farming point of view and from scientific literature; 2) monitoring of existing farm weed management strategies; and 3) initiating small-scale field trials. Apart from standard scientific review methods, project researchers have been collecting data on farmer weed management practice using semi-formal interviews (over fifty done) and collating and presenting them as case study data. This has dovetailed with monitoring the impact of management operations and rotations on dock populations on ten farms in addition to setting up some simple small-scale trials on mechanical dock control as compared with the farmers' current practice. In addition, background information on topics of interest arising from the on-going work is being compiled and will be fed back to the relevant groups in order to stimulate further activities as the project progresses. An integral part of the project is communication of information between stakeholders, and the project has specifically addressed this theme at all meetings and public events. Although many farmers have expressed a desire for information as leaflet summaries, it is intended to hold much of the information publicly on a website (HDRA, 2003) which will also provide ample opportunity for feedback on all topics and issues arising during the project. Eventually, all project working documents will be held on the website but will also be available in leaflet form.

FUTURE WORK

From the perspective of a systemic approach to weed management the project has began to place 'weeds' in the context of organic farming systems, mainly from the perspectives of farmers and researchers. It is hoped that these approaches will eventually draw in wider range of stakeholders, and that problems and solutions will begin to flow from the processes being used to engage the various stakeholders. This soft systems approach is more likely to lead to the development of relevant sustainable organic farming practices that meet the needs for practical weed management on the part of farmers, to provide stimulating and engaging research projects for researchers, as well as going some way to fulfilling the aspirations of other stakeholders in the organic movement in general.

REFERENCES

BARBERI P. (2001) Weed management in organic agriculture: are we addressing the right issues? *Weed Research,* 42, 177-193.

DAVIES G. and GIBBON D. (2004) Systems thinking in organic research: does it happen? In: Hopkins A. (Ed.) *Organic Farming: Science and Practice for Profitable Livestock Production* (Joint BGS/ AAB/ COR Conference, 20-22 April 2004). Occasional Symposium of the British Grassland Society No. 37, pp. 216-219.

GIBBON D. (2002) Systems thinking, interdisciplinarity and farmer participation: essential ingredients in working for more sustainable farming systems. *Proceedings of the UK Organic Research 2002 Conference, 26-28 March 2002,* Aberystwyth, pp 105-108.

GEROWITT B., BERTKE E., HESPELT S.-K., and TUTE C. (2003) Towards multifunctional agriculture- weeds as ecological goods? *Weed Research*, 43, 227-235.

GRUNDY A. and TURNER B. (2002) Horticultural weed control in organic systems- a modelling approach. *Proceedings of the UK Organic Research 2002 Conference, 26-28 March 2002,* Aberystwyth, pp 295-298.

HDRA (2003). HDRA Organic Weed Management Site < http://www.organicweeds.org.uk>

KOLB D. A (1984). Experiential learning: experience as the source of learning and development. Eaglewood Cliffs, NJ: Prentice-Hall.

KRISTIANSEN P. (2000) Work with nature for effective weed control. In: Horsley, P. (Ed.) *The Organic Alternative: the complete guide to organic farming,* pp. 80-84, Kondinin: Cloverdale.

LIEBMAN M. and DAVIS A.S. (2000) Integration of soil, crop and weed management in low-external-input farming systems. *Weed Research,* 40, 27-47.

MARSHALL E.J.P., BROWN V.K., BOATMAN N.D., LUTMAN P.J.W., SQIURE G.R. and WARD L.K. (2003) The role of weeds in supporting biological diversity within crop fields. *Weed Research,* 43, 1-13.

PACKHAM R. (2003) Concepts behind the farming systems approach. *1st National Australian Farming Systems Conference, 7-10 Sept 2003, Toowoomba, Queensland.* pp 26- 48.

Plant Breeding for Agricultural Diversity

S.L. PHILLIPS and M.S. WOLFE
Elm Farm Research Centre, Hamstead Marshall, Nr Newbury, Berkshire, RG20 0HR, UK

ABSTRACT
Plants bred for monoculture require inputs for high fertility, and to control weeds, pests and diseases. Plants that are bred for such monospecific communities are likely to be incompatible with the deployment of biodiversity to improve resource use and underpin ecosystem services. Two different approaches to breeding for agricultural diversity are described: (1) the use of composite cross populations and (2) breeding for improved performance in crop mixtures.

INTRODUCTION
Monocultural plant communities dominate modern agriculture. Monocultures are crops of a single species and a single variety; hence the degree of heterogeneity within such communities is severely limited. The reasons for the dominance of monoculture include the simplicity of planting, harvesting and other operations, which can all be mechanized, uniform quality of the crop product and a simplified legal framework for variety definition.

Monocultural production supports the design of crop plants from conceptual ideotypes. The wheat plant ideotype is a good example of a plant designed for monoculture. Wheat plants that perform well in monoculture interfere minimally with their neighbours under high fertility conditions, where all ameliorable factors are controlled. The aim of this design is to provide a crop community that makes best use of light supply to the best advantage of grain production (Donald, 1968). This design has produced wheats with a high proportion of seminal roots, erect leaves, large ears and a relatively dwarf structure. This 'pedigree line for monoculture' approach is highly successful, but it has delivered crop communities that do best where light is the only, or the main, limiting factor for productivity: therefore the products of this approach to breeding require inputs to raise fertility, and to control weeds, pests and diseases. This breeding effort, coupled to the increasing convenience of monoculture, now dominates modern farming but the restrictions involved have led some people to question the value of this approach to farming and breeding.

THE ECOLOGICAL ROLE OF AGRICULTURAL DIVERSITY
Darwin had a seamless view of population biology, evolutionary biology and ecosystem processes. The advantages of such a view are now being realized. For instance, Tilman (2001) points out two key findings: (1) that a greater number of terrestrial plant species can lead to greater ecosystem productivity and resource use and (2) that greater diversity can lead to greater ecosystem predictability and temporal stability. This links two key concepts: that diversity can underpin productivity and the stability of productivity; and that diversity underpins ecosystem functioning and therefore the ecosystem services required for sustainability.

Biodiversity in agroecosystems provides ecosystem services beyond the production of food, fibre, fuel and income. Altieri (1999) suggests that examples of ecosystem services include the recycling of nutrients, control of local microclimate, regulation of local hydrological processes, regulation of the abundance of undesirable organisms, and detoxification of noxious chemicals. In addition, agrobiodiversity supports above- and below-ground trophic levels. For instance, Marshall *et al.* (2003) suggest that many arable weed species support a high diversity of insect species, that in turn support several bird species; indicating that weeds have roles within agroecosystems of supporting biodiversity more generally. The restricted biodiversity associated with monocultural plant communities limits the ecosystem services of those production systems, one simple example being the use of herbicides to eliminate all weeds within the crop.

Furthermore, Altieri (1999) recognizes the ecological sensitivity of monospecific communities, stating that nowhere are the consequences of biodiversity reduction more evident than in the realm of agricultural pest management. Altieri explains that the reasons are complex, but he sums up the problem as the loss of inherent self-regulation. Amongst the options that Altieri suggests for utilizing biodiversity to limit pest problems are high crop diversity through mixing crops in time and space, the presence of tolerable levels of specific 'weed' species and the deployment of genetic diversity by using variety mixtures or multilines.

The challenge is to link all of these ecological concepts by moving away from monoculture with its biological simplification and consequences for ecosystem services, while simultaneously providing functionality to the diversity incorporated within the crop community.

EVOLUTIONARY THEORY IN PLANT BREEDING

The neo-Darwinian view of the process of evolution describes four basic components: (1) the initial generation of variability, (2) the exchange of DNA between chromosomes (recombination), (3) differential reproduction and (4) isolation in space and time. Simmonds (1962) points out that the modern breeding of crops, (or mankind's pursuit of adaptation of plants to monocultural cropping systems), has seen steps (1) and (2) as important steps only in the initial phases of generating new varieties, but (3) and (4) have come to dominate the process of plant breeding. This breeding effort produces (especially for inbreeding species) homozygous varieties that are adapted (perform well) to the conditions under which they were isolated or selected by the breeder. This process is distinct from providing crop genotypes or populations with a degree of adaptability. Adaptation is the property of a genotype which permits survival under selection; adaptability is the property of a genotype or population of genotypes which permits subsequent alteration of the norms of adaptation in response to changed selection pressures. Hence there has been a tendency to eliminate variability and adaptability in crop varieties and populations, and to pursue the notion that strictly uniform crop populations, adapted to a specific set of circumstances, is a universal ideal.

Suneson (1956) described a "new" method of plant breeding. He suggested that it is important to recognize the value of evolutionary fitness in plant breeding. Evolutionary change is based upon the interaction between populations and their environments, where environmental interactions are both abiotic and biotic. Suneson describes a process of assembly of seed stocks with diverse evolutionary origins, recombination by hybridization, the bulking of F_1 progeny, and subsequent prolonged natural selection for mass sorting of the progeny in successive natural cropping environments. Therefore, Suneson promoted a method of plant breeding that moves away from the notion of strictly uniform crop populations and towards populations with a high level of heterogeneity, thereby repeatedly harnessing all components of the neo-Darwinian view of evolution.

COMPOSITE CROSS POPULATIONS

The bulk population breeding method described by Suneson (1956) depends upon the nature and outcome of mass trials by artificial and natural selection acting on a heterogeneous mixture of competing genotypes. This is distinct from pedigree selection schemes where early and continuous individual selection begins in the F_2 generation. In composite crosses a large number of carefully chosen varieties are intercrossed and all the hybrids are bulked together for propagation. The basic idea of the composite populations is that the introduction of genetic diversity may a) allow the isolation of superior individual lines in a cost effective manner,and b) that diverse populations may offer better performance than pure lines. These lines and populations are the result of adaptation to those selection pressures imposed during the breeding process, both natural and artificial, providing improved fitness to given environmental conditions. Indeed, the evidence from barley composite crosses, is that directional and stabilizing selection over a number of years tends to produce agronomically superior crops.

Composite cross populations also provide the option of farmer participation in the process of selection; this may be important since low input production systems are difficult to characterize, and all require slightly different emphases in the interactions between crop vigour, disease resistance, weed resistance, fertility scavenging and pest resistance. But also, importantly, with populations or lines with a broad genetic base there should be some capacity for a genetic response to selection, or adaptability. There are advantages to producers of providing a compromise between adaptation and adaptability, especially for low input production systems, because predicting the range and intensity of limiting factors year on year is impossible.

EFRC, in collaboration with the John Innes Centre, have produced six composite cross populations that are growing in trials across the UK. This work is part of a DEFRA-funded project 'Generating and evaluating a novel genetic resource in wheat in diverse environments.'

BREEDING COMPONENTS FOR MIXTURE PERFROMANCE

Following an effective crop rotation, the simplest step forward for introducing diversity into cropping systems is to grow variety mixtures, followed by species mixtures. However, components are very rarely selected for performance in variety mixtures. Decisions on mixture composition are often based on yield in monoculture. But the underlying assumption may be incorrect: yielding ability, which is required for high monocultural performance, is not necessarily the same as competitive ability in mixtures (Hill, 1996).

Hill (1996) calls for an approach to breeding for mixtures that selects for good general and specific combining abilities with other varieties/species. Hill suggests two possible strategies, either focusing on one component of a mixture only, 'the passive approach', or focusing on both components using alternating cycles of selection, 'the active approach'. In the passive approach a number of genotypes from one component could be assessed in all possible binary combinations against a set of testers drawn from the other component. In the active approach the roles of tester and tested are reversed in alternate cycles of selection. The active approach permits a degree of coevolution of components; the passive approach should deliver varieties that are better suited to mixing with a particular crop type than varieties bred for monocultural communities. Breeding for mixture performance is especially important for organic or low input systems where the predictability of important variables is less certain than in conventional systems, and therefore the need for a crop community to buffer against the risks of these variables is more important.

CONCLUSIONS

Monocultural plant communities have different demands from plant breeding than cropping systems based on diversity. The ability of diverse cropping systems to provide inherent buffering against both biotic and abiotic variables without resorting to synthetic inputs is clear. As a consequence, the potential for biodiverse cropping systems to contribute to important ecosystem processes is greater than for monocultural communities. Composite cross populations are a way of producing crop communities with a higher degree of heterogeneity, as is breeding varieties for good ecological combining abilities in mixtures. However, any adoption of breeding for agricultural diversity requires shifts in legal and administrative frameworks and an improvement in the market acceptability of heterogeneous crops.

REFERENCES

ALTIERI M.A. (1999) The ecological role of biodiversity in agroecosystems. *Agriculture, Ecosystems and Environment,* 74, 19-31.

DONALD C.M. (1968) The design of a wheat ideotype. In: Finlay K.W. and Shepherd K.W. (Eds) *Proceedings of the Third International Wheat Genetics Symposium*, pp. 377-387, Canberra: Butterworths.

HILL J. (1996) Breeding components for mixture performance. *Euphytica,* 92, 135-138.

MARSHALL E.J.P., BROWN V.K., BOATMAN N.D., LUTMAN P.J.W., SQUIRE G.R. and WARD L.K. (2003) The role of weeds in supporting biological diversity within crop fields. *Weed Research,* 43, 77-89.

SIMMONDS N.W. (1962) Variability in crop plants, its use and conservation. *Biological Reviews,* 37, 442-465.

SUNESON C.A. (1956) An evolutionary plant breeding method. *Agronomy Journal,* 48, 188-191.

TILMAN D. (2001) An evolutionary approach to ecosystem functioning. *Proceedings of the National Academy of Science, USA,* 98 (20), 10979-10980.

Compost Teas - A Simple Disease Control Solution for Organic Crops?

A.M. LITTERICK[1], C.A. WATSON[1], P. WALLACE[2] and M. WOOD[3]

[1]SAC, Craibstone Estate, Aberdeen, Scotland, AB21 9YA, UK
[2]Enviros Ltd, 20-23 Greville Street, London, EC1N 8SS, UK
[3]Soil Science Department, The University, PO Box 233, Reading, RG6 6DW, UK

ABSTRACT

Compost teas and extracts are defined and their effects on the health of crops are examined with examples from recent scientific literature. Available information on the safety and efficacy of compost teas is discussed and priorities for research to develop safe effective compost teas for use in crop production systems are outlined.

INTRODUCTION

Sprays based on compost extracts have been used for hundreds of years, but interest in them waned when pesticides became available in the twentieth century, since pesticides tend to give better, more reliable control of most foliar diseases. However, the recent increase in sustainable and organic farming and problems relating to pesticide use has led to renewed interest in compost extracts and teas. Most work to develop improved methods for preparation and use of compost extracts and teas has been done in the United States and much of this has been done by commercial companies rather than independent institutes. The work and the key findings from it are discussed in this paper.

DEFINITION OF COMPOST TEAS AND COMPOST EXTRACTS

Exact definitions of compost teas and compost extracts vary. The term "compost tea" is used here to describe the product of recirculating water through loose compost or a porous bag or box of compost suspended over or within a tank with the intention of maintaining aerobic conditions. It is important to distinguish between compost teas prepared using aerated and non-aerated processes; therefore the terms aerated compost tea (ACT) and non-aerated compost tea (NCT) will be used in this paper to refer to the two dominant compost fermentation methods. ACT will refer to any method in which the water extract is actively aerated during the fermentation process. NCT will refer to methods where the water extract is not aerated or receives minimal aeration during fermentation apart from during the initial mixing. The term "compost extract" has been used in the past to define water extracts prepared using a very wide range of different methods (Scheuerell and Mahaffee, 2002). For the purposes of clarity in this paper, the term compost extract refers to the filtered product of compost mixed with any solvent (usually water), but not fermented.

POTENTIAL EFFECTS OF COMPOST TEAS

Compost extracts/teas have been shown to help prevent or control a wide range of foliar diseases in glasshouse and field grown edible and ornamental crops (Scheuerell and Mahaffee, 2002). Examples of diseases controlled in this way are shown in Table 1.

Control has not been achieved with all pathogens in all tests and efficacy of compost teas varies depending on the crop and experimental system.

Most of the published evidence to demonstrate control of foliar disease concerns NCTs or compost extracts. There is a shortage of data on the efficacy of ACT's and few studies have compared the efficacy of ACTs and NCTs in controlling foliar diseases. The few trials which have been carried out have shown that the effects of ACTs vary considerably. For example, no effect of ACT applications on early blight of tomato was observed; lettuce drop (several pathogens) was reduced in a summer but not a spring crop; post-harvest fruit rot of blueberries was significantly reduced, but this was offset by reduced yields. The impact of ACTs on plant health and crop yield can therefore be crop specific and may depend on the experimental system and environmental conditions. General statements about the efficacy of ACTs cannot be made.

Compost teas are also being widely advertised and used on both organic and conventional farms in the United States as an inoculant to restore or enhance soil microflora (www.attra.ncat.org). However, very little work has been done to establish the effectiveness of teas used in this way or to quantify the benefits if there are any. NCTs have been shown to reduce the impact of seedborne pathogens when used as a seed treatment (Tränkner, 1992). Compost teas have also been shown to affect soilborne pathogens *in vitro*, but it is well known that successful disease control *in vitro* does not always translate to field conditions. Recent work has shown that fusarium wilt of pepper (*F. oxysporum* f.sp. *vasinfectum*) and cucumber (*F. oxysporum* f.sp. *cucumerinum*) was controlled by drenching NCT on to soil under greenhouse conditions (Ma *et al*., 2001).

Table 1. Examples of proven plant disease suppression following application of compost teas or extracts made from known feedstocks

Target pathogen	**Common name**	**Crop**	**Principal feedstocks in compost**
Botrytis cinerea	grey mould	strawberry bean	MSW 'tea'
Erisyphe graminis	powdery mildew	barley wheat	MSW 'tea'
E. polygoni	powdery mildew	phaseolus bean	MSW 'tea'
Sphaerotheca fuliginea	powdery mildew	cucumber	MSW 'tea'
Phytophthora infestans	potato blight	tomato potato	MSW 'tea'
Venturia inaequalis	apple scab	apple	SMS 'tea'

NB.: MSW = municipal solid waste; SMS = spent mushroom substrates

MODE OF ACTION OF COMPOST TEAS

Compost teas sprayed on to plant leaves act on the leaf surface. The principal active agents in compost teas are bacteria in the genera *Bacillus* and *Serratia,* and fungi in the genera *Penicillium* and *Trichoderma*, although other genera are involved (Brinton *et al*., 1996). It is thought that compost extracts/teas act in three main ways: through inhibition of spore germination, through antagonism and competition with pathogens and through induced resistance against pathogens. The efficacy of compost extracts/teas depends on live microorganisms. Sterilized or micron filtered compost extracts have usually been shown to have significantly reduced activity against test pathogens.

PREPARATION AND USE OF COMPOST TEAS

The production of ACTs and NCTs involves compost being fermented in water for a period of 1 to 5 days. Both methods require a fermentation vessel, compost, water, incubation and filtration prior to application. Nutrients may be added prior to or following fermentation, and additives or adjuvants may be added prior to application. Several companies now sell fermentation units that produce aerobic compost tea by suspending compost in a fermentation vessel and aerating, stirring or recirculating the liquid.

Several hundred commercial growers including producers of ornamentals, field vegetables and glasshouse crops in the United States and the Netherlands are making compost teas and spraying them on to foliage as plant strengtheners and/or to help prevent foliar disease. However, compost teas are very little used and little understood as yet in the UK. At present, around thirty UK growers of ornamental crops and a small number or organic producers are experimenting with compost teas on their own holdings.

POTENTIAL PROBLEMS WITH THE USE OF COMPOST TEAS

The main potential problem with compost teas (apart from reports of variable efficacy in controlling plant disease) is the concern that fermenting compost could potentially support the growth of human pathogens. For example, Welke (1999) detected faecal coliform and salmonella populations in the source compost, the NCT fermentation and on samples of broccoli and leek growing in a field and sprayed with the NCT. Pathogens have been shown to grow during the production of both ACTs and NCTs. However, the indications are that pathogen growth is not supported when ACTs or NCTs are prepared without fermentation nutrients (Scheuerell and Mahaffee, 2002). Further work is required to ensure that the production and use of compost teas and extracts can be guaranteed not to propagate and spread human pathogens onto food intended for human consumption.

THE FUTURE FOR COMPOST TEAS

There is mounting pressure on UK organic farmers to increase both crop yield and quality in order that they can maintain and improve their place in the European marketplace. There is some evidence that the use of compost extracts/teas can help them do this through direct and indirect control of pests and diseases.

Considerable work is required to ensure predictable disease suppression and control from organic residues including composts and compost teas for different crops in different climates and soils. Much of the current work has been done, or is being done in the United States on different crops and in very different climates, soils and farming systems from those in the UK. It will be necessary to develop the techniques and protocols successfully developed in other countries for use in UK organic farming systems.

A great deal of the recent work on composts has been carried out using feedstocks which are not readily available in the UK. Research is required to assess the quality and disease suppressive properties of composts made from feedstocks which are cheap and readily available to UK farmers.

Work relating to compost teas is still in the early stages, although some consultants and farmers (both conventional and organic) in the United States are claiming disease control when using them. Work is required to identify the key active microorganisms in compost teas/extracts and to develop production processes to ensure that they exist in appropriate numbers. Application technology, which has been developed mainly to ensure optimal application of pesticides must be adapted for use with compost teas. An improved understanding of the mode(s) of action of compost teas may also allow the combination of other natural products and biological agents to treat organic crops. Work on compost teas is continuing rapidly in the United States. Much of the information relating to current preparation and application methods for compost extracts/teas is available free on the internet. Again however, considerable work is required to develop and adapt the techniques currently used in the United States for use on UK organic crops.

ACKNOWLEDGEMENTS

The authors of this report have received financial support from the Scottish Executive Environment and Rural Affairs Department (SEERAD) and the Department for the Environment, Food and Rural Affairs (DEFRA, Project No. OF0313)

REFERENCES

BRINTON W.F., TRÄNKNER A. and DROFFNER M. (1996) Investigations into liquid compost extracts. *Biocycle* 37 (11), 68-70.

MA L.P., QIAO X.W., GAO F. and HAO B.Q. (2001) Control of sweet pepper fusarium wilt with compost extracts and its mechanism. *Chinese Journal of Applied and Environmental Biology,* 7, 84-87.

SCHEUERELL S. and MAHAFFEE W. (2002) Compost tea: principles and prospects for plant disease control. *Compost Science and Utilization* 10, 313-338.

TRÄNKNER A. (1992) Use of agricultural and municipal organic wastes to develop suppressiveness to plant pathogens. In: Tjamos E.S., Papavizas G.C. and Cook R.J. (Eds.) *Biological Control of Plant Diseases*, pp. 35-42, New York: Plenum Press.

WELKE S. (1999) Effectiveness of compost extracts as disease suppressants in fresh market crops in BC. *Organic Farming Research Foundation. Grant Report* 19-31, Santa Cruz, California, USA.

Varieties and Integrated Pest and Disease Management for Organic Apple Production in the UK

S. CUBISON[1] and J. CROSS[2]
[1] Henry Doubleday Research Association, Ryton Organic Gardens
Coventry, CV8 3LG, UK
[2] Horticulture Research International, East Malling, West Malling,
Kent, ME19 6BJ, UK

ABSTRACT

Over recent years, consumer demand for organic apples in the UK has increased. This demand is expected to continue to rise, yet over 90% of current supplies are imported. The volume of UK production is currently very small and totally inadequate to meet the rising demand. Current methods of organic apple production are unsatisfactory with yields low and erratic and quality poor. This is mainly due to pest and disease problems and a lack of suitable resistant varieties. The aim of this five-year Horticulture LINK project is to develop an effective Integrated Pest and Disease Management (IPDM) programme for organic apple production, using cultural pest and disease control methods and optimizing the use of organically permitted plant protection products. Suitable apple varieties for organic production will also be identified. This paper discusses the initial challenges and progress of the project to date. Preliminary results suggest that cultural practices and programmes of copper and sulphur sprays are currently the only effective means of controlling the diseases scab and mildew on disease-susceptible varieties. Further results will become available on completion of the project in March 2005.

INTRODUCTION

Whilst overall consumption of apples in the UK has remained static, consumer demand for organic apples has risen over the last decade, with this trend set to continue for the foreseeable future. However, this demand is currently met largely through imports and the volume of UK-grown fruit is very small, (Firth and Lennartsson, 1999). The majority of organic fruit sold in the UK (both home grown and imported) is marketed through the major supermarket retailers and it is here that the best opportunity for expansion and growth exists. However, if this is to be achieved, new strategies to facilitate the production of organic apples need to be developed.

Organic apple growing is one of the most technically challenging areas of production within the organic sector. Inadequate methods of pest and disease control are the main reason for the poor performance of UK crops. Apples are subject to attack from a wide range of highly damaging pests and diseases. The diseases scab (*Venturia inaequalis*) and mildew (*Podosphaera leucotricha*) are particularly debilitating and can severely reduce tree growth, yield and quality. Damaging insect pests include moth species such as the codling moth (*Cydia pomella*), fruit tree tortrix (*Archips podana*), summer fruit tortrix (*Adoxophyes orana*), winter moth (*Operophtera brumata)* and the apple sawfly (*Hoplocampa testudinea*). Aphids can also be a significant problem – notably the rosy apple aphid (*Dysaphis plantaginea*) which attacks young shoots in early spring leading to severely distorted and curled leaves and puckered, malformed fruits. The green apple aphid (*Aphis pomi*) and rosy leaf curling aphid (*Dysaphis devecta*) can also be troublesome.

In 1999, we commenced a five-year research project ken to develop strategies for reducing damage from pests and diseases in order to assist the production of high-quality organic apples in UK. Now in its final full year (2004), the project consists of four component objectives:

1) to evaluate and refine an innovative IPDM programme
2) identify 4-6 disease-resistant varieties of high fruit quality with a range of seasons (storage potentials) and markets (dessert, culinary, juicing) which are suitable for organic production in the UK.
3) to identify effective controls for the diseases scab and mildew
4) to identify effective controls for rosy apple aphid.

The project is overseen by a consortium of industry representatives with an interest in organic apple production. These include an established organic topfruit grower, a successful conventional topfruit grower with a trial organic orchard, the two largest marketing organizations for English topfruit (OrchardWorld and WorldWide Fruit), a major processor (Fourayes Farms Ltd), two major multiple retailers (Sainsbury's and Waitrose), the Horticulture Development Council (formerly the Apple and Pear Research Council), and the East Malling Trust. The research is being undertaken by Horticulture Research International and the Henry Doubleday Research Association, blending conventional and organic expertise and experience.

PROGRESS TO DATE

Integrated Pest and Disease Management Programmes

The research into IPDM programmes has been undertaken in two very different orchards, representative of the two extremes faced by growers - a mature organic orchard planted with the disease-susceptible variety Fiesta (Red Pippin), and a new organic orchard planted with two disease-resistant varieties, Pinova and Topaz. A number of experimental approaches are used to control the main disease problem – scab, including the destruction of leaf litter in the autumn, together with applications of copper pre bud burst and a sulphur spray programme. A similar approach is used for mildew with the removal of shoots infected with primary mildew and a sulphur spray programme. The effect of this programme in controlling sooty blotch, which has proved to be a considerable problem, is also being evaluated. The experimental approaches against pests involve granulovirus for codling moth, Bt (*Bacillus thuringiensis*) against tortrix and winter moth, predator refugia and physical destruction of infested shoots against rosy apple aphid and pyrethrum against blossom weevil.

The new organic orchard was initially planted with bench-grafted trees of three disease susceptible and five disease resistant varieties. Establishment was hampered by an unusually severe drought. The bench-grafted trees were grubbed and the orchard replanted with well-grown 2-year-old nursery trees of the varieties Pinova and Topaz on MM106 rootstocks in spring 2002. Mulching with composted green waste was provided to aid establishment.

Preliminary results from the established organic Fiesta orchard

Plot yields from the experimental IPDM programme in the mature organic Fiesta orchard were compared with plots of the grower's existing organic pest and disease control programme and untreated control plots. The experimental programme showed a large increase in yield compared with the grower's programme and an untreated control from 2000-2003 (Table 1) and a slightly better gradeout was obtained. However, in reality, yields during 2000-2002 were very low compared with those achieved in conventional orchards and the gradeout was very poor. A higher yield was achieved during 2003.

Table 1. Yield and gradeout in the organic Fiesta orchard, 2000-2003

	Yield (t/ha) per treatment		
	Experimental	**Growers**	**Untreated**
2000			
Class 1	3	1.5	1.0
Class 2	7	5.5	1.5
Outgrade	3	3	0.8
2001			
Class 1	2	5	2.5
Class 2	5.5	3	3.5
Outgrade	10	4.5	2.5
2002			
Class 1	3.5	1	0
Class 2	4	0.8	0.2
Outgrade	1.5	0.2	0
2003			
Class 1	10.9	7.6	5.3
Class 2	9	8.9	8.2
Outgrade	6.4	9.2	5.4

Scab is difficult to control and was one of the main reasons for the poor yields during 2000-2002, confirming that highly scab-susceptible varieties such as Fiesta are unsuitable for organic production. A much higher yield was obtained in 2003 due to the very dry spring and low incidence of scab. Partial control of scab was achieved through early copper spays followed by sulphur. A pre bud-burst copper spray has proved particularly effective. Mildew was well controlled by sulphur sprays and removing shoots infected by primary mildew. Apple blossom weevil (*Anthonomus pomorum*) has also been a major cause of yield loss and sprays of pyrethrum have proved to be the only effective option at present. Sooty blotch on fruit has also been a severe problem at the site although well-pruned trees reduce the problem and the fungus can be reduced by pre bud-burst copper sprays and early full rate sulphur sprays.

Suitable varieties

The second objective of the project is to identify 4-6 varieties of low disease susceptibility which will produce fruit that meets the full range of market requirements, providing high levels of consumer satisfaction whilst performing well in organic production. A world-wide search was undertaken to identify potential varieties. Initially, 160 promising resistant (or partially resistant) varieties were assessed by specialists for fruit quality. This selection process resulted in a short list of 27, which were then propagated via bench grafting and the resulting trees planted in a replicated field trial on registered organic land. The agronomic performance and fruit quality of each variety is being assessed, with storage tests carried out during the final two years of the project. The most suitable 4-6 varieties will be selected and recommended to growers on completion of the project.

Controls for scab and mildew

The third objective for the project has been the identification of effective organic fungicidal controls for scab and mildew. This is also involving an evaluation of products and methods to accelerate leaf rotting, thus minimizing infection from overwintered spores. Nine organic products for which claims of effectiveness have been made overseas or in the UK were tested in glasshouses to a standard test and compared with copper and sulphur standards. Although one product, Ulmasud B reduced sporulation of mildew colonies, the alternative products tested have demonstrated little effectiveness against scab and mildew. To date, sulphur and copper remain the only effective treatments against mildew and scab respectively. Products high in nitrogen have so far proved most effective in accelerating leaf rotting.

Control of rosy apple aphid

The search for effective treatments for rosy apple aphid has involved the testing of six treatments, alongside Derris and untreated controls, in replicated experiments in commercial apple orchards. Two strategies have been trialled: a) control when colonies develop after hatching in the spring, the normal time to control this pest, and b) applications in the autumn against migrant offspring before they lay winter eggs. All materials tested have been ineffective against established colonies in the spring but the autumn strategy is very promising. A high degree of control has been obtained with a single application of a conventional insecticide applied in early October and work is ongoing to test organically acceptable treatments at this time. It may be that the aphid is more vulnerable at this time than in the spring.

ACKNOWLEDGEMENTS

This work is funded by the Department for Environment, Food and Rural Affairs.

REFERENCES

FIRTH C. and LENNARTSSON M. (1999). Economics of organic fruit production in the UK. MAFF project no. OF0151. Coventry: HDRA, Ryton Organic Gardens.

The Effects of Conversion to Organic Field Vegetable Production on the Populations of Two Perennial Weeds, Couch Grass (*Elytrigia repens*) and Creeping Thistle (*Cirsium arvense*)

J.E. ADAMS, P.D. SUMPTION and R.J. TURNER
IOR-HDRA, Ryton Organic Gardens, Coventry, CV8 3LG, UK

ABSTRACT

Couch grass (*Elytrigia repens*) and creeping thistle (*Cirsium arvense*) are invasive and problematic weeds that, if established, can cause serious impacts on the agronomic and financial sustainability of organic horticultural systems. This study asked 'are couch and creeping thistle populations increasing in organic field vegetable systems post-conversion?' Data collected in the DEFRA-funded 'Conversion to Organic Field Vegetable Production' project (OF0126 T and OF0191), between 1996 and 2002 from two research stations and nine commercial farms in the UK were used in an attempt to answer this question. Field walk data were analysed that included occurrence of couch grass and creeping thistle, both quantitatively and qualitatively. There was little evidence of an increase in populations of couch grass and creeping thistle on these sites during and after conversion. Couch was more widespread in its occurrence than creeping thistle and on two farms this has had an agronomic and economic impact. Creeping thistle has occurred more sporadically and only on two farms occurred at levels where it could be considered a problem, even here it does not appear to be increasing year on year.

INTRODUCTION

The profitability and sustainability of organic systems could be threatened by increased populations of perennial weeds (HDRA, 2001). Couch grass (*Elytrigia repens*) and creeping thistle (*Cirsium arvense*) are two such weeds that are highly competitive, and can significantly reduce crop yields, slow or impede harvest, and contaminate seed grain crops, thereby reducing economic returns. Both spread vegetatively via underground rhizomes or fibrous taproots, respectively, and can regenerate from small root fragments. Creeping thistle is difficult to manage in organic systems. Arable rotations, particularly those without grazed leys can provide ideal conditions for perennial weeds to thrive (Cormack, 2002).

HDRA have been leading the DEFRA-funded 'Conversion to Organic Field Vegetable Production' project which has monitored two research stations and nine commercial farms converting from conventional rotations to an organic system which includes organic field vegetables. Four farms have been monitored since 1997 and a further six farms since 1999. The farms, which range in size from 20 ha to 1900 ha, with target conversions of between 4 ha and 700 ha, are on varied soil types and geographical locations. In 2002 they grew between 2 and 115 ha of organic vegetables. In 2001 the project reported that creeping thistle and couch grass were the most widespread perennial weeds on the reference farms (HDRA, 2001). This study aimed to identify whether or not the populations of couch and creeping thistle had increased on reference farms of the conversion project during the monitoring period (1996-2002) and if so, to identify possible reasons for any changes.

MATERIALS AND METHODS

HDRA staff and Organic Advisory Service advisers have visited the farms in the project on average six times per year through and after conversion. Information has been gathered through a combination of farmers' records, interviews with farmers and detailed field assessments. Weeds have been monitored through regular crop walks on the ten reference farms. Using quadrat samples (10 quadrats placed in each crop), occurrence of individual weeds was recorded along with total weed cover as a percentage.

Data were analysed from monitoring visits between June and September. Data, recorded as percentage occurrence in ten quadrats in a field, were summarized for couch grass and creeping thistle, for each year on each farm. Factors such as different farm weeding practices, farm size, soil type, topography, climate, crop rotations and crop competitiveness were considered to try and explain population changes in couch grass and creeping thistle over the six-year project. The data were displayed graphically on a farm by farm basis.

RESULTS AND DISCUSSION

Couch grass was not observed on any of the farms in the first year of conversion, but in the last full year of monitoring it was observed on 6 of the 10 farms. This may be partially due to the difficulty of identifying couch in grass/clover leys at the start of conversion. While it is difficult to say with certainty that it is increasing, with levels fluctuating between years, it has been a significant problem on two farms. On farm 5 it was not observed in the first year of organic cropping but was a major problem in 2001 and 2002, with occurrence in up to 80% of quadrats in some fields (Figure 1).

Figure 1. Percentage occurrence of couch grass over time at Farm 5

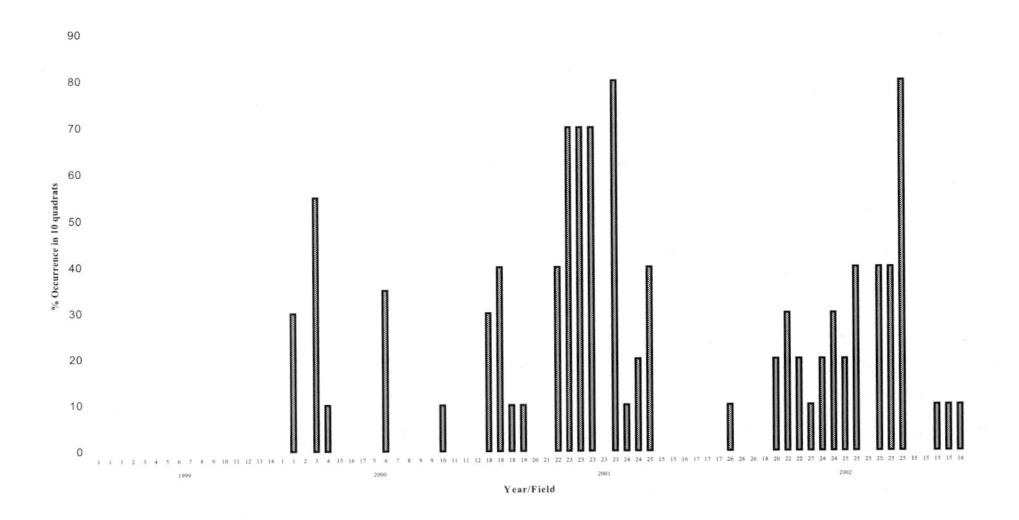

Both of the farms with couch problems were arable conversions on black Fen soils, and which had had to take land out of production and use a bastard fallow and cultivation to reduce weed populations. This obviously had incurred an economic cost. On one farm a 15 ha field was fallowed, the total cost of cultivations was £216/ha plus the loss of revenue from the vegetable crop that could have been grown. The average gross margin for vegetables on that farm was £2,750/ha. If this strategy had not been adopted then there would have been the risk of a low or negative margin in the subsequent crop, due to higher costs of weeding and harvesting.

Creeping thistle only appears to be a problem weed on two farms, where it was present at the start of conversion (50% occurrence on one farm and up to 90% occurrence on the other) and levels have been variable since but always recorded as present at the farms.

The changes in occurrence of couch grass and creeping thistle across the conversion period did not follow a definite trend. Some of this fluctuation can be related to the effectiveness of the weeding strategies in general. This in turn can be related to weather, farmer experience of using new or unfamiliar weeding techniques and equipment, and also missing weeding windows through the pressure of time, often after a rapid expansion in organic cropping area. The weather has a great influence on weed levels and the effectiveness of weeding strategies. In a wet year, such as 2000/01, a greater level and number of weeds occurred, whereas in 2002/03,which was a drier season and easier for weed management, fewer weeds occurred overall. Similarly, soil types such as black Fen soils have a higher inherent weed pressure; therefore farms with these soil types had a higher percentage of weed cover overall making comparisons of different weeding practices rather difficult. The larger farms with extensive arable rotations had more flexibility to change rotations when problems such as couch infestation occurred, and thus choose to grow the higher value vegetable crops on cleaner land. The importance of keeping the land clean therefore assumes greater importance for smaller more intensive growers.

The conversion project has monitored farms through conversion and their first organic rotation to produce case study type data on each farm scenario. Information has been collected on a range of aspects at each visit, not specifically weeds. If more time had been available a specific monitoring of weeds could have included, for example, counting couch shoots or even recording individual weed numbers rather than simply presence or absence in a quadrat. This would have produced a fuller data set from which to draw conclusions. It is possible that now farms have been identified with a particular perennial weed problem that future recording methods could assess those weeds in more detail as a separate study.

CONCLUSIONS

While couch grass and creeping thistle were recorded on all farms throughout the conversion period, occurrence was sporadic. Consequently, from the data collected it was concluded that over the six-year conversion project 'Conversion to organic field vegetable production' (OF126T and OF0191) that couch grass and creeping thistle populations had not increased, when information from all farms were considered. Certainly two farms have seen an increase in couch grass to a level whereby they had to introduce a bastard fallow into their rotation to control the weed.

Further monitoring to follow trends over the subsequent rotations are therefore needed and the new three-year DEFRA-funded project 'Sustainable organic vegetable systems network' will use and build on methods and approaches developed, and data collected during the conversion project, thus enabling the continued monitoring of these two perennial weed populations.

REFERENCES

CORMACK W.F. (2002) Effect of mowing a legume fertility-building crop on shoot numbers of creeping thistle (*Cirsium Arvense* (L.) Scop.) In: *Proceedings of the UK Organic Research 2002 Conference*, Aberystwyth, 2002, pp225-228.

HDRA (2001) *Conversion to Organic Field Vegetable Production Report*. Ryton, UK: HDRA.

Can N Use and Farm Income be Optimized for Organic Field Vegetable Rotations in Europe?

U. SCHMUTZ[1], C. FIRTH[1], F. RAYNS[1] and C. RAHN[2]
[1]HDRA-IOR, Ryton Organic Gardens, Warwickshire, CV8 3LG, UK
[2]HRI Wellesbourne, Warwickshire, CV35 9EF, UK

ABSTRACT

Most fresh organic vegetables are produced in intensive rotations, which rely heavily on large inputs of nitrogen to maintain the yield and quality of produce demanded by customers. Field vegetable crops often use nitrogen inefficiently and may leave large residues of nitrogen in the soil after harvest, which can lead to damage to soil, water and air quality. The four-year project EU-ROTATE_N "Development of a model-based decision support system to optimize nitrogen use in horticultural crops rotations across Europe" aims to reduce some of these problems. The project, led by HRI Wellesbourne, started in January 2003 and involves seven research organizations from countries in northern, central and southern Europe. Work includes the evaluation of the effects of varying levels of N supply on both product quality and farm income for organic and conventional rotations, as well as case studies for the evaluation of agricultural strategies with respect to N losses and economics for vegetable crops in Europe. This paper describes the work carried out at HDRA which focuses on farm economics and organic field vegetable rotations.

INTRODUCTION

Most fresh vegetables, organic and conventional, are produced in intensive rotations with high financial inputs and outputs; there are consequently high production risks involved. Economic penalties for crops failing to meet market criteria are so high relative to the cost of nitrogen (N) input, that growers can be tempted to adjust their N applications as insurance, even significantly above the predicted optimum. While this is particularly true of conventional systems, organic farms which are seeking to meet supermarket specifications may face similar temptations in their use of permitted inputs. The need to include fertility-building phases within organic crop rotations may also add to the economic pressure on growers, causing high opportunity costs and forcing the other crops in the rotations to obtain high returns from the land to pay for fixed costs.

IMPLICATIONS OF ORGANIC PRODUCTION STANDARDS FOR NITRATE LEACHING POTENTIAL

Environmental problems associated with N leaching in conventional agriculture are widely recognised (e.g. Strebel *et al.*, 1989). External costs of nitrates in drinking water originating from UK agriculture are estimated to be £16m/year (Pretty *et al.*, 2000). In contrast, up to 50% lower nitrate leaching rates per hectare have been measured in organic systems (Stolze *et al.*, 2000). If, however, nitrate leaching is expressed per unit output of grain or milk, the performance of organic systems is similar to conventional systems (Stolze *et al.*, 2000). The rotational approach and restrictions of N inputs within organic farming systems tends to lead to a lower overall risk of nitrate leaching. However, there are some problems, like the management of residual N from fertility building leys or the use of increased amounts of organic fertilizer, e.g. chicken manure.

The EU regulation (ECC 2092/91, 1991-2003) limits nitrogen input from manure to 170 kg N/ha/year across the whole farm; however, there are no specific recommendations for vegetable production. Member states may establish lower limits, taking into account the characteristics of the area concerned, the application of other nitrogen fertilizers to the land and the nitrogen supply to the crops from the soil. The UK Standards (UKROFS, 2001) mention this possibility but do not make use of it. Private standards in the UK (e.g. Soil Association Certification Ltd, 2002) do not mention supplementary N fertilization or environmental problems with N leaching in the horticultural production section of their standards. The general section of the standards state that the maximum permitted application is 250 kg N/ha for field crops as organic manure application. Higher rates may be permitted in protected cropping. Because organic N fertilizers are often used well before harvest, applications of manures may occur at a time of low crop demand and high leaching potential. Organic Farmers & Growers standards in the UK (Organic Farmers & Growers, 2001) suggest acceptable rotations with a nitrogen balance. One example is the following rotation where the two-year white clover/ryegrass ley followed by winter wheat has a significant potential for leaching much of the rotational N surplus:

5-year suggested rotation	**kg N/ha/year**
White clover/Ryegrass	0
White clover/Ryegrass	400
Winter wheat	-64
Potatoes	-126
Spring Barley & Clover undersown	30
Balance	**240 kg N/ha**

In Germany the Bioland Association restricts N applications in organic vegetable production to 110 kg N/ha in the field and 330 kg N/ha in the greenhouse (Bioland, 2002). Biokreis, another German sector body, specifies that only 110 kg N/ha can be applied as an average over the whole vegetable rotation; if more than 50% of the N is from manure then 140 kg N/ha can be used. In Switzerland to qualify for direct ecological payments, each main vegetable (including members of the same family) has to be followed by at least a 2-year break (Swiss Confederation, 2003).

From the above limited consideration of organic production standards in Europe we conclude that potential environmental problems are possible even where farmers follow EU organic standards. Possible problems with N leaching are not addressed in detail by the production standards and there is no explicit control of the potential for N leaching. There is also scope to further optimize N use in field scale vegetable production both to reduce environmental impact and to make efficient use of this valuable N resource in organic horticulture.

THE APPROACH TAKEN BY THE RESEARCH PROJECT EU-ROTATE_N

The four-year project EU-ROTATE_N "Development of a model-based decision support system to optimize nitrogen use in horticultural crops rotations across Europe" is funded by the European Commission within the Fifth Framework Programme. It is led by HRI Wellesbourne and started in January 2003. The modelling approach builds on the previous N fertilizer prediction models N_ABLE and WELL_N (Rahn *et al.*, 1996). The model includes generalized relationships for growth and its dependence on plant size, N content and temperature, the development of roots, N uptake, N release from soil organic matter and incorporated crop residues, evapotranspiration, soil water content, and leaching. The model currently operates on a daily time-step and provides simulations for growth and N requirements for more than twenty different vegetable crops.

The model is, however, based on the simulation of the N dynamics in single cropping seasons and does not include crop rotations or cover crops; it also makes no assessment of economic implications of crop management. The model is adapted to UK or central European conditions and does not include Scandinavian or Mediterranean climates or crops. On-going research therefore involves countries from northern, central and southern Europe and includes work packages on the evaluation of the effects of varying levels of N supply on product quality and farm income, the development and testing of a new decision support system for crop rotations in Europe, and case studies for the evaluation of agricultural strategies with respect to N losses and economics. Other work includes root growth modelling and the simulation of the influence of irrigation and freeze-thaw cycles on nitrogen dynamics. The result will be a model of N dynamics suitable for use in both conventional and organic field vegetable rotations and during conversion.

ONGOING RESEARCH AT HDRA

HDRA is responsible for compiling a database, containing selected economic data for all major field vegetables across the different European climate zones. The database is almost complete and includes prices, variable costs, and gross margin data for conventional and organic field vegetables.

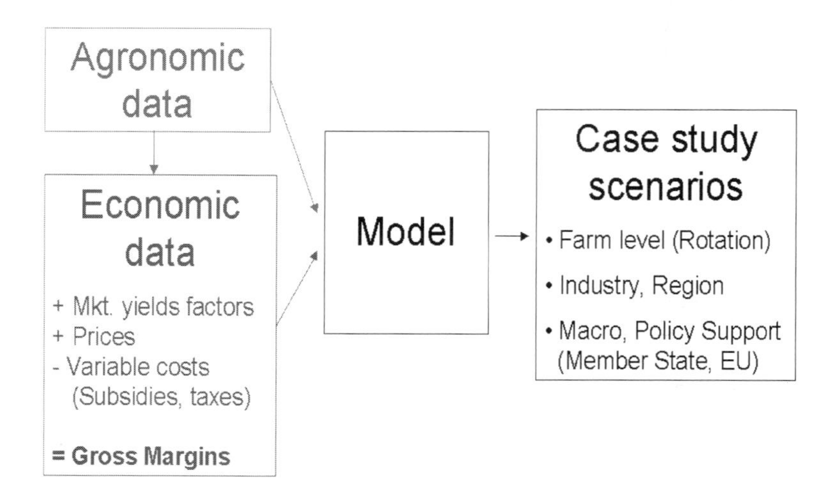

Other crops, often found in organic or conventional field vegetable rotations, such as cereals or fertility-building legumes are also included. There are large differences in the amount of economic information available for each country. Once complete, the database will be linked with yields generated from the agronomic model, it will then be possible to examine the effect of varying the supply of N on the economics of vegetable rotations. The effect of sub-optimal nitrogen supply on marketable yields and management practices such as the inclusion of fertility-building crops in organic rotations will also be simulated. It is necessary to convert the dry matter yield produced in the agronomic part of the model into marketable yield used in the economic model. Different factors and algorithms for this conversion are currently being researched to allow the data to be used in both economic and agronomic models.

In a different work package, the HDRA soil science team is sourcing information to enable simulations of N accumulation and release from green manure cover crops in the model. Field studies have also begun to validate the current model predictions of N release from the addition of organic manures to soils in organic systems.

EXPECTED OUTPUTS - CALL FOR PARTICIPATION

Upon completion and following validation of the model, we aim to examine the effects of using various strategies to optimize the use of N and to minimise nitrate leaching within the rotation on farm economics, for example through the use of winter cover crops in both conventional and organic rotations. Farmers, advisers or researchers are welcome to contact us and to contribute any ideas about which rotations and strategies are likely to be adopted by farmers and hence should be tested. In future, the model could also be used to scale up data to consider impacts for a borehole catchment (Vinten and Dunn, 2001), a Nitrate Vulnerable Zone, a vegetable producing region, or even for a EU member state. This would allow the implications of adopting various N management strategies to be evaluated on a wider scale, at farm and the broader macro-economic level.

REFERENCES

BIOLAND (2002) *Bioland Organic Standards.*. Mainz, Germany: Bioland.

CONVEDERATIO HELVETICA (2003)*Direktzahlungsverordnung SR 910.13*, The federal authorities of the Swiss Confederation.

ECC 2092/91 (1991-2003) *Council Regulation (EEC) 2092/91 on organic production of agricultural products and indications referring thereto on agricultural products and foodstuffs. With annex and 2003 amendments*. Brussels: EC European Commission, EU European Union. s.

ORGANIC FARMERS & GROWERS (2001) *Control Manual*. Shrewsbury, UK.

PRETTY, J. N., BRETT, C., GEE, D., HINE, R. E., MASON, C. F., MORISON, J. I. L., RAVEN, H., RAYMENT, M. D. and Van Der BIJL, G. (2000) An Assessment of the Total External Costs of UK Agriculture. *Agricultural Systems*, 65: 113-136.

RAHN C, GREENWOOD D.J & DRAYCOTT A. (1996) Prediction of nitrogen fertilizer requirement with the HRI WELL_N computer model. *Progress in Nitrogen Cycling* (eds. O Van Cleemput et.al.). 255-258.

SOIL ASSOCIATION CERTIFICATION LTD (2002) *Organic standards and certification*. Bristol.

STOLZE M., PIORR A., HÄRING A. and DABBERT S. (2000) *The Environmental Impacts of Organic Farming*. University of Hohenheim. Stuttgart, Germany. 6. 127.

STREBEL O., DUYNISVELD W. H. M. and BOTTCHER J. (1989) Nitrate pollution of groundwater in western Europe. Agriculture, *Ecosystems & Environment*, 26: 189-214.

UKROFS (2001) *UKROFS Standards for organic food production*. DEFRA UK, UKROFS United Kingdom Register of Organic Food Standards. London.

VINTEN A. J. A. and DUNN, S. M. (2001) Assessing the effects of land use on temporal change in well water quality in a designated nitrate vulnerable zone. *The Science of the Total Environment*, 265: 253-268.

WEB SOURCES: www.hri.ac.uk/eurotate, www.hdra.org.uk/research

Effect of Within-Crop Diversity on Colonization by Pest Insects

S. FINCH and R.H. COLLIER
Horticulture Research International, Wellesbourne, Warwick CV35 9EF, UK

ABSTRACT
Cauliflower host plants were surrounded by three species of marigold (*Tagetes* spp.) and five other species of aromatic plants to determine whether aromatic plants were more effective than non-aromatic plants at disrupting egg-laying by pest insects. The "model" test insect was the cabbage root fly (*Delia radicum*) and the non-aromatic plants tested included common garden bedding plants and weeds. Of the 24 plant species tested, 20 disrupted egg-laying by the fly. Aromatic plants were no more disruptive than non-aromatic plants. The disruptive effect varied from one plant species to another and depended upon the architecture of the non-host plants. In the current tests, the least disruptive plant was common fumitory (36% reduction in egg numbers) and the most disruptive was fat hen (82% reduction). Plants of cineraria, alyssum and lobelia, that appeared grey, white and blue, respectively, were not disruptive. To be disruptive the non-host plants must be close to the cauliflower plants, equivalent to growing in the "within row" area, and be green. Contrary to popular belief, the pungent chemicals found in aromatic plants did not deter the pest insects from landing or repel them after landing. The disruption by all non-host plants is caused simply by the presence of additional green surfaces.

INTRODUCTION
Many researchers have shown that the numbers of pest insects found on crop plants are reduced considerably when the crop is allowed to become weedy, when it is intercropped with another plant species or when it is undersown with a living mulch (Finch and Collier, 2000). It has been suggested that the non-host plants "disrupt" the pest insects from finding their host plants by physical obstruction or by visual camouflage. It has also been suggested that aromatic plants should be more disruptive than non-aromatic ones. This arises because some researchers believe that the odours given off by aromatic plants can "mask" those of the host plant so that the host-plants are never found, or are sufficiently pungent to deter the searching insects either from landing on, or staying on, the leaves of aromatic plants (see Finch and Collier (2000) for additional references).

In Finch and Collier (2000) we concluded, from experiments using real and artificial plants, "it is just the number of green surfaces surrounding a host-plant that disrupts host-plant finding and colonization by plant-feeding insects". Therefore, in this paper we try to answer the question "Are the aromatic plants, e.g. marigolds (*Tagetes* spp.), that are used often by organic growers as "companion plants" more effective at disrupting insect behaviour than non-aromatic plants such as weeds?"

MATERIALS AND METHODS
The flies were produced in the Insect Rearing Unit at Wellesbourne and were tested when 5-6 days old, the time at which they are ready to lay eggs. The host-plants were 20 cm tall cauliflower plants.

Experiments were done in field cages (4.2m x 2.2m x 1m high) divided into 32 plots, each measuring 0.5m x 0.5m. Each cage enclosed 24 species of non-host plant and 8 check (bare soil) treatments. A potted cauliflower was "planted" in the centre of each plot and then surrounded by four plants of one of the 24 test species. The non-host plants included aromatic plants, garden bedding plants, companion plants, one vegetable plant and weeds (see Table 1 for details). The reasons for selecting the various plants are given in detail in Finch *et al.*(2003).

One hundred female flies were released into each cage and allowed to lay eggs for 2 days. Each test was repeated simultaneously in three cages and was repeated on five occasions during the season. The numbers of eggs laid around the plants surrounded by non-host plants were compared with those laid around the plants surrounded by bare soil.

Direct observations were also made to determine whether the flies were deterred or repelled by any of the plant species. This was done by releasing flies into a 106cm x 106cm x 90 cm high laboratory chamber and recording the numbers of landings made on the plants when presented alone (no-choice situation) or alongside a cauliflower plant (choice situation). The times the flies remained on the various leaf surfaces were also recorded to determine whether some leaf surfaces do actually repel flies.

RESULTS

In the field-cage experiments (Table 1), the least disruptive plant was common fumitory (36% reduction in egg numbers) and the most disruptive was fat hen (82% reduction). With the exception of thyme (42%), which was relatively low growing, the other four aromatic plants reduced egg-laying on average by about 56%. The average reduction recorded for the six disruptive bedding plants (62%) and the three companion plants (62%) were also of the same order. The weeds were the group that showed the greatest variation, as the reduction from common fumitory, greater chickweed and scentless mayweed averaged 39%, compared with 70% and 82% for rosebay willow-herb and fat hen, respectively. In the current tests, four plants, namely geranium with the green foliage, carrot, rosebay willow-herb and fat hen, reduced egg numbers by 70% and more.

In the laboratory observation experiments (Table 1, column 4), the numbers of flies that landed on the various plant types were similar ($P = 0.05$). There was also no evidence that the flies were deterred from landing on the plants in either the "no-choice" (18 plants tested – see Table 1) or the "choice" (only three plants tested) situations. In addition, the time ± SE spent on each leaf of the cauliflower host plant in the "no-choice" experiment was 1.3 m (77±18 s), whereas the mean time spent on leaves of non-host plants ranged from 3.1 m (188±51 s) on African marigold to 6.7 m (403±68 s) on the geranium with the reddish leaves. In the "choice" experiment, the flies stayed for 1.4 m (82±13 s) on the leaves of the cauliflower plants compared with 7.6 m (454±66 s), 6.1 m (400±58 s) and 4.1 m (245±50 s) on the leaves of African marigold, the reddish-leaved geranium, and tagetes, respectively.

DISCUSSION

We showed earlier (Finch and Collier, 2000) that undersowing cauliflower plants with clover reduced the numbers of eggs laid, to a greater or lesser extent, by all of the eight pest insect species that were included in the tests. Hence, undersowing disrupts host-plant finding by all species; it does not deter some species and favour others.

Table 1. The numbers of cabbage root fly eggs recovered from around cauliflower plants, each surrounded by four plants of one of 24 non-host plant species, expressed as a percentage reduction of the numbers of eggs recovered from around cauliflower plants surrounded only by bare soil. The final column shows the time (in m) the flies remained on a leaf before becoming active again.

Latin name	Common English name	% reduction in number of fly eggs	Time (min) spent on leaf of plant
Aromatic plants			
Thymus vulgaris	thyme	42	4·1
Mentha piperita x *citrata*	lavender mint	57	4·8
Rosmarinus officinalis	rosemary	58	3·2
Mentha pulegium	penny-royal	60	-
Helichrysum bracteatum	curry plant	63	4·6
Bedding plants			
Cineraria maritima	cineraria	21 NS	4.8
Lobelia erinus	lobelia	22 NS	5·6
Lobelia maritima	alyssum	24 NS	5·5
Pelargonium x *hortorum (red)*	geranium	56	6·7
Nicotiana x *sanderaea*	tobacco plant	57	3·5
Dahlia variabilis (red)	dahlia	60	3·8
Antirrhinum majus	antirrhinum	62	3·3
Dahlia variabilis (green)	dahlia	65	5·6
Pelargonium x *hortorum (green)*	geranium	70	5·5
Companion plants			
Tagetes patula	French marigold	58	4·4
Tagetes tenuifolia	tagetes	58	4·8
Tagetes erecta	African marigold	68	3·1
Vegetable plant			
Daucus carota ssp. *sativus*	carrot	75	4·8
Weeds			
Sallopia convolvulus	black bindweed	18 NS	-
Fumaria officinalis	common fumitory	36	-
Stellaria media	greater chickweed	38	-
Tripleurospermum inodora	scentless mayweed	44	-
Chamaenerion angustifolium	rosebay willow-herb	70	-
Chenopodium album	fat hen	82	6·2
Bare soil	cauliflower	0	1·3

The current tests were done in part to show whether other non-host plants were as disruptive as clover. The results from the current experiments show that all of the non-host plants that were green, except the low growing black bindweed, also disrupted host-plant finding, and several were as disruptive as clover (based on assessments in previous experiments). The three other plants that were not disruptive were cineraria, which had grey foliage, and alyssum and lobelia, which flowered profusely during the tests, and so appeared white and blue, respectively.

To be disruptive, the non-host plants have to be relatively robust and tall (Finch *et al.*, 2003), which helps to explain why carrot and fat hen were so effective. In some seasons, however, carrot could lose its effectiveness if not irrigated, as its fleshy leaves tend to droop and expose the cauliflower plants during dry periods. The weeds, common fumitory, greater chickweed and scentless mayweed were disruptive initially but lost their impact during later tests. This occurred because these plants were true ephemerals and had started to senesce during some of the later exposures. They could still be of use, however, as the period during which insects need to be deterred from crops can be relatively short in some instances. It is important to remember, however, that we have only considered 24 plant species in the current experiments and there are hundreds that could be tried, now that we have confirmed that the disruption is caused simply by the presence of additional green surfaces.

Originally, we had thought that the leaves of aromatic plants, such as mint, thyme and marigolds, would have been coated with deterrent chemicals. Therefore, we expected the flies to leave some leaves within seconds of landing. However, the relatively long periods of time (3-7 min.) the flies stayed on the leaves of the 18 non-host plants (Table 1) indicated that they were not being deterred or repelled by the chemicals present on, or in, the leaf surfaces of even the highly aromatic plants. It appears that the flies become relatively inactive after landing on any plant leaf surface and are only stimulated to become more active when the leaf contains those chemicals, in this case the glucosinolates, which characterize their host plants.

For a more detailed description of how to overcome some of the practical problems of intercropping, we recommend the papers of Theunissen (e.g. Theunissen *et al.,* 1995) and for a general overview of the subject the paper of Finch and Collier (2003).

ACKNOWLEDGEMENTS

We thank Defra for supporting this work and Vladimir Kostal, Manuela Kienegger, Kate Morley, Helen Billiald and Marian Elliott for their contributions.

REFERENCES

FINCH S and COLLIER R.H. (2000) Host-plant selection by insects – a theory based on 'appropriate/inappropriate landings' by pest insects of cruciferous plants. *Entomologia experimentalis et applicata*, 96, 91-102.

FINCH S and COLLIER R.H. (2003) Insects can see clearly now the weeds have gone. *Biologist,* 50, 132-135.

FINCH S., BILLIALD H and COLLIER R.H. (2003) Companion planting - do aromatic plants disrupt host-plant finding by the cabbage root fly more effectively than non-aromatic plants? *Entomologia experimentalis et applicata*, (in press).

THEUNISSEN J., BOOIJ C.J.H and LOTZ L.A.P. (1995) Effects of intercropping white cabbage with clovers on pest infestation and yield. *Entomologia Experimentalis et Applicata*, 74, 7-16.

Seed Spacing and Treatment for Organically Grown Sugar Beet

W.F. CORMACK[1], P.J. JARVIS[2] and P.M.J. ECCLESTONE[2]
[1]ADAS Terrington, Terrington St Clement, King's Lynn, Norfolk, PE34 4PW, UK
[2]British Sugar plc, Oundle Road, Peterborough, PE2 9QU, UK

ABSTRACT

Seed spacing and treatment were compared for organically grown sugar beet in nine experiments, over three years, on a range of soil types. Yields were generally good, with five experiments yielding over 8 t/ha of sugar. Sowing at 9 cm rather than the standard 17.5 cm spacing increased plant population at five sites, but increased yield at only one, even where populations were well below the recommended 75,000/ha. However, individual plants grew larger and so could cause increased losses in machine-harvested commercial crops. Advantage™ primed seed gave a greater sugar yield at only one site but experience in non-organic crops is that it gives faster, more even emergence, allowing an earlier start to mechanical weed control, essential in organic crops. In wet conditions on heavier soils, slug grazing caused severe plant loss. The lack of an effective commercial field-scale organic control is a major constraint to organic beet growing on heavier soils.

INTRODUCTION

Arable cropping has been shown to be economically and technically viable under organic management (Cormack, 1999). However, in the late 1990s, despite strong growth, the market for organic arable crops was limited to cereals, pulses and potatoes. Sugar beet was, and remains, a major part of the rotation of many Eastern Counties arable farms, and the lack of a market for organic sugar beet was seen as a key constraint to conversion. The production of sugar from organically grown beet would also allow British Sugar plc to meet their market demand for organic sugar and organic animal feeds from home-grown sources and so replace imports and help protect the UK share of the EU sugar quota. Little work has been done or published on organic sugar beet production. Swedish work indicates that sugar yields of over 9 t/ha are possible (Larsson, 1999). The first UK organically grown commercial sugar beet crops were harvested in 2001. To gain information on how best to manage these crops, a programme of experiments was funded by the British Beet Research Organization. The objectives of the work were to identify challenges for organic beet production on different soil types in the UK and develop initial agronomic strategies to meet these challenges. The scope of the work included crop establishment, weed control, and avoidance of virus infection. This paper presents results of the first three of four years of crop establishment experiments.

MATERIALS AND METHODS

Sites were selected to give a range of soil types (Table 1) but choice was limited by the low numbers of arable organic farms in the Eastern Counties with experience of growing sugar beet. Experimental design was a factorial combination of two seed spacings and two seed treatments in randomized blocks with four replicates. Seed was either sown 'to a stand' at the standard 17.5 cm spacing, or sown at 9 cm spacing and hand-thinned to standard spacing. Seed was pelleted and was either untreated or primed (Advantage™) (Anonymous, 2003a). Plot size was 6 rows by 12 m with an assessment and harvest area of 3 rows by 9 m. Row spacing was 50 cm. Seedbed preparation and drilling was carried out according to the normal commercial practice for the sites. The cultivars were Jackpot in 2000 and 2001, and Stallion in 2002. All sites were on organic registered land. Weed control was by a combination of tractor hoeing between the rows, and hand hoeing within the row. The aim was to simulate a commercial operation with 2 to 4 tractor hoeings and 2 hand hoeings. No fertilizers or pesticides were applied to any site apart from a foliar application of manganese made at Lode in response to visual symptoms of manganese deficiency. Harvest was using a three-row plot harvester, apart from experiments at West Acre and Terrington, which were hand-harvested due to particularly small and large sized beet, respectively. All beet were taken to Rothamsted Research-Broom's Barn for weighing and quality analysis. Data were analysed by Analysis of Variance.

Table 1. Experiment site details

Expt.	Year	Location	Soil type	Previous crop	Sown	Harvested
1	2000	Blankney, Lincolnshire	Clay loam over limestone	Spring barley	4 May	1 Nov.
2	2000	Lode, Cambridgeshire	Organic sandy loam	Broccoli	20 April	25 Oct.
3	2000	West Acre, Norfolk	Sandy loam	Winter wheat	18 April	24 Oct.
4	2001	Blankney, Lincolnshire	Clay loam over limestone	Winter wheat	3 May	20 Nov.
5	2001	Gayton, Norfolk	Sandy loam	Vetch	1 May	15 Nov.
6	2001	ADAS Terrington	Silty clay loam	White clover	2 May	23 Nov.
7	2002	Blankney, Lincolnshire	Clay loam over limestone	Lucerne	12 April	25 Nov.
8	2002	Bury St Edmunds, Suffolk	Silty clay loam	Clover	11 April	19 Nov.
9	2002	ADAS Terrington	Silty clay loam	White clover	16 April	24 Oct.

RESULTS AND DISCUSSION

Soil fertility was reasonable apart from the two light land sites at West Acre and Gayton, and the Bury St Edmunds site, which all had K Indices of 1, and Na of less than 10 mg/l (Table 2). These were well below the 40 mg/l threshold for Na application (Anonymous, 2003b), and in the absence of K fertilizer, yield was probably limited by nutrient supply on these experiments. Soil organic matter was also very low at the West Acre and Gayton sites, probably more a reflection of the soil texture than management.

Sowing of the non-organic beet crop starts in early March, but this was considered to be too early for an organic crop grown without the protection of pesticides incorporated in the seed pellet. The aim was to sow the organic crops in mid-April when germination and emergence would be swifter.

Table 2. Pre-sowing soil analysis.

Expt.	pH	P mg/l	K mg/l	Mg mg/l	OM %	Na mg/l	SMN kg/ha
1	7.9	25 (2)	161 (2)	80 (2)	4.3	18	210
2	7.0	19 (2)	67 (1)	74 (2)	13.5	77	224
3	7.2	46 (4)	142 (2)	48 (1)	1.2	5	171
4	7.8	20 (2)	196 (2+)	90 (2)	4.1	11	130
5	7.8	17 (2)	99 (1)	29 (1)	1.8	4	148
6	7.8	17 (2)	166 (2-)	185 (4)	2.3	12	105
7	7.8	15 (1)	134 (2-)	141 (3)	4.2	21	165
8	7.5	10 (1)	113 (1)	40 (1)	2.8	9	116
9	7.9	25 (2)	316 (3)	148 (3)	4.0	11	161

All three crops established well in 2000. In 2001, beet again emerged well in moist and warm conditions. The crops at Blankney and Gayton established successfully, but grazing by slugs caused almost total plant loss at Terrington leading to re-sowing on 24 May. In 2002, soil conditions were excellent for sowing in mid-April, but heavy rain later in the month resulted in very wet soils and plant loss from slug grazing at Terrington and Bury St Edmunds. Advantage™ seed gave a greater plant population at West Acre. At 17.5 cm spacing, only two sites achieved the target 75,000/ha plant population (Anonymous, 2003b). Closer seed spacing increased plant population at five of the nine sites but overall only four achieved 75,000/ha (Table 3).

Table 3. Established plant population '000/ha.
(NS = not significant, ** = P<0.01, *** = P<0.001)

Expt.	Seed spacing 17.5 cm	Seed spacing 9.0 cm		Seed treatment Standard	Seed treatment Advantage™	
1	88.7	138.0	***	113.0	113.7	NS
2	54.4	56.7	NS	57.6	53.5	NS
3	41.4	65.2	**	42.1	64.4	**
4	76.9	81.3	NS	78.7	79.4	NS
5	49.7	52.8	NS	49.2	53.3	NS
6	65.1	55.9	NS	60.6	60.5	NS
7	72.2	106.7	***	91.5	87.5	NS
8	51.9	80.6	***	64.4	68.1	NS
9	27.5	40.9	***	34.1	34.3	NS

In 2000, difficulty in controlling rampant weed growth on the organic sandy loam at Lode, and low fertility and competition from perennial weeds at West Acre, probably restricted growth and reduced yield potential (Table 4). Despite the late sowings in 2001, all three sites yielded greater than 8 t/ha of sugar. Seed spacing affected sugar yield only at West Acre in 2000 despite some very marked differences in plant population. The ability of individual plants to compensate is illustrated by a yield of over 9 t/ha from only 40,000plants/ha at Terrington in 2002, and suggests that in most situations, sowing at the normal spacing of 17.5 cm is sufficient for organic crops. However, there may be problems in machine harvesting of very large beet in such low populations. Advantage™ seed gave a greater sugar yield at Blankney in 2000, but had no effect otherwise. This reflects the generally good conditions from the late sowings. The expected faster and more even emergence from primed seed would be more likely to lead to a yield difference in the poorer growing conditions of earlier sowings. However, the other benefits of faster and even emergence, particularly in allowing early mechanical weed control, essential in organic crops, make it a useful part of organic beet agronomy.

Table 4. Sugar yield t/ha. (NS = not significant, * = $P<0.05$)

Expt.	Seed spacing			Seed treatment		
	17.5 cm	9.0 cm		Standard	Advantage™	
1	7.26	6.82	NS	6.63	7.44	*
2	5.54	6.02	NS	5.87	5.69	NS
3	3.93	4.28	*	3.99	4.21	NS
4	8.37	8.33	NS	8.38	8.32	NS
5	8.21	8.81	NS	8.14	8.87	NS
6	9.49	9.43	NS	9.40	9.52	NS
7	10.10	10.59	NS	10.45	10.24	NS
8	7.27	7.87	NS	7.27	7.87	NS
9	8.79	9.42	NS	9.12	9.09	NS

CONCLUSIONS

Seed spacing for organically grown sugar beet need not be, in most cases, closer than the 17.5 cm used in conventional crops. However, closer spacing may be necessary on soils prone to slug activity and to avoid harvesting problems with over-large beet. Advantage™ primed seed increased yield at only one site, but the faster and more even emergence, not recorded here, but an accepted benefit in non-organic crops, should allow earlier mechanical weed control. In wet conditions on heavier soils, slug grazing caused severe plant loss. The lack of an effective commercial field-scale organic control is a major constraint to organic beet growing on heavier soils.

REFERENCES

ANONYMOUS (2003a) http://www.germains.com/advantage/

ANONYMOUS (2003b) Sugar Beet: a Grower's Guide. http://www.rothamsted.bbsrc.ac.uk/broom/grguide/

CORMACK W.F. (1999) Testing a stockless arable rotation on a fertile soil. In: Olesen J.E., Eltun, R., Gooding M.J., Jensen E.S. & Köpke U. (Eds.) *Designing and Testing Crop Rotations for Organic Farming, Proceedings from an International Workshop,* Denmark, June 1999, DARCOF Report no.1, pp 115 -123.

LARSSON H. (1999) Experiments with leguminous crops in a stockless organic farming system with sugar beets.. In: Olesen J.E., Eltun, R., Gooding M.J., Jensen E.S. & Köpke U. (Eds.) *Designing and Testing Crop Rotations for Organic Farming, Proceedings from an International Workshop,* Denmark, June 1999, DARCOF Report no.1, pp 311-317.

Estimating Nitrogen Fixation by Fertility-Building Crops

S.P. CUTTLE[1] and G. GOODLASS[2]
[1]IGER, Plas Gogerddan, Aberystwyth, Ceredigion SY23 3EB, UK
[2]ADAS, High Mowthorpe, Duggleby, Malton, North Yorkshire YO17 8BP, UK

ABSTRACT
Information from a literature review and simple models are being used to develop improved guidelines on the use of legumes to supply N in organic farming rotations. Although published information indicates a wide range of fixation values for different legumes (60 - 500 kg N/ha/year), the quantity of N fixed by a particular species is often closely related to the crop yield. This can be used as the basis of a simple on-farm estimation of N fixation after adjusting for factors, such as soil N status, which influence the proportion of N derived from the atmosphere. Existing models of N fixation can be used to extend information from a limited number of studies to a wider range of managements. For managing crop rotations, information about how much of this fixed N (and other forms of soil N) will be available to following crops is more useful than simple estimates of N fixation *per se*. For non-annual crops this requires the use of models to estimate N losses and crop removals during the fertility-building phase. Modelling the subsequent mineralization of N, requires information about the quality (C:N ratio) of the legume residues.

INTRODUCTION
Estimates of the quantity of N fixed by legumes and information about the subsequent mineralization of soil N are essential for the planning and successful operation of crop rotations on organic farms. As part of a current study (www.organicsoilfertility.co.uk), we have conducted a literature review of the quantities of N fixed by legumes and of the factors influencing fixation and the availability of N to following crops. Information from this review and simple models of N fixation and N mineralization are being used to develop improved guidelines for farmers on the use of fertility-building crops in organic rotations.

QUANTITIES OF N FIXED BY LEGUMES
A wide range of legumes are grown, or could potentially be grown, on organic farms in the UK. There is an extensive literature about N fixation in the more commonly grown crops such as white and red clover, peas and field beans. There is relatively little published about less common species or, where information is available, it generally refers to climatic conditions very different from those in the UK. A wide range of fixation values have been measured for various legumes; Table 1 lists the ranges for a number of legume crops grown under climatic conditions broadly similar to those in the UK. These have been adjusted to include an estimate of the fixed N contained in the plant roots. Although the wide ranges for individual crops make prediction difficult, some of the variation can be readily explained; for example, by differences in the proportion of clover in grass/clover mixtures. Similarly, some low values refer to catch crops that had been grown for less than a full season. Fortunately, there is often a close relationship between the quantity of N fixed by a particular legume and the crop yield, with higher yielding

crops fixing more N. This can be a convenient starting point for estimating fixation. Estimates based on yield also take account of the effects of growing conditions that would be difficult to quantify otherwise. A poorly performing crop, whether due to disease, inadequate P or K, drought, etc., will fix less N and this will be reflected in the yield. Similarly, where soil conditions directly affect fixation activity, for example because of ineffective rhizobia in the soil, this will be reflected in reduced yields. Estimates based on clover yields also take account of the generally lower proportion of clover in grazed than in cut swards. On commercial farms, N fixation by grain and silage crops can be estimated from harvested yields; information about yields is less readily available for grazed or green manure crops. For grass/clover swards and other legumes grown with a companion crop, an estimate of the proportion of legume in the total yield is also required. Visual estimates of the proportion of clover in mixed swards are particularly unreliable. As with other aspects of farm management, the difficulty of obtaining the on-farm data necessary is a major obstacle to accurate estimation of N fixation.

Table 1. Ranges of values from the literature for the amount of N fixed by various legumes and of the fixed N remaining in the soil after harvest (all adjusted to include an estimate of fixed N in roots).

Crop	N fixed (kg/ha/year)	N remaining after harvest (kg/ha/year)
White clover/grass (cut)	68 - 438	26 - 166
White clover/grass (grazed)	66 - 262	51 - 204
Red clover (cut)	169 - 456	59 - 159
Vetch (cut & mulched)	101 - 235	96 - 223
Lucerne (cut)	126 - 502	41 - 162
Lupin (grain)	121 - 251	91 - 161
Forage peas	80 - 284	45 - 115
Field beans (grain crop)	200 - 377	95 - 167

Where yield information is available, estimates of N fixation are still subject to uncertainties about the relative proportions of plant-N derived from the atmosphere and that taken up from the soil. Legumes will utilize more soil N and obtain a smaller proportion of their N via fixation when grown in soils containing high levels of available N. Published values for the percentage N derived from the atmosphere in white clover are commonly in the range 70 - 99%, though values as low as 44% have been reported for grazed pastures (Ledgard, 2001). Although grain legumes are considered to obtain a smaller proportion of their N by fixation, the overall range is similar to that for clovers (<40 - 95%) but with a smaller proportion of values in the upper part of the range. Much of this variation can be explained by differences in soil N status. In most organic rotations, legumes are grown following a fertility-depleting phase when soil N reserves are low and mineral-N concentrations are unlikely to be sufficient to inhibit fixation. However, concentrations will be higher where manures have been applied (especially poultry manures, pig and cattle slurries with high contents of readily available N).

Similarly, where legumes are cut and mulched, the fresh crop residues will rapidly decompose and add to the mineral-N in the soil. Mineral-N contents will also be relatively high following cultivation of soils with a long history of grassland cropping. These effects have to be taken into account when predicting the amount of N fixed but are difficult to quantify. This is particularly so where legumes are grown with a non-fixing companion crop, such as in grass/clover swards or where peas are grown with cereals. The non-fixing species would be expected to utilize much of the available N in the soil and thus reduce inhibitory effects on fixation. There is some evidence that fixation in mixtures is less affected by additions of N than where the legume is grown alone.

Most studies have only measured fixed N in the above-ground parts of plants. The additional contribution of fixed N in the roots and stubble below cutting height is of particularly importance in those circumstances where much of the above-ground crop is harvested and removed from the field. Values in the literature suggest that the quantity of fixed N measured in the harvested parts of white clover swards should be multiplied by between 1.23 and 1.65 to obtain the total N fixation (e.g. Jørgensen and Ledgard, 1997; Korsaeth and Eltun, 2000). This correction factor is influenced by cutting height and age of the sward.

NITROGEN AVAILABILITY TO FOLLOWING CROPS

Estimates of N fixation alone are of limited practical value for planning crop rotations. A series of workshops with farmers confirmed that what is most required is information about the amount of N that will be available to following crops. This requires information about the amount of N remaining in the soil after the legume crop and also about the nature of the crop residues and how rapidly they are mineralized. As shown in Table 1, quantities of fixed N remaining after harvest can be very variable and in many cases are only poorly related to the total amount of N fixed. This is largely due to differences in management. Where much of the above-ground plant is harvested and removed, only the N remaining in roots, stubble and crop residues contributes to the soil N pool. Some studies with grain legumes have reported net negative N balances, where the N removed in the crop exceeds the amount fixed; however, balances under UK conditions generally appear to be positive. In contrast to harvesting crops, grazing retains much more of the fixed N in the field. About 80% of the N consumed by grazing livestock is returned directly to the soil in excreta, resulting in a much greater build-up of soil N. However, these estimates must also take account of the large losses that can occur from urine patches. Similarly, cutting and mulching returns almost all of the fixed N to the field, though as noted above, mineralization of high-N residues may inhibit the amount of N fixed. Although harvesting of grain legumes or silage crops will remove much of the fixed N from the field, this will be retained if the crop is fed to livestock on the same farm. The fixed N, minus losses during housing, storage and spreading, will ultimately be returned to the soil in manure.

Estimating the N available to following crops is particularly difficult for perennial legume crops that accumulate N over a number of years. For example, estimates for grass/clover leys need to take account of the amount of N removed by harvesting or grazing each year, N losses during the ley and the recycling and transfer of fixed N to the grass component of the sward.

DEVELOPING GUIDELINES FOR FARMERS

Given the complexity of the factors affecting N fixation and the fact that commercial farmers do not generally have accurate input data on which to base predictions, any guidelines are at best an approximation. Within these constraints, the current project aims to summarize the available information in an accessible form and to make it relevant to the farmer's particular situation. To make the information directly applicable to the user's farm, simple models have been used to extend the data from published studies to a wider range of soil types, climates and managements.

The model by Korsaeth and Eltun (2000) has been used to estimate N fixation for a range of legumes, based on potential fixation, total yield, proportion of legume in mixtures and the amount of mineral N supplied in manure. It has also been modified by incorporating a simple mineralization function to allow for the effect of N mineralization from residues where crops are cut and mulched. The model suggests that N mineralization from the mulch reduces fixation by less than 10% when compared with cutting and removing herbage. Nitrogen losses and the accumulation of soil N in grass/clover leys have been estimated using the NCYCLE model (Scholefield *et al.*, 1991) with the outputs adjusted to more accurately represent a clover-based sward. This has been used to compare N losses, crop removals and the resulting N accumulation under cutting and grazing managements. For example, the model indicates that a cut grass/clover ley on a moderately freely draining loam of intermediate N status and with an annual (above-ground) fixation of 100 kg N/ha, would accumulate 28 kg soil N/ha in the first year. In comparison, a grazed sward with a similar N input would accumulate about 110 kg/ha.

To be able to predict the pattern of N release to following crops, it is also necessary to have information about the quality of the N residues from the legume crop: for example, the C:N ratio and partitioning of N between fast, medium and slow mineralizing pools.

ACKNOWLEDGEMENTS

This work was funded by the Department for Environment, Food and Rural Affairs. IGER is supported by the Biotechnology and Biological Sciences Research Council.

REFERENCES

JØRGENSEN F.V. and LEDGARD S.F. (1997) Contributions from stolons and roots to estimates of the total amount of N_2 fixed by white clover (*Trifolium repens* L.). *Annals of Botany*, 80, 641-648.

KORSAETH A. and ELTUN R. (2000) Nitrogen mass balances in conventional, integrated and ecological cropping systems and the relationship between balance calculations and nitrogen runoff in an 8-year field experiment in Norway. *Agriculture, Ecosystems and Environment*, 79, 199-214.

LEDGARD S.F. (2001) Nitrogen cycling in low input legume-based agriculture, with emphasis on legume/grass pastures. *Plant and Soil*, 228, 43-49.

SCHOLEFIELD D., LOCKYER D.R., WHITEHEAD D.C. and TYSON K.C. (1991) A model to predict transformations and losses of nitrogen in UK pastures grazed by beef-cattle. *Plant and Soil*, 132, 165-177.

Systems Thinking in Organic Research: Does it Happen?

G. DAVIES[1] and D. GIBBON[2]
[1]HDRA-IOR, Ryton Organic Gardens, Coventry, CV8 3LG, UK
[2]RULIVSYS, Lower Barn, Cheney Longville, Shropshire, SY6 8DR, UK

ABSTRACT
Systems thinking has developed in tune with the attempt to define and model complex behaviours in both physical and biological systems. Many tools have been developed that enable complex systems to be described. Such tools have enormous potential for use in understanding organic farming systems, which are complex social-economic-ecological systems. Agricultural research in general, including research on organic farming, has tended to oversimplify complex situations, and thus has generally delivered solutions that are often only partial, or sometimes positively damaging, to the systems into which they have been applied. The aims of the organic agricultural movement would seem to be more sympathetic to the application of an holistic overview of the research and development process, and to developing a 'systems or systemic approach' to problems at several levels of aggregation. However, a more systematic inclusion of systems thinking in current organic research and development programmes will have to be actively encouraged if this potential is to be fully realized.

INTRODUCTION
Biological sciences, and in particular, agricultural science, have tended to lag behind many areas of thinking in systems research, as exemplified by some branches of the physical, ecological or social sciences. In this paper we examine some thoughts on systems thinking and what this might imply for research into organic agricultural research and development programmes. We will also ask whether much current research is really addressing relevant issues, and discuss some alternative methodologies that might better address the research agenda for the organic agriculture movement.

SYSTEMS
Systems and complexity are intertwined as concepts in the popular scientific literature. Whilst the subject area is not without polemic, some consensus has emerged. It is widely recognized that systems are characterized by elements linked within a boundary by flows of material or information. Systems show emergent properties arising from the total relationships between the elements, which are different from the sum of the individual element outputs. Systems have boundaries, but these depend on perspective and the purpose for which they are needed (e.g. at level of field, farm, water catchment area, region). It therefore depends on the point of view of the observer where the boundaries are drawn. Systems may be self-organizing and adaptive, i.e. detect and react to external or internal stimuli (through feedback loops). In effect, systems are models that can be used to give meaning to a situation (OU, 2003). The extent to which, and how, this might mirror 'reality' or 'truth' has been, and will remain, a contentious philosophical debating point. However, failure to appreciate them can often lead to focusing on systems emergent properties, rather than on an understanding of underlying structural causes, which in turn can lead to ineffective or damaging actions. This has been especially true of much recent controversial conventional agricultural research (e.g. GMOs, pesticides) and

needs to be avoided while developing organic systems.

Systems can be characterized in different ways, but can be broken down broadly into 'tame' and 'wicked' systems (Madron and Jopling, 2003); the former characterized by more linear complexities and consequently often be solved by tackling the sub-components, analogous to repairing a car. The latter involve non-linear complexities and not easily broken down into sub-components without loosing the overall perspective, more akin to dealing with road congestion. We argue that farming systems are 'wicked' in this sense and more suited to systems thinking approaches (e.g. Seppänen, 2002) rather than more traditional sub-component research programmes (e.g. Delate, 2002).

FARMING SYSTEMS

Farming systems are complex systems arising from an interaction of social, ecological and physical factors. They can be classified as wicked or non-linear in the sense that: (1) there are often no definitive statements of the problems possible (i.e. they are embedded in an interlocking system of constraints and issues and the problem is only really understood by developing and testing interim solutions); (2) there is no objectively right answer or solution (i.e. there are many people who have a stake in the problem to be solved and this makes the process social rather than purely technical; stakeholders will often accept the most promising sounding solution); (3) the constraints on the solution change over time (politics or money or relevance); (4) the problem solving process is often constrained by resources (time, effort, money).

In such a context it becomes difficult to develop and apply meaningful research programmes in the traditional reductionist sense of identifying a problem, formulating and testing hypotheses, analysing the data and proposing a solution. In fact, solving complex problems usually involves jumping back and forth between the various stages outlined above to understand and analyse the problem and refine solutions within some (often arbitrary set of) limits (e.g. time, resources, etc.). Kolb (1984) has generally explained this as a series experiential learning cycles of active experimentation that gradually broadens areas of knowledge through experience and critical reflection.

SYSTEMS THINKING

Systems thinking has arisen as a means of understanding and managing these complex situations (Checkland, 1981). In agriculture, these methods have become collectively known as farming systems research (and extension). It has emerged from a general consensus that a strictly reductionist perspective, examining a narrow range of factors (e.g. curing a pest problem on an individual crop), is better replaced by a more constructive approach that takes a larger number of factors into consideration (e.g. integrated pest management). More integrated management programmes have often resulted in a 'hard systems' approach to defining farming systems (using tools and models to analyse inputs, outputs and controls in a system defined by the researcher, e.g. models of nutrient or energy flows), which in turn is increasingly giving way to a more 'soft systems' approach (farming as a human activity system defined by the participants) (Gibbon, 2002). Packham (2003) gives an overview of the concepts behind farming systems research. Farming systems, and especially soft systems approaches, rely on a toolkit of methods that can be used to gain perspectives on a system from many differing points of view. They are often characterized by cycles of thinking, action and learning; a participatory approach to analysing and solving problems, which can include, amongst many methods, surveys, interviews, mapping, focus groups and case studies (see IDS, 2003). These would imply that researchers would need to employ a range of skills and tools complementary to those they usually use in their research programmes, more akin to those used by social scientists and geographers. However, it should also be pointed out that the well-tested scientific methods generally used in agricultural research are still necessary and vital elements of much research.

CURRENT ORGANIC AGRICULTURAL RESEARCH

Organic agriculture practitioners in general, and researchers in particular, have always maintained that they take a holistic view of agricultural processes, and that these are in some sense tailored to fit into natural ecological cycles (agroecology). At the same time there has been increasing pressure to at least take on board participatory research methods in research programmes. However, most organic research programmes, as currently reported (Table 1), are very similar to conventional or traditional research studies, in the sense that they appear to be developing a technological or other aspect of organic systems in relative isolation and are largely researcher driven. Some of the studies take a more system wide approach, breaking farm systems into subcomponents (especially ecological or economic ones). The hard systems studies seem to veer towards a modelling approach with a transfer of technology, as opposed to a consultative approach. The soft systems projects involve at least some level of farmer consultation and feedback (e.g. workshops, expert groups). Based on this analysis it would seem that organic research projects in the UK still bend towards a technical development or at best a hard system approach. It should be conceded that at least some of the focused research programmes might be reintegrated into organic farming systems at several different levels, although the mechanisms for achieving this are often not explicitly considered as part of the projects.

Table 1: Classification of Organic Projects (DEFRA, 2003)

Project Type	No	Subject areas
Technology	28	crop/ animal production, companion cropping, storage, variety selection, IPM, rotation, weeds, seeds, shelf life
Agroecology	3	biodiversity, fruit production, pest management
Marketing	7	marketing, economics
Hard Systems	21	nutrient flows, fertility, sustainability, conversion, animal production, weed control
Soft Systems	8	animal health, improving N use, weeds, cereal varieties
Other	19	databases, reviews, policy
No data	2	

DISCUSSION

While some systems thinking appears to be entering into the approaches used by organic researchers there is scope for considerable improvement. We suggest that researchers could educate themselves as to the value of systems approaches and take responsibility for integrating their studies into wider and higher order perspectives. However, the social context in which the research is carried out remains largely unchanged (Edwards-Jones, 2001). Professional obstacles were also highlighted by Collinson (2001), whilst Riley and

Fielding (2001) also point out some of the statistical complications faced by researchers.

Notwithstanding this, the advantages of increased systemic approaches would be a broad development of the organic movement that maintains a dialogue and communication between all stakeholders and practitioners, a factor that has bedevilled conventional farming. Strong linkages and open communication will not only help to develop markets for organic produce but will also serve to enhance linkages within organic systems that will add benefit to them, through for example, the development of both production and ecological services. These linkages will also naturally encourage a diverse and sustainable livelihoods approach to rural development and ensure organic agriculture is a strong component of such development. At the least the 'organic movement' should raise the level of debate and discussion about the implications of the types of research undertaken on behalf of organic farming systems. Failure to do so would condemn organic agriculture to repeat the mistakes of the dominant conventional research paradigm. Finally, there is nothing new about this proposal, as farming systems approaches have been an integral and valued part of many research systems worldwide for nearly 40 years (Collinson, 2000). We feel that the time is appropriate for a review of research strategy and the acceptance of systemic thinking as a valid element within all organic research. Immediate practical steps might be a series of seminars through the COR network, and establishment of a core group of systems researchers to develop guidelines and programmes on participatory, and systemic learning.

REFERENCES

CHECKLAND P.B. (1981) *Systems Thinking , Systems Practice.* NY: John Wiley and Sons.

COLLINSON M. (2000) *A History of Farming Systems Research* . FAO and CABI.

COLLINSON M. (2001) Institutional and professional obstacles to a more effective research process for smallholder agriculture. *Agricultural Systems*, 69, 27-36.

DEFRA (2003) DEFRA science and research projects; subject policy area organic farming <http://www2.defra.gov.uk/research/project_data/>. Accessed 25 September 2003.

DELATE K. (2002) Using an agroecological approach to farming systems research. *HortTechnology*, 12(3), 345-354.

EDWARD-JONES G. (2001) Should we engage in farmer-participatory research in the UK? *Outlook on Agriculture*, 30, 129-136.

GIBBON D. (2002) Systems thinking, interdisciplinarity and farmer participation: essential ingredients in working for more sustainable farming systems. in *Proceedings of the UK Organic Research 2002 Conference , 26th -28th March,* Aberystwyth. pp 105-108.

IDS (UNIVERSITY OF SUSSEX) (2003) Participation Resource Centre. <http://www.ids.ac.uk/ids/particip/information/index.html>. Accessed 27th October 2003

KOLB D. A. (1984). Experiential learning: experience as the source of learning and development. Eaglewood Cliffs, NJ: Prentice-Hall.

MADRON R. and JOPLING J. (2003) The web of democracy. *New Internationalist*, 360, 16-19.

OU (2003) CENTRE FOR COMPLEXITY AND CHANGE Sysweb Home Page. < http://systems.open.ac.uk/page.cfm>. Accessed 6 Oct. 2003

PACKHAM R. (2003) Concepts behind the farming systems approach. in *1st National Australian Farming Systems Conference*, Toowoomba, Queensland 7-10th Sept. 2003, pp26- 48.

RILEY J. and FIELDING W.J. (2001) An illustrated review of some farmer participatory research techniques. *Journal of Agricultural, Biological, and Environmental Statistics*, 6, 5-18.

SEPPÄNEN L (2002) Creating tools for farmers' learning: an application of developmental work research. *Agricultural Systems*, 73, 129-145.

Vegetable Variety Testing for Organic Farming Systems: are Growers' Needs Being Met?

G. DAVIES , S. HARLOCK and P. SUMPTION
IOR-HDRA, Ryton Organic Gardens, Coventry, CV8 3LG, UK

ABSTRACT
Organic vegetable production is an important component of diverse types of organic farming systems. Vegetables are a mainstay of fresh organic produce; they are marketed under many different schemes, either directly to the consumer through various types of community agricultural schemes, or through wholesale trade, or multiple retail outlets. Current trends in varietal development and testing, coupled to recent changes in organic standards, could limit the future development of diverse vegetable cropping systems. The issues centre around a general lack of knowledge on the performance of unfamiliar or untested organic varieties, a lack of confidence in organic seed, especially as regards seed-borne disease, and variety trialling programmes that centre on producers selling into the multiple retail trade. A more regionally based approach to variety testing on growers' holdings, together with structured information-sharing systems, would enable growers to make better and more informed choices to enable flexible cropping systems with a diverse mix of species and varieties to be developed.

INTRODUCTION

Organic vegetables are produced in a wide range of organic farming systems ranging from intensive stockless production through to mixed enterprise systems. They form an important component of organic purchases and are marketed in diverse ways, with 'quality' (taste, texture), 'freshness' and perceived 'health benefits' being important considerations for purchasers (Soil Association, 2002a). Indeed, organic principles aspire to deliver a wide range of environmental and healths benefits to consumers and, in effect, address many of the external costs of 'conventional' intensive agriculture. As such, organic systems aim to be diverse and sustainable. In order to satisfy both principles and demand, organic vegetable production should be moving towards local and seasonally based production and marketing systems. The aims of this paper are to present a brief overview of organic vegetable trialling in the UK, to discus some of the implications, and make some recommendations for moving forwards.

VARIETIES FOR ORGANIC SYSTEMS

Variety choice is arguably one of the more important facets of a grower's production practice, and vegetable variety trialling can potentially provide valuable information that would allow growers to choose varieties suitable for their production needs and market outlets. Here we argue that organic farmers currently depend on varieties supplied by conventional plant breeders for conventional production systems. However, conditions in organic farming systems can be very different, and desirable characteristics for organic varieties are often different to those for conventional ones (Lammerts Van Bueren *et al.*, 2002). Organic growers generally need varieties adapted to lower , and organic, inputs, which implies varieties with root systems able to cope with fluctuating nitrogen dynamics and microbial interactions in organic soils. They also need varieties with plant architecture suitable for suppressing weeds, with general field resistance to pests and diseases, and with a level of yield stability among other traits. Whilst some growers will need uniformity to supply larger markets, it is likely that in future, and to satisfy organic aspirations, growers will be seeking non-standard varieties which promote diversity in their production systems, and aid local marketing schemes, both through novelty and by extending seasons. However, as in all biological systems, trade-offs will ultimately be necessary, for example, Lindig-Cisneros *et al.* (2002) showed domestication of beans led to a reduction in secondary compounds with antifungal properties produced by the plants.

VARIETY EVALUATION FOR ORGANIC SYSTEMS

Generally, variety trialling in the UK, whether organic or conventional, aims to evaluate varieties for suitability for marketing under the EU list system and/or to compare varieties that are commercially available under this legally codified system. There is also a requirement for organic growers to use organically certified seed; generally understood to mean seed produced from mother plants grown under organic conditions over at least one generation, or two growing seasons in the case of perennial plants (Soil Association, 2002b). For instance, current NIAB organic trials aim to address both these issues by evaluating varieties that are available as organic seed and to identify varieties that would be useful in organic systems so that organic seed can be produced.

There are three principal actors in varietal evaluation in the UK: seed companies, research institutes and growers. Seed companies run privately funded organic breeding and evaluation networks which are often geared towards the larger markets, i.e. standard varieties for the multiple retail sector, although there is also an arguably more diverse, though smaller, market for amateur growers. Various research institutes also carry out variety trialling work, either consumer based for the amateur market (e.g. RHS), or more formal bodies, such as agricultural colleges and NIAB. These are often funded through the government or crop levy bodies, whilst the latter is also responsible for maintaining the national seed list. Research institutes undertake some breeding work but seed companies usually lead near-market research. Finally, growers have developed many informal testing systems on their holdings, some of which can be quite sophisticated, but remain largely undocumented in a formal sense.

ARE GROWERS NEEDS BEING MET?

We have argued that organic vegetable production systems should, as far as possible, be local and seasonal in order to satisfy both demand and organic principles, and that varietal choice will be an important component in developing such systems. Information on varietal characteristics is therefore vital in order to allow growers to improve their production systems. The main current source of varietal information is through seed catalogues, composed of information from seed company testing programmes. These are often backed up by open days to which growers and the press are invited. Research institutes publish trial results, though in some cases a subscription is necessary to access them and they may also run open days to disseminate trial results. Growers often cite other growers as valuable sources of information on varieties, though diffusion of grower experience can be slow or erratic. Some recent projects (e.g. COSI, 2003) have attempted to aid this process by publishing trial results (from research institutes or growers' own trials) on-line and encouraging growers to provide feedback.

Despite this, there is a feeling, especially amongst growers, that there is a dearth of information on varietal performance under organic conditions. Growers tend to prefer to use varieties that they have tested and know perform well in their production systems and meet the specifications of their market outlet. Lack of information on equivalence can lead to a lack of confidence in adopting new varieties. Many growers also feel that the results of current variety trials might not be directly relevant to them as there are many factors that might not be applicable to their particular production system (e.g. soil type, climatic conditions, water regime etc.). Coupled to this there is also a worry about lack of information concerning seed-borne disease and the implications for organic vegetable production (Borgen, 2002). Consequently, there is a feeling that there is insufficient diversity or quantity of organic vegetable seed available both to satisfy demand for current production systems and to develop new ones.

FUTURE DIRECTIONS

In summary, vegetable variety breeding, selection and dissemination are, in effect, market led in the UK. Independent comparison trials are run by research institutes but there is probably now less public variety testing (Williams and Roberts, 2002), one reason being that they are often seen as problematic from a 'scientific' point of view (Cantliffe, 2002). Diffusion of results is often through national open days, organized by seed companies or research institutes, at which growers can meet to exchange information, and from which the agricultural press often draw and publicize information. We would argue that these programmes are not serving organic vegetable growers well in terms of developing their farming systems to meet the aspirations of both organic standards and consumers. In particular, there is a danger that growers will actually reduce the diversity of varieties grown in their systems as current research programmes address a narrow segment of the market and organic standards are tightened to enforce use of organic seed. The issue of confidence in organic seed also needs to be addressed.

We recommend a change in emphasis in some current programmes to reorientate the focus of varietal evaluation. In particular, growers would seem to value practical discussion of variety performance in the field, above other means of dissemination. It is also well recognized that regional and local differences are perceived as being important. In this case, participatory approaches to varietal evaluation, well established in other countries, could serve to address both these issues (e.g. for potatoes (Ruissen, 1999; Speiser and Tamm, 2000) and rice (Joshi and Witcombe, 2002)). Participatory trials could be run on a semi-formal basis, preferably on growers' holdings but also using demonstration farm networks to underpin work. Frequent open days run by grower groups would score varieties for desirable characteristics and provide feedback to other groups. Researchers could be used to facilitate the process and other stakeholders such as multiple retailers could be involved in assessing variety characteristics.

Dissemination of information is an important component of any evaluation programme. We have already recommended that researchers facilitate information exchange among growers. Internet-based information systems along the lines of the COSI website (http://www.cosi.org.uk) would seem to be the most practical method of reaching a wide audience, especially if strongly underpinned by participatory varietal evaluation programmes. Zhang *et al.* (2002) describe a web-based database approach for official variety trials (in soybean in this case). Williams and Roberts (2002) also point out the importance of presenting data that growers want, not necessarily yield data only, but data that enables them to put the trials in context, e.g. soil type data, crop spacing and grade out specifications to emphasize a few factors. Variety trial results should be presented in standard formats to aid this process, but with ample scope to include rapid feedback. Consideration of funding mechanisms might also be important to sustain these efforts.

Adding value to vegetable production is likely to be important, especially for smaller scale producers. In such cases producing novel cultivars, and stressing the health angle to eating fresh and local food are likely to be important. Velasques *et al.* (2003) promote the idea of providing new research impetus by focusing on such (up-to-the-minute) subject areas, as does Maynard (2002), who promotes the idea of increasing the scope of variety trials to include other aspects of production practice such as evaluating weeding regimes. This could also include encouraging seed swap mechanisms, and the evaluation and propagation of heritage or heirloom varieties. These aspects of trialling would dovetail neatly with the participation of growers in variety evaluation discussed above and aid growers in developing their systems to meet both consumer demand and organic aspirations whilst developing viable businesses.

REFERENCES

BORGEN A. (2002) Control of seed borne diseases in organic cereals and legumes. In: *Proceedings of the 4th ISTA - PDC seed health symposium: Healthy seeds, the basis for sustainable farming*. ISTA. p. 18.

CANTLIFFE D.J. (2002) Vegetable variety trials: an administrative point of view. *HortTechnology,* 12(4), 584-586.

JOSHI K.D. and WITCOMBE J.R. (2002) Participatory varietal selection in rice in Nepal in favourable agricultural environments- a comparison of two methods assessed by varietal adoption. *Euphytica,* 127, 445-458.

LAMMERTS VAN BUEREN E.T., STUICK P.C. and JACOBSEN E. (2002) Ecological concepts in organic farming and their consequences for an organic crop ideotype. *Netherlands Journal of Agricultural Science,* 50, 1-26.

LINDIG-CISNEROS R., DIRZO R. and ESPINOSA-GARCIA F.J. (2002) Effects of domestication and agronomic selection on phytoalexin antifungal defence in *Phaseolus* beans. *Ecological Research,* 17, 315-321.

MAYNARD D.N. (2002) Enhancing the scope of vegetable cultivar evaluation in Florida. *HortTechnology,* 12(4), 560-561.

RUISSEN T. (2000). Participatory research - potato late blight as a case in Norwegian organic agriculture. In Allard G., David C. and Henning J. (eds) *Organic Agriculture Faces its Development;the future issues*, Volume Les Collques , 95, pp. 380-384. INRA: Paris.

SOIL ASSOCIATION (2002a) Organic Food and Farming Report 2002. Bristol: Soil Association.

SOIL ASSOCIATION (2002b) Soil Association standards for organic farming and production. Revision 14. Bristol: Soil Association.

SPEISER B. and TAMM L. (2000) "Participatory research"- a promising approach to the introduction of new, blight tolerant potato varieties in practice. In Alfodi T., Lockeretz W. and Niggli U. (eds) *Proceedings 13th International IFOAM Scientific Conference*. FiBL:Basel, Switzerland. pp. 231.

VELASQUEZ C., EASTMAN C. and MASIUNAS J. (2003) Value-added vegetable cultivars: an assessment of farmers' perceptions. *HortTechnology,* 13(3) 518-521.

WILLIAMS T.V. and ROBERTS W. (2002) Is vegetable variety evaluation and reporting becoming a lost art? An industry perspective. *HortTechnology,* 12(4), 553-559.

ZHANG L., QI W., SU L., WHISLER F. (2002) Deltasoy- an internet based soybean database for official variety trials. *Agronomy Journal,* 94, 1163-1171.

When is a Weed not a Weed? Pest Control in Organic Farming

G. DAVIES and R. GUNTON
HDRA-IOR, Ryton Organic Gardens, Coventry, CV8 3LG, UK

ABSTRACT

Weeds can reduce crop yields by competition. However, many studies have suggested that weeds can also reduce insect pest damage to crops. We review a variety of explanations for "protective effects" of weeds and cite evidence to show which ones may be most important. Organic farmers could maximize marketable yield through an integrated approach to weed/pest management by modifying basic cultural practices. The topic requires further research and organic farmers could contribute with realistic validation trials of weed/pest management regimes.

INTRODUCTION

Both weeds and attacks by insect herbivores inflict yield loss on crops, and their control can be expensive to the farmer. Relatively few studies to date have examined how these two effects are linked or asked what the economic implications are. There is a considerable body of evidence that non-crop vegetation in fields can reduce herbivore populations and damage to crops. In the light of this, we can ask questions about the protective effects of weeds as compared to their competitive costs and about how to optimize crop yield. These types of questions can be especially useful in organic systems, which promote a holistic, whole-ecosystem approach to crop health. This brief review draws on the large body of work relevant to this topic but does not provide an exhaustive list of references due to limitations on space. In does, however, highlight where further work should be directed in order to improve cultural practices.

MECHANISMS FOR PROTECTIVE EFFECTS

Two hypotheses, stated by Root (1973), are commonly invoked to explain the phenomenon of reduced herbivore damage in a weedy culture. The "resource concentration hypothesis" considers effects on herbivore behaviour (colonization and emigration rates), while the "enemies hypothesis" considers mortality (by predation/parasitism) of herbivores. These general hypotheses have been broken down into more detailed mechanisms (that work both to increase and to decrease herbivore damage in various ways) by various authors; we present here a working classification of these mechanisms for how protective effects might arise in order to develop the following discussion.

Deterrence /attraction hypotheses suggest that weeds could keep herbivores away from crops by releasing repulsive odours (Dimock and Renwick, 1991), modifying the microclimate (Tahvanainen and Root, 1972) or attracting herbivores to a crop by providing complementary resources (such as alternative host species) (Schellhorn and Sork, 1997).

"Concealment" hypotheses state that herbivores are less likely to find stands of host plants that are diversified with weeds. Effects include forming physical barriers, camouflaging a crop against a (bare-earth) background (Finch and Collier 2003), or masking olfactory stimuli from the crop.

"Distraction" effects occur where weeds are colonized by herbivores *instead of* the

crop, either because they prefer the weed or because they reach the weed before the crop.

The *"bottom-up effect"* (Hooks and Johnson, 2003) is due to crops becoming less attractive to herbivores as competition with weeds reduces their vigour. Recent evidence suggests that some crop plants' chemical defence mechanisms against herbivores may be compromised by competition (Cipollini and Bergelson, 2002).

"Natural enemies" (*"top-down"*) hypotheses usually state that more-diverse communities provide more of the habitat requirements of insects that feed on herbivore species (Sengonca *et al.*, 2002), although weeds could also attract predators away from herbivores on crop plants by providing alternative resources.

There is no general consensus on which of these hypotheses provide the most realistic description of pest-host relationships. The current balance of evidence suggests that: (1) the "resource concentration hypothesis" is probably more important than the "enemies hypothesis" (Andow, 1991); (2) weedy culture is more effective where monophagous herbivores are prevalent or where candidate natural enemies take the form of predators rather than parasites (Andow, 1991); (3) visual concealment is probably more important than olfactory concealment or deterrence (Finch and Collier, 2003); (4) distraction effects can work (Kloen and Altieri, 1990); and (5) bottom-up effects need further investigation, as they could undermine the benefits of weedy culture. In fact, different mechanisms are likely to work in different systems, and probably at different stages in the unfolding and complex arena of pest-host interaction.

USING WEED CONTROL TO BEST ADVANTAGE

In the light of this, several aspects of cultural strategies would seem to provide opportunities to manipulate crop-weed balance so that herbivore damage may be reduced and marketable yield increased. These are addressed below, with reference to the putative mechanisms for protective effects outlined above.

Timing

Studies on critical period have shown that, provided weeding is carried out at a certain stage or period during crop development, weeds do not inflict significant yield loss. If "camouflaging" a crop during its early growth can effectively reduce colonization by herbivores, then an initial weed flora that develops and is destroyed at the start of the critical period could satisfy the interests of both weed control and pest defence. The effectiveness of this will depend on the colonization habits of prevalent herbivore species. On the other hand, if weeds growing between crop plants play a role at later stages in crop growth (e.g. by forming a visual barrier or attracting natural enemies), delaying weed establishment until the end of the critical period could reduce these benefits. In these situations, the costs of weedy culture may inexorably outweigh the benefits.

Many herbivore species seek new host plants at predictable times during the season. It might therefore be possible to synchronize weed flushes with pest peaks. Similarly, if the local weed flora happens to be dominated by a species that serves as an alternative host for a prevalent herbivore (e.g. charlock for flea beetle), a "trap crop" of weeds could be allowed to develop. Disturbance of established weed communities at appropriate times, may also encourage predators to move into a crop. Some weed species are more useful than others. It is possible to influence the species composition of vegetation that develops on cultivated land by the time of year when cultivation operations are carried out (Altieri and Whitcomb, 1979). Over time, a farmer might encourage marginal land to develop a flora that is most effective at protecting crops from pests – either directly, or by contributing to the seed bank of adjacent fields.

Spatial arrangement

The "camouflage" effect requires weeds to grow within and between rows. Taller weeds confined to corridors where weed control is relaxed, however, might produce the "visual barrier" effect. In turn, "natural enemy" effects may be obtained from weeds at still greater remoteness. "Distraction" effects might require only a border of the alternative host around the field margin (as with conventional trap cropping). The efficacy of these strategies will depend on the behaviour of the relevant predator, parasitoid and herbivore species. It should be borne in mind that the spacing of crop plants is more easily controlled than that of weeds and some thought might need to be given to what arrangement of weeds and crop plants could promote "concealment" effects.

Other cultural issues

The weed flora can be manipulated by sowing living mulches of selected species around a crop. This has an economic cost, but could help bring persistent weeds under control, while volunteer plants of the species that were sown may regenerate in subsequent years and continue to influence the flora. Where a protective effect depends heavily on the presence of certain species, sowing may be the only way to obtain them in sufficient abundance. Choice of crop varieties may have a role to play but it is unclear what balance of competitiveness as compared with innate herbivore resistance might be used to maximize yield.

CAN IT BE ECONOMICAL?

Different crops have different economics, and biological yield loss is not equivalent to economic yield loss. Unfortunately, there is a lack of studies on economic outcomes of weedy culture and, to date, only a few experiments have obtained net increases in biological yield (e.g. Altieri *et al.*,1977; Ryan *et al.*,1980; Hooks and Johnson, 2002). Some experiments have found no significant reduction in yield, using either weeds or living mulches (e.g. Andow *et al.*, 1986; Kemp and Barrett, 1989; Rodenhouse *et al.*, 1992; Schellhorn and Sork, 1997). Most of these studies have used either brassica varieties or soybeans – both high-value crops where protective effects could pay off.

Additional costs of exploiting weedy culture may come from altered management regimes. For example, they may include obtaining new equipment (to perform different weeding operations). In addition, the costs of weedy culture may increase if, for example, weather conditions interfere with intended management regimes. Leaving weed plants to develop to more advanced stages might increase the risk of increasing weed seedbanks.

At this point, two mitigating factors are worth noting. First, weeds growing among crops can have other benefits. They may reduce soil erosion and compaction, while enhancing levels of organic matter, water penetration and nutrient retention. They may also moderate temperature and mediate mycorrhizal effects. Second, if a useful weed flora can be fostered on field margins and other unproductive land, some of the benefits of weedy culture may be obtained with zero yield loss from competition.

CONCLUSIONS

Ultimately, the benefits of weedy culture depend on pest pressure. Being largely unpredictable, this adds an uncertainty factor. It is therefore important to take a long-term, whole-system view and ask whether allowing weeds a freer reign is compatible with sustaining weed management on the whole farm. Research results suggest that there is potential for weeds to play a role in protection against pests. The generalizations in this paper have been necessitated not only by a shortage of relevant experimental studies, but also by the nature of the problem: a three-way interaction between crops, weeds and pests in an ephemeral environment is bound to be subject to great variability and be sensitive to local conditions. Greater attention to weedy culture has the potential to benefit organic farmers, but it will require individual farmers to experiment and contribute to pools of local knowledge, as well as support from the agroecology research community. The best way forward may be to invite farmers into active participation in agroecology research.

REFERENCES

ALTIERI M.A., SCHOONHOVEN A. and DOLL J.D. (1977) The ecological role of weeds in insect pest management systems: a review illustrated with bean (Phaseolus vulgaris) cropping systems. *Pest Articles and News Summaries,* 23, 195-205.

ALTIERI M.A. and WHITCOMB W.H. (1979) Manipulation of insect populations through seasonal disturbance of weed communities. *Protection Ecology,* 1, 185-202.

ANDOW D.A. (1991) Vegetational diversity and arthropod population response. *Annual Review of Entomology,* 36, 561-586.

ANDOW D.A., NICHOLSON A.G., WIEN H.C., and WILSON H.R. (1986) Insect populations on cabbage grown with living mulches. *Environmental Entomology,* 15, 293-299.

CIPOLLINI D.F. and BERGELSON J. (2002) Interspecific competition affects growth and herbivore damage of Brassica napus in the field. *Plant Ecology,* 162, 227-231.

DIMOCK M.B. and RENWICK J.A.A. (1991) Oviposition by field popoulations of *Pieris rapae* (Lepidoptera: Pieridae) deterred by an extract of wild crucifer. *Environmental Entomology*, 20, 802-806.

FINCH S. and COLLIER R. (2003) Insects can see clearly now the weeds have gone. *Biologist,* 50, 132-135.

HOOKS C.R.R. and JOHNSON M.W. (2002) Lepidopteran pest populations and crop yields in row intercropped broccoli. *Agricultural and Forest Entomology,* 4, 117-126.

HOOKS C.R.R. and JOHNSON M.W. (2003) Impact of agricultural diversification on the insect community of cruciferous crops. *Crop Protection,* 2, 223-238.

KEMP J.C. and BARRETT G.W. (1989) Spatial patterning: impact of uncultivated corridors on arthropod populations within soybean agroecosystems. *Ecology,* 70, 114-128.

KLOEN H. and ALTIERI M.A. (1990) Effect of mustard (Brassica hirta) as a non-crop plant on competition and insect pests in broccoli (Brassica oleracea). *Crop Protection,* 9, 90-96.

RODENHOUSE N.L., BARRETT G.W., ZIMMERMAN D.M. and KEMP J.C. (1992) Effects of uncultivated corridors on arthropod abundances and crop yields in soybean agroecosystems. *Agriculture, Ecosystems and Environment,* 38, 179-191.

ROOT R.B. (1973) Organization of a plant-arthropod association in simple and diverse habitats: the fauna of collards (*Brassica oleracea*). *Ecological Monographs,* 43, 95-124.

RYAN J., RYAN M.F., and MCNAEIDHE (1980) The effect of interrow plant cover on populations of the cabbage root fly Delia brassicae. *Journal of Applied Ecology,* 17, 31-40.

SCHELLHORN N.A. and SORK V.L. (1997) The impact of weed diversity on insect population dynamics and crop yield in collards, *Brassica oleracea* (Brassicaceae). *Oecologia*, 111, 233-240.

SENGONCA C., KRANZ J. and BLAESER P. (2002) Attractiveness of three weed species to polyphagous predators and their influence on aphid populations in adjacent lettuce cultivations. *Journal of Pest Science,* 75, 161-165.

TAHVANAINEN J. and ROOT R.B. (1972) The influence of vegetational diversity on the population ecology of a specialized herbivore, *Phylotreta cruciferae* (Coleoptera:Chrysomelidae). *Oecologia,* 10, 321-346.

Sustaining Plant Available Soil Phosphorus in Broad-Acre Organic Cropping in Australia

JEFFREY EVANS and LATARNIE MCDONALD
NSW Agriculture, Agricultural Institute, Private Mail Bag,
Wagga Wagga, New South Wales 2650, Australia.

ABSTRACT

Plant available soil phosphorus concentrations are deficient (< 5 mg/kg; Olsen P) on some organic farms in the cereal cropping region of southern Australia and may herald an emerging widespread problem that will threaten the opportunity to further develop a sustainable, productive organic grain industry. Reactive rock phosphate has poor agronomic effectiveness under the rainfall regime of much of this cropping region. Two field trials are investigating the effect on available P of the co-application of reactive rock phosphate and elemental sulphur, using acid production from oxidation of the sulphur to facilitate release of orthophosphate from the rock phosphate. At one site, Olsen P has increased from a background level of 4 mg/kg to 11 mg/kg, and at the other site, from 8 mg/kg to 18 mg/kg. These increases have occurred within 3 years of soil application of ground reactive rock phosphate intimately mixed with ground sulphur and incorporated with soil.

INTRODUCTION

Soils on the Australian continent are generally deficient in phosphorus (P) for modern agricultural crops. Consequently Australian agriculture has a history of extensive soil application of fertilizer P, commonly superphosphate, dating from the 1950s. In the early stages of converting from conventional to organic cropping, it is highly likely that organic grain yields are supported by accumulated soil reserves of P from this previous superphosphate history. However, an analysis of grain yields and crop nutrient content from an organic farm with a 25-year history of organic farming (Derrick, 1996) indicated yields well below those of a neighbouring conventional farm, associated with marginal tissue P concentrations. Were this deficiency characteristic of maturing organic grain enterprises, efforts to grow a significant, sustainable domestic and export organic grain industry will be jeopardized.

Rock phosphate is an allowable P fertilizer for organic farming. However, previous research on the agronomic effectiveness of reactive rock phosphate (Bolland *et al.*, 1988; Bolan *et al.*, 1990) reported that its efficacy was largely limited to soils with pH ($CaCl_2$) < 5.5 in environments with annual rainfall > 800 mm. The rainfall limitation excludes the majority of the southern Australian cropping zone. Sulphur addition to rock phosphate granules may be used to increase acidity of the granule to facilitate release of orthophosphate, though this was not effective in temperate Australia (Swaby, 1975). However, it may be hypothesised that RPR-S reaction is compromised by the large particle size of granules. RPR-S granules of 0.2 mm – 2 mm were effective in potted soil held at field capacity (Rajan, 1983).

The first aim of the current study was to assess the plant available P status of cereal cropping soils on organic farms. The second aim was to assess the role of ground elemental sulphur (S) in assisting dissolution of ground reactive rock phosphate to increase plant available soil P in the southern Australian cereal belt.

MATERIALS AND METHODS

Soil P status on organic cereal farms

Surface soil (0-10 cm) was collected from fields on several organic grain-producing farms in central and southern New South Wales. All farms had a history of organic farming exceeding 5 years. Plant available soil P was estimated by the Olsen P method; a P concentration of < 15-20 mg/kg was taken as being too low for meeting potential grain yield (Holford *et al.*, 1985; Hedley *et al.*, 2001).

Field trials with rock phosphate and sulphur

Two field trials were established in 2001. One trial (site 1: lat. 34°82′S; long. 147°74′E) was located on a sandy loam of pH 4.4 ($CaCl_2$) in an area of mean annual rainfall of 532 mm. Rainfall in 2001 was 422 mm, and in 2002, 420 mm. The second trial (site 2: lat. 34°35′S; long. 146°90′E) was located on a red loam of pH 5.4 ($CaCl_2$) in an area of mean annual rainfall of 489 mm. Rainfall in 2001 was 380 mm, and in 2002, 251 mm.

At site 1, soil treatments included: nil, granulated reactive phosphate rock (RPR) (500 kg/ha), or ground RPR (500 kg/ha). Each of these treatments was split for nil or ground S (500 kg/ha). In June 2001, plots were sown to various pasture mixes, or to wheat under-sown with pasture. The pasture species were allowed to regenerate in 2002. Soil was cultivated in autumn of 2003 and sown to wheat.

At site 2, soil treatments included, nil, ground RPR (500 kg/ha), or ground RPR + ground S (500 + 500 kg/ha). In July 2001, the plots were sown to pasture species, to wheat under-sown with clover, or left to volunteer species. Pasture growth was either green-manured or cut and removed; grain was harvested. In 2002, following cultivation in autumn, the plots were sown to vetch (*Vicia sativa*). Vetch was also established into cultivated soil in 2003.

At each site, all soil treatments were included for each pasture treatment and all treatments were represented in 3 blocks. The RPR had 13% P with 1% citric acid soluble P. When combined with S, the RPR and S were thoroughly mixed prior to spreading; all soil amendments were incorporated into the surface 10 cm in May 2001 (site 1) and June 2001 (site 2). Plant available soil P (Olsen P) and soil pH (1:5, $CaCl_2$) were estimated before soil amendments, and post soil amendments: in August of 2001 (both sites), in April of 2002 and 2003 (site 1), or May 2002 and June 2003 (site 2). Each estimate is a mean of at least 9 replicates of a soil treatment.

RESULTS AND DISCUSSION

Soil P status on organic cereal farms

Table 1 indicates that plant available P was frequently well below the critical level.

At the concentrations indicated, yield levels of 50% of potential, or less, would be expected, which is consistent with related farm yields. The data suggest a chronic lack of attention to managing P balance.

Table 1. Plant-available soil P levels (Olsen P; mg/kg) on organic farms in south-eastern Australia.

Soil	Olsen P	Soil	Olsen P
1	5.5	9	4.3
2	1.5	10	3.0
3	4.4	11	3.4
4	3.8	12	1.8
5	3.1	13	2.1
6	13	14	1.6
7	1.4	15	2.5
8	2.8	16	3.1

Field trials with rock phosphate and sulphur

To date, there have been no significant effects of the various pasture treatments on available P at either site; hence, soil treatment effects are compared across pasture species. Soil treatment effects on available P (0-10 cm) are shown in Table 2 (site 1) and Table 3 (site 2).

At site 1, by August 2001, and relative to background levels, Olsen P had not been affected significantly by soil amendment with either granulated RPR or ground RPR, irrespective of co-application with S. In the absence of S, similar was true for the subsequent years of 2002 and 2003. Co-application of ground RPR + ground S had significantly increased Olsen P by April 2002 and the effect was sustained into June 2003. RPR + S increased Olsen P to concentrations at, or exceeding, the critical value. Soil pH in the RPR + S plots declined from pH = 4.4 to pH = 4.1 at the third year. Olsen P level has also increased in the 10 – 20 cm zone at April 2003 (data not shown).

Table 2. Influence of soil amendments applied in late May 2001 on Olsen P (mg/kg) at site 1 over 3 years. Values in brackets are the standard errors.

	Nil RPR		Ground RPR		Granulated RPR	
	Nil S	+S	Nil S	+S	Nil S	+S
August 2001	8.9	8.3	10.1	8.9	8.5	8.0
	(0.85)	*(0.54)*	*(0.6)*	*(0.53)*	*(0.31)*	*(0.34)*
April 2002	8.3	9.3	9.9	17.5	7.8	9.1
	(0.94)	*(0.68)*	*(0.60)*	*(0.78)*	*(0.30)*	*(0.43)*
April 2003	7.4	9.8	9.5	15.2	8.0	9.7
	(0.51)	*(0.50)*	*(0.81)*	*(0.61)*	*(0.5)*	*(0.44)*

At site 2, relative to background levels, soil amendment with ground RPR has had no significant effect on Olsen P over the 3 years of testing. Co-application of RPR + S though, had increased Olsen P by August 2001, and sustained significantly higher increases in Olsen P over the succeeding 2 years. Soil pH has declined from 5.4 to 5.0 in the RPR + S treatment. Most response to the increased soil concentrations of available P following soil amendment with RPR + S has occurred in background gramineous species and wheat. At site 1 this has led to greater incidence of acid tolerant *Vulpia spp.* on this treatment and, at site 2, to pasture dominated by annual ryegrass (*Lolium spp.*).

Table 3. Influence of soil amendments applied in June 2001 on Olsen P (mg/kg) at site 2 over 3 years. Values in brackets are the standard errors.

	Nil RPR	Ground RPR	
	Nil S	Nil S	+S
August 2001	4.6	5.4	6.9
	(0.20)	*(0.19)*	*(0.28)*
May 2002	3.8	4.0	8.9
	(0.13)	*(0.16)*	*(0.48)*
June 2003	5.4	5.8	11.0
	(0.14)	*(0.31)*	*(0.43)*

CONCLUSIONS

Very low concentrations of plant-available soil P (Olsen P) in soils collected from a sample of 'old' organic farms practicing broad-acre cropping may indicate a widespread problem of P deficiency that could endanger the growth of a sustainable productive organic grain industry in SW Australia. However, field trials indicate that acid produced during the microbially-mediated oxidation of elemental S, may increase the availability of P from RPR in the Australian southern cropping zone. Ground fertilizer amendments appear to be critical to the success of this strategy. Analyses will continue in the current study to quantify residual RPR and S levels; this and further allied research is needed to identify the minimal fertiliser rates required to amend severely P-degraded soil, and to identify maintenance P requirements for a range of soils and rainfall zones.

REFERENCES

BOLAN N.S., WHITE R.E. and HEDLEY M.J. (1990) A review of the use of phosphate rocks as fertilizers for direct application in Australia and New Zealand. *Australian Journal of Experimental Agriculture*, 30, 297-313.

BOLLAND M.D.A., GILKES R.J. and D'ANTUONO M.F. (1988) The effectiveness of rock phosphate fertilizers in Australian agriculture: a review. *Australian Journal of Experimental Agriculture*, 28, 83-90.

DERRICK J.W. (1996) *A comparison of agroecosystems: organic and conventional broad-acre farming in south east Australia*. Ph.D. Thesis, Australian National University.

HEDLEY M.J., MORTVEDT J.J., BOLAN N.S. and SYERS K. (2001) Phosphorus fertility management in agroecosystems. In: *Phosphorus in the Global Environment* (Chapter 5, 42 pp). www.icsu-scope.org.

HOLFORD I.C.R., MORGAN J.M., BRADLEY J. and CULLIS B.R. (1985) Yield responsiveness and response curvature as essential criteria for the evaluation and calibration of soil phosphate tests for wheat. *Australian Journal of Soil Research*, 23, 201-209.

RAJAN S.S.S. (1983) Effect of sulphur content of phosphate rock / sulphur granules on availability of phosphate to plants. *Fertiliser Research*, 4, 287-296.

SWABY R.J. (1975) Biosuper-biological superphosphate. In: K.D.McLachlan (Ed.). *Sulphur in Australasian Agriculture*, pp 213-20, Sydney University, Sydney.

Potato Cyst Nematode Populations and Spatial Distribution: Temporal Variation within a Stockless Organic Rotation

W.F. CORMACK[1] , A.E. RIDING [2]and W.E. PARKER[2]
[1]ADAS Terrington, Terrington St Clement, King's Lynn, Norfolk, PE34 4PW, UK
[2]ADAS Woodthorne, Wergs Road, Wolverhampton, WV6 8TQ, UK

ABSTRACT
The distribution of potato cyst nematodes (PCN) was mapped annually, in winter, from December 1998 to December 2001 within five plots of an organic stockless rotation at ADAS Terrington. Conversion to organic methods was started in 1990 and was completed in 1995; the land is currently certificated by the Soil Association. Each 2 ha plot is under a different phase of a five-year rotation (clover, potatoes or calabrese, winter wheat, spring beans, undersown spring cereal). The five plots were divided into sub-plots of approximately 625 m^2 (25 x 25 m). The corners of each sub-plot were geo-referenced using a GPS (Global Positioning System) to record actual location and to aid relocation. On each sampling occasion, 30 soil cores were taken to 15 cm depth and bulked to give one representative sample for each sub-plot. PCN cysts were extracted and egg counts made using standard assessment methods. Data were analysed and spatial distribution plotted within a GIS. A discrete (patchy) spatial distribution was observed for PCN counts which remained fairly static throughout 1998 to 2001. There was no evidence of a significant multiplication of PCN following potatoes. However, the growing of non-PCN resistant cultivars may have resulted in a more significant increase in PCN.

INTRODUCTION
Potato cyst nematodes (PCN; *Globodera rostochiensis* and *G. pallida*) are the most important pests of potato in the UK, and are a potential threat to the sustainability of organic potato production in the UK. Within organic rotations, PCN management is largely achieved by rotation (PCN populations decline in the absence of a host-crop) and the use of PCN-resistant cultivars. *G. pallida* is now widespread in the UK (Minnis *et al.*, 2002), and only a few commercially-acceptable potato cultivars have (at best) partial resistance to *G. pallida*. An understanding of the effect of an organic rotation on the longer-term population dynamics of PCN is therefore essential.

This work was part of a larger project carried out at ADAS Terrington (Kings Lynn, Norfolk) to evaluate the sustainability of a stockless organic arable rotation in terms of nutrient supply, perennial weed control, pests and diseases (Cormack, 1999). The core of the project was an unreplicated system study with field-scale plots to allow meaningful study of potential crop protection issues and give confidence to farmers that the system could work on a farm scale. Although problems with PCN had not been encountered during the conversion to organic farming at the ADAS Terrington site, PCN are endemic in the local area, and the development of a serious PCN problem at the Terrington organic site would necessitate a major change in crop rotation and so affect profitability. The incidence and distribution of PCN was therefore mapped at the site over a four-year period.

MATERIALS AND METHODS

The distribution of PCN within the five plots of the stockless organic rotation at ADAS Terrington was mapped annually, in winter, between December 1998 and December 2001. Each of the five plots was divided into sub-plots of approximately 625 m^2 (25 x 25 m). The corners of each sub-plot were geo-referenced using a GPS to record actual location and to aid relocation. On each sampling occasion, 30 soil cores were taken to 15 cm depth from each sub-plot and bulked to give one representative sample. Soil samples were analysed for PCN cysts and eggs using the Fenwick can technique (Southey, 1986).

RESULTS AND DISCUSSION

The PCN population levels observed over the four cropping years covered by the sampling remained at low levels (Table 1). Variations in the percentage of plots infested, and PCN eggs/g of soil in individual plots, were within the range of sampling errors normally encountered when assessing low PCN populations. Critically, there was no evidence of a significant multiplication in the PCN population in plots that had been cropped with potatoes immediately prior to a sampling occasion (Table 2).

Table 1. Percentage of sub-plots infested and the range of PCN populations found on each sampling occasion.

Measure of infestation	Sampling date			
	Dec 1998	Jan 2000	Jan 2001	Dec 2001
% with no cysts found	87.9	92.4	87.3	83.4
% with very low PCN levels (non-viable cysts)	5.7	2.5	5.1	9.5
% with low PCN levels (viable cysts)	6.4	5.1	7.6	7.0
Range of eggs/g soil	0.03 – 4.1	0.05 – 4.6	1.0 – 7.0	1.0 – 5.0

None of the PCN populations found were sufficiently numerous or viable for a PCN species test to be carried out. The use in all plots of the potato variety Santé, which has full resistance to both yellow PCN (*Globodera rostochiensis*) and partial resistance to white PCN (*Globodera pallida*), would have contributed significantly to suppressing the build-up of PCN populations.

Small differences were seen in the PCN counts between different plots (Table 2 and Figure 1); plot 4 consistently had a higher population than the other four plots. A map showing the change in the spatial distribution of the cysts (Figure 1) indicates that from 1998 to 2001 there were distinct, and largely temporally persistent, foci of PCN in the eastern part of plot 4 and the north-west and south-west corners of plots 1 and 2 respectively. These infested areas probably reflected the general location of initial foci of PCN infestation rather than an impact of the recent cropping history.

Figure 1. Spatial differences in PCN counts between plots, 1998-2001.

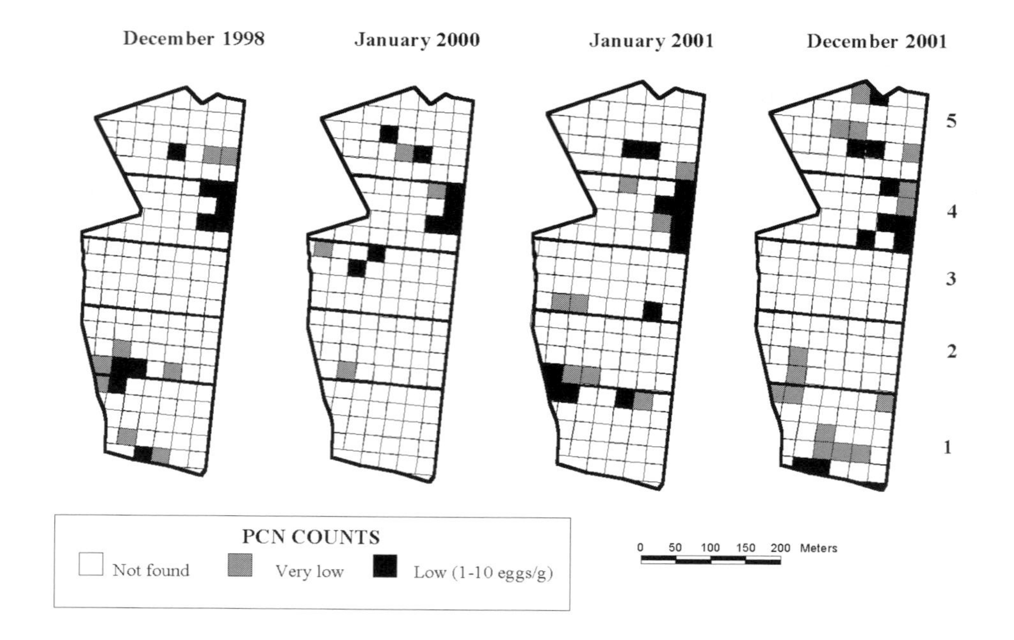

These results indicate that PCN population is not changing significantly under the current rotational frequency of potatoes (one year in five) whilst the resistant variety Santé is grown. However, the use of non-PCN resistance varieties may well have resulted in more significant population build-up and future work should consider the impact of variety choice on the sustainability of organic potato cropping on PCN-infested land. It should be noted in particular that the highest PCN populations tended to occur on sampling occasions following a potato harvest year, indicating some small PCN multiplication had occurred under the potato crop. On close rotations (less than 1 in 5) over a period of 20 years, such low populations can build up to unmanageable levels (e.g. Trudgill *et al.*, 2003) unless the use of resistant varieties is integrated into the management strategy.

Table 2. Percentage of sub-plots containing viable PCN cysts for each sampling occasion in relation to potato cropping history. (bold figures indicate samples taken in the winter immediately following a potato crop)

Plot	Last potato harvest	% sub-plots with viable PCN cysts			
		Dec 1998	Jan 2000	Jan 2001	Dec 2001
1	1998	**6.3**	0.0	9.4	9.4
2	2000	6.5	0.0	**3.2**	0.0
3	1999	0.0	**6.1**	3.0	0.0
4	2001	19.2	15.4	19.2	**19.2**
5	1997	2.9	5.7	5.7	8.6
Mean	-	6.4	5.1	7.6	7.0

ACKNOWLEDGEMENTS

We are grateful to Defra for funding this research and to all ADAS staff involved in the field and laboratory work.

REFERENCES

CORMACK W. F. (1999) Testing a Stockless Arable Rotation on a Fertile Soil. *Proceedings of an International Workshop on Designing and Testing Crop Rotations for Organic Farming, Borris, Denmark, June 1999.* Danish Research Centre for Organic Agriculture (DARCOF) Report 1/1999, pp. 115-123.

MINNIS S. T., HAYDOCK P.P.J., IBRAHIM S.K. and GROVE I.G. (2002) Potato cyst nematodes in England and Wales – occurrence and distribution. *Annals of Applied Biology,* 140, 187-195.

SOUTHEY J.F. (1986) *Laboratory methods for work with plant and soil nematodes.* London: Her Majesty's Stationery Office.

TRUDGILL D.L., ELLIOTT M.J., EVANS K. and PHILLIPS M.S. (2003) The white potato cyst nematode *(Globodera pallida*) – a critical analysis of the threat in Britain. *Annals of Applied Biology,* 143, 73-80.

Potato Varieties for Organic Production in Northern Ireland

C.E HALL,[1]E.M. WHITE, [1]L.R. COOKE, [2] G. LITTLE, [2] and A.R. SAUNDERS [3].
[1] Northern Ireland Plant Testing Station, Department of Agriculture and Rural Development, 50 Houston Road, Crossnacreevy, Castlereagh, BELFAST, BT6 9SH
[2] Applied Plant Science Division, Department of Agriculture and Rural Development, Newforge Lane, BELFAST, BT9 5PX
[3] Greenmount Campus, College of Agriculture, Food and Rural Enterprise, Department of Agriculture and Rural Development, ANTRIM, BT41 4PU

ABSTRACT
Organic potato production requires varieties combining good blight resistance, agronomic characters and consumer acceptability. Results are presented from field plot trials of selected varieties in an organic situation. Blight incidences were low and yields were acceptable in both years of the trial. However defects reached substantial amounts in some varieties. Supporting non-organic trials are reported on research to assess varietal performance under high blight pressure. In both sets of trials some cultivars exhibited excellent foliage blight resistance, but initial results also showed that some varieties exhibited good resistance to tuber blight despite having foliar infection. It is concluded that further work is needed over a number of seasons to evaluate blight resistance, effectiveness of fungicide treatment, consumer characteristics and husbandry requirements of varieties which may suit organic production in Northern Ireland.

INTRODUCTION
The Northern Ireland weather favours late blight, caused by *Phytophthora infestans*, in most years, and organic potato production requires varieties combining good blight resistance, agronomic characters and consumer acceptability. To aid this selection process, DARD's Plant Testing Station, Crossnacreevy, is carrying out field plot trials of selected varieties in association with Glens of Antrim Potatoes on the Greenmount Campus organic unit.

In supporting trials, DARD's Applied Plant Science Division, Newforge, is, in a non-organic situation, assessing varietal performance under high blight pressure through exposure to unsprayed plants inoculated with blight, and evaluation in copper oxychloride-treated and untreated plots.

CROSSNACREEVY TRIALS
MATERIALS AND METHODS
The potato varieties grown in randomized blocks with three replicates (2002, plots 4 drills x 50 tubers; 2003, plots 4 drills x 20 tubers) under organic management at Greenmount in 2002 and 2003 were Lady Balfour (2003 only), Milagro, Milva (2003 only), Orla, Remarka, Santé, Valor (2002 only). The objective was to produce maximum yields of 45-65 mm tubers by removing the haulms at the optimal stage.

Foliage blight was monitored throughout the season and copper oxychloride was applied as required. After defoliation and harvest, tubers were assessed for tuber blight, yield, defects, and cooking characteristics.

RESULTS

In 2002 Milagro had no foliage blight at all while Santé and Remarka were severely infected at haulm removal in mid-August. In 2003 blight was not detected on any of the varieties despite conducive weather conditions during early growth.

Marketable yields (45-85 mm) in 2002 (only healthy tubers were retained, rotted tubers were not weighed) varied between 8.0 t/ha in Valor and 20.2 t/ha in Milagro (Table 1). Marketable yields in 2003 were much higher, irradiance being much higher, and varied between 24.6 t/ha in Milagro and 33.0 t/ha in Milva. Defects accounted for between 5 and 23% of total yield in 2002 and between 7 and 25% in 2003, Milagro being worst in both

Table 1. Yield (t/ha) and defects (as % of total yield) of potato varieties under organic management

	2002				*2003*			
		Yield (t/ha)		*Defects**		*Yield (t/ha)*		*Defects**
	Total	*45-85mm*	*45-65mm*		*Total*	*45-85mm*	*45-65mm*	
Lady Balfour		Not in trials			42.0	30.7	27.7	14.4
Milagro	30.1	20.2	15.4	23.0	35.4	24.6	19.6	24.7
Milva		Not in trials			45.5	33.0	31.3	11.2
Orla	27.0	16.11	5.9	5.7	43.9	29.9	26.9	10.8
Remarka	20.1	12.1	11.8	9.1	35.8	24.9	18.4	24.2
Sante	26.1	15.7	15.7	5.4	40.6	31.7	27.4	6.8
Valor	18.0	8.0	8.0	5.2		Not in trials		

* % of total yield

CONCLUSIONS - CROSSNACREEVY TRIALS

Blight incidences were low and yields were acceptable in both years of the trial. However, defects reached substantial amounts in some varieties.

NEWFORGE TRIALS

MATERIALS AND METHODS

Trials were carried out in 2002 and 2003 at the Agriculture & Food Science Centre, Belfast to assess varietal performance under high blight pressure.

In 2002, eight cultivars (Milagro, Navan, Orla, Remarka, Santé, Valor, Axona and Tominia) were used and each was evaluated in both copper oxychloride-sprayed plots (5 applications of 1.25 kg copper/ha, 4 July - 29 August) and also in unsprayed plots. A split-plot design was used with fungicide treatments as main plots and cultivars as sub-plots in four randomized blocks. The unit plot consisted of two rows of eight tubers. Groups of four sub-plots were separated by single rows of unsprayed cv. Désirée in which plants were inoculated (24 July) with Northern Ireland isolates of *Phytophthora infestans*. Foliage blight was assessed weekly from 24 July to 4 September when foliage was destroyed. Tubers were lifted on 2 October; the yield from each plot was graded and assessed for tuber blight incidence.

In 2003, nine cultivars (Milagro, Milva, Orla, Lady Balfour, Remarka, Santé, Mira, Tominia and Axona) were evaluated in unsprayed plots only. A fully randomized block design was used with four replicates. The unit plot was three rows of 10 tubers and groups of three plots were separated by single rows of cv. Désirée inoculated (16 July) with Northern Ireland *P. infestans* isolates. Foliage blight was assessed twice weekly from 29 July to 9 September. The haulm was destroyed on 10 September and tubers lifted 24-25 September. Yield has not been assessed at the time of writing, but results are presented in the poster.

Also in 2003, additional plots of the above nine cultivars, plus Home Guard and Up-to-Date, were maintained free of foliage blight and used to provide tubers for direct assessment of tuber blight susceptibility. Two replicates of 20 tubers were lifted on 26 August and immediately inoculated by spraying with a sporangial/zoospore suspension of *P. infestans*. The tubers were incubated at 15°C and assessed for tuber blight incidence after two weeks.

RESULTS

Foliage infection 2002

Foliage blight reached *c.* 90-100% in the untreated plots of Navan, Orla, Valor and Santé by 4 September (Table 2). The untreated Remarka and Milagro developed significantly less foliage infection (75 and 63%, respectively on 4 September) and no blight was seen in Axona and Tominia. The ranking order was similar for the copper oxychloride-treated plots with the most severely infected cultivars (Navan, Orla, Valor and Santé) having 50-60% infection by 4 September.

Yield and tuber blight 2002

The extremely wet season resulted in a lot of bacterial soft rot in tubers, which made it difficult to ascertain how much rotting was initiated by tuber blight. The results should therefore be interpreted with caution. Copper oxychloride treatment gave little increase in yield in most varieties. Valor, Orla and Remarka produced the greatest yield of healthy tubers with very few rotted ones, despite Orla and Valor being among those with the most foliage blight.

Foliage blight 2003

By early September, the foliage of cvs.Orla, Santé, Milva and Remarka had been almost completely killed by blight. Infection of Milagro and Lady Balfour was significantly delayed and less, while no blight was observed in plots of Axona, Mira and Tominia (Table 2).

Yield and tuber blight 2003

Results are not available at the time of writing, but are presented in the poster.

Direct tuber inoculation 2003

Direct tuber inoculation gave the expected results for the susceptible standards Home Guard and Up-to-Date. Lady Balfour proved highly susceptible to tuber infection and *c.* 50% of tuber of cvs. Milva and Remarka were infected. Cvs.Santé and Milagro were less susceptible and only 10% of Orla tubers were infected. Only one tuber of cvs. Mira and Tominia and none of Axona developed blight (Table 2).

The results emphasize the need to assess tuber as well as foliar susceptibility when selecting cultivars suitable for organic production.

Table 2. Foliage blight development in 2002 and 2003 field trials and tuber blight in directly-inoculated tubers

Cultivar	Foliage blight (%)			Tubers infected (%)
	4 Sept 2002		5 Sept 2003	
	untreated	CuOCl-treated	untreated	
Lady Balfour	nt	nt	74.2	75.0
Milagro	62.5	4.7	85.4	32.5
Milva	nt	nt	100.0	52.5
Navan	99.4	50.0	nt	nt
Orla	97.5	61.2	99.6	10.0
Remarka	75.0	18.1	98.8	50.0
Santé	88.1	50.0	99.6	40.0
Valor	92.5	50.0	nt	nt
Axona	0.0	0.0	0.0	0.0
Mira	nt	nt	0.0	2.5
Tominia	0.0	0.0	0.0	2.5
Home Guard	nt	nt	nt	100.0
Up-to-Date	nt	nt	nt	100.0
L.S.D. ($P<0.05$)	17.06		6.50	12.48

nt = not tested

CONCLUSIONS - NEWFORGE TRIALS

In 2002, some cultivars developed less foliage blight than others when untreated, though blight levels were still high. Copper oxychloride had some effect on reducing foliage blight in 2002, though considerable infection was still evident.

In the directly-inoculated tubers in 2003 considerable variation in foliage blight infection was shown between cultivars, with some exhibiting extremely low levels of infection.

OVERALL CONCLUSIONS – BOTH TRIALS

Both trials showed that some cultivars exhibited excellent foliage blight resistance. However, initial results from both trials also showed that some varieties exhibited good resistance to tuber blight despite having foliar infection. This emphasizes the need to assess tuber as well as foliar susceptibility when selecting cultivars suitable for organic production.

Further work is needed over a number of seasons to evaluate blight resistance, effectiveness of fungicide treatment, consumer characteristics and husbandry requirements of varieties which may suit organic production in Northern Ireland.

Changes in Abundance and Diversity of the Weed Seedbank in an Organic Field-Scale Vegetable System: from Conversion through the First Course of a Rotation

R.J. TURNER
IOR-HDRA, Ryton Organic Gardens, Coventry, CV8 3LG, UK

ABSTRACT
Changes in the abundance and diversity of the weed seedbank was monitored over the course of conversion and first sequence of cash crops in an organic arable and vegetable rotation at HRI-Wellesbourne, UK. Three areas with different levels of fertility building and cropping length were investigated. There was an increase in viable weed seed numbers ranging from 54-495% in all areas at the end of one rotation. The rotations with the greatest proportion of cereals showed the highest increase in viable weed seed numbers. Two rotations maintained a constant (+/- 2 species) total weed species diversity over the course of the rotation. One rotation increased total weed diversity by eight species after the final crop with increases seen after the first and second cereal crops.

INTRODUCTION
The seedbank can be defined as 'the reserves of viable seeds present in the soil and on its surface' (Roberts, 1981). The weed seedbank is said to be the 'memory of the land' and its species abundance and diversity will reflect the previous cropping history (Buhler *et al.*, 1997). It will also contain the key to potential future weed problems. Weeds are seen as a major barrier for those considering conversion to organic farming (Davies *et al.*, 1997). Information and documentation on what happens to the weed seedbank during this change of farming system is required.

This paper presents the preliminary results of a six-year study investigating the changes in the weed seedbank during conversion from conventional arable to an organic system including field vegetables in the rotation. The weed seedbank of a field (Hunts Mill, Wellesbourne, Warwickshire, UK) has been monitored in three (0.8 ha) experimental areas (4-6) between 1996-2001, to establish if any of the practices or rotations used in conversion affect the composition of the weed seedbank. Results are presented in terms of changes in weed seed numbers abundance and diversity for an entire area at start and end of fertility building, then after the first, second and third crop. The results must be considered in the context that data have been summarized so there is no direct replication of rotations or statistical analysis and is therefore a descriptive document. Diversity data are presented in terms of the total weed seedbank which may mask/buffer the shorter term effects on the viable weed seedbank of the processes of conversion. This study is part of a DEFRA-funded project investigating 'Conversion to organic field vegetable production' (OF0126T). Further details are available from HDRA (2000).

MATERIALS AND METHODS
Three rotations are compared. The initial fertility building period and length of rotation varied; area 4, a 30-month grass/clover ley (5 years); area 5, an 18-month grass/clover ley (4 years) and area 6, a 6-month vetch crop (3 years) (Table 1). The three areas were further divided into six strips that have had slightly different rotations (e.g. area 4 1999, 3

strips were cropped with potatoes and 3 with cabbage) but for the purpose of this study data from a whole area have been combined. The basic rotations are given in Table 1.

Table 1. Rotations for areas 4, 5 and 6.

	1996	1997	1998	1999	2000	2001
Area 4	G/C ley			P/Ca	O/C/L	Sp. barley
Area 5	G/C ley		Sp. barley	O/C/L	Sp. barley	
Area 6	Vetch	Sp. barley	O/C/L	Sp. barley		

KEY: G/C= Grass/clover, P=Potatoes, Ca=Cabbage, O=Onions, C=Carrots, L=Leeks, all barley under sown with clover or grass/clover

Samples were taken at the start of conversion and after each crop. Each area (100.0 m x 80.0 m) was divided into six strips (16.7 m x 80.0 m). These strips were further sub-divided into quarters (16.7 m x 20.0 m) producing 24 samples per area. An auger (3 cm diameter) was used to take ten soils cores to a depth of 10 cm in each quarter. The soil was then air dried and analysed in 200 g sub sample portions using the wet sieving, flotation and filtering extraction methods as described by Roberts and Ricketts (1979). The numbers of 'apparently' viable and non-viable seeds per sub sample were identified and recorded.

RESULTS AND DISCUSSION

Weed species abundance in the soil seedbank

The average viable and total number of weed seeds per kg of dry soil is shown in Table 2. The table also shows the percentage that the viable weed seeds comprise of the total numbers. The baseline sample at the start of conversion and the sowing of the fertility building crops in August 1996 shows the total number of weeds to be very similar (range 106-109 per kg dry soil in the three areas).

Table 2. Average viable and total weed seed numbers per kg dry soil, and viable numbers expressed as a percentage of the total seeds at five sampling dates.

Area	Start of fertility build			End of fertility build			End of crop 1			End of crop 2			End of crop 3		
	Viable	Total	%	Viable	Total	%	Viable	Total	%	Viable	Total	%	Viable	Total	%
4	13	109	12	9	124	7	15	72	21	13	53	25	20	65	31
5	16	106	15	18	109	17	25	115	22	41	126	33	95	134	71
6	8	106	8	7	133	5	9	127	7	16	103	16	20	103	19

The rotation in area 4 was a 30-month grass/clover ley, two vegetable crops and a spring barley. There was a gradual increase in viable seed numbers (54% rise between start and end) and a decline in the total weed seedbank (40%) over the course of the 5 year rotation. Viable weed seed numbers declined during the grass clover ley and this the longest rotation had the smallest rise in viable seeds.

Area 5 had the highest number of viable weeds seeds (double that in area 6) from the start and throughout the rotation. Crop sequence was 18-month grass/clover ley, spring barley, vegetables and a second spring barley. The viable seed population rose considerably, by 495%, and the total by 27% between the start and end of the 4-year rotation. This is similar the findings of Albrecht and Sommer (1998) who recorded a

427% rise in total weeds germinated from the seedbank in conversion over a 3-year arable rotation. The main increase in seeds had been after the underwsown cereal crops. Work by Sjursen (2001) in Norway had found the lowest seedbank levels after the first crop, a cereal after a 3-year ley. The main increase in this study was after the second cereal where bird damage in the crop had left open canopy areas and weeds, particularly *Matricaria* spp (mayweed), had flourished.

Area 6 had the shortest (3 years) and most intensive cropping sequence, a 6-month vetch, spring barley, vegetables, and spring barley showed a gradual increase in viable seeds (150% rise) and a similar level of total weed seed number (3% decline). This more frequently cultivated area had the potential to stimulate germination and decrease the weed seedbank although any weeds left to mature would soon counterbalance the seed reduction (Roberts, 1981). The fallow periods (16% over the rotation) may have increased viable seed shed (Zwerger *et al.*, 1993).

In all three rotations the fertility building phase kept weed seed numbers at a low level. The 30-month grass/clover ley (area 4) and the 6 month overwinter vetch (area 6) actually showed a decrease in viable seeds showing the importance of leys and dense suppressing cover crops on weeds. This is consistent with work by Younie *et al.* (2002) that found the higher the proportion of grass/clover ley in the rotation the lower the weed seedbank levels.

The rotations with the greatest proportions of cereals (5 and 6) had the highest increase in viable weed seeds; this was consistent with Albrect and Sommer (1998) and Sjursen (2001) and supports the general conclusion that increased proportions of annual cropping increase the weed seed bank.

Weed species diversity in the soil seedbank

Table 3 shows the total species diversity at intervals in the rotation. At the start of conversion total species diversity was 15 in area 4 and 6, and 13 in area 5. During the course of the rotation area 4 and 6 remained fairly stable (+/- 2 species). Area 5 ranged between 9-21 species dependent upon cropping, the greatest diversity recorded after the second cereal crop at the end of the rotation.

Table 3. Total weed species diversity in the seedbank at five sampling dates.

Area	Start of fertility build	End of fertility build	End of crop 1	End of crop 2	End of crop 3
4	15	13	14	16	14
5	13	12	19	9	21
6	15	14	13	15	15

Area 4 showed little change in species composition, *Chenopodium album* (fat hen) and *Viola arvensis* (pansy) occupied 85% or more of the seedbank through the rotation, dominance of fat hen actually increasing to 70% by the end. At the start, area 5 was dominated by fat hen and pansy (together 81% of the seedbank), at the final sample date the species were much more equally distributed, Mayweed now being the most abundant seed (30%) and six species making up 85%. In area 6, fat hen and *Papaver* spp. (poppy) comprised 85% of the seedbank, this dominance reduced to 66% by the end.

The seedbank in these areas is typical of those associated with cropped systems of Northern Europe, a few species dominating, and a wider range making up a low proportion (Roberts, 1981; Chauval *et al.*, 1989). Over the time period monitored there has been no specific shift in weed type. The grass clover phase has not favoured annual grass weeds as suggested by Roberts and Chancellor (1986) and Davies *et al.*, (1997). When only viable weed species diversity and separate rotation replications are considered a clearer pattern may emerge.

The greatest diversity and more equitable seedbank has been seen in area 5 as a consequence of a bird damaged barley crop allowing establishment and seed shed of a range of weed species.

CONCLUSIONS

A rise in viable weed seed numbers and the proportion of the weed seedbank occupied by viable weed seeds has been shown to increase following conversion to an organic arable/field vegetable system after one rotation. Cereal crops had greatest effect on viable seed increase, but also increased the diversity and reduced the dominance of the seedbank by individual weeds. Further analysis and statistical testing of the data set is required.

ACKNOWLEDGEMENTS

We thank DEFRA for funding this research (Projects OF0126T and OF0191).

REFERENCES

ALBRECHT H. and SOMMER H. (1998) Development of the arable weed seed bank after the change from conventional to integrated and organic farming. *Aspects of Applied Biology*, 51, 279-288.

BUHLER D.D., HARTZLER R.G. and FORCELLA F. (1997) Implications of weed seedbank dynamics to weed management. *Weed Science*, 45, 329-336.

CHAUVAL B., GASQUEZ J. and DARMENCY H. (1989) Changes of the weed seed bank parameters according to species, time and environment. *Weed Research*, 29, 213-219.

DAVIES D.H.K., CHRISTAL A., TALBOT M., LAWSON H.M. and McN WRIGHT G. (1997) Changes in weed population in the conversion of two arable farms to organic farming. *Proceedings of the Brighton Crop Protection Conference 1997* 9C-2 pp 973-978.

HDRA (2000) Conversion to Organic Field Vegetable Production Report. Coventry: HDRA.

ROBERTS H.A. and RICKETTS M.E. (1979) Quantitative relationship between the weed flora after cultivation and the seed population in the soil. *Weed Research* 19, 269-275.

ROBERTS H.A. (1981) Seed Banks in Soils. *Advances in Applied Biology*, 6, 1-54.

ROBERTS H.A. and CHANCELLOR R.J. (1986) Seed banks of some arable soils in the English Midlands. *Weed Research* 26, 251-257.

SJURSEN H. (2001) Change of the weed seed bank during the first complete six-course crop rotation after conversion from conventional to organic farming. *Biological Agriculture and Horticulture* 19, 71-90.

YOUNIE D., TAYLOR D., COUTTS M. MATHESON S., WRIGHT G. and SQUIRE G. (2002) Effect of organic crop rotations on long-term development of the weed seedbank. *Proceedings of the UK Organic Research 2002 Conference , 26th -28th March,* Aberystwyth. pp 215-220.

ZWERGER P., LECHNER M. and HURLE K. (1993) Does rotational fallow cause weed problems in subsequent crops. *Proceedings of Brighton Crop Protection Conference 1993* 4B-5, pp 299-304.

Organic Carrot Production: a Weed Management Perspective

R.J. TURNER[1] and A.C. GRUNDY[2]
[1]IOR-HDRA, Ryton Organic Gardens, Coventry, CV8 3LG, UK
[2]HRI-Wellesbourne, Warwick, CV35 9EF, UK

ABSTRACT
Weed management is a major technical constraint to organic carrot production. Weed control is notoriously difficult in this relatively uncompetitive, but high value, cash crop. Two field experiments (1999 and 2000) were carried out to evaluate weed management strategies in maincrop carrots. Trial treatments included cultural, mechanical and thermal control methods. The trial site had a low inherent weed population and weather conditions favoured the carrot crop over the weeds in these two seasons. Results suggested that pre-emergence flaming was a cost effective technique and that weeding only once at three or five weeks after 50% emergence was sufficient to prevent yield loss due to weeds. No significant difference between weeding implements in terms of carrot yield could be confirmed. A separate investigation (2001) was made into the efficacy of a steerage hoe, brush weeder and flame weeder on three weed types at two growth stages. Steerage hoeing proved to be the most effective weed control method when considered in terms of the three weed species tested. The flame weeder was effective on *Chenopodiun album* and *Stellaria media,* but had virtually no effect on *Poa annua*. There was some re-growth after flaming, particularly of *S. media*. The brush weeder was not effective when used in this hard dry soil. There was little effect of growth stage on weeder efficacy at the two relatively large growth stages tested for each species.

INTRODUCTION
Carrots are a particularly difficult crop to manage in terms of weed control (Litterick, 1999). The crop is sensitive to poor seedbed conditions, slow to germinate and only reaches canopy closure towards the end of the season (Peacock, 1991; Tamet *et al*., 1996; Baumann, 2001). Carrots form an important component of organic crop rotations as a high value cash crop and are promoted as a healthy dietary component (Radics *et al*., 2002). The organic sector is gradually increasing their retail share of the market. In the 2000/01 season, 457,300 tonnes of carrots were produced in the UK (DEFRA, 2002); the Soil Association estimated that 5,000 tonnes were retailed organically (Soil Association, 2001). In 2002, 9,000 tonnes of British organic carrots were marketed; an increase of 44% on the previous year and a market value of £2,475,000. This represented only 65% of total sales, with the remainder of the crop being imported (Firth *et al*., 2003), highlighting the opportunity for expansion of the home-grown British market.

Weed control is a major concern of organic vegetable growers and a research priority for the UK organic governing body. There has been little published work on practical weed control programmes for organic growers and this study aims to address weed control strategies in maincrop carrots. Organic growers have to rely on a sound rotation and effective cultural measures to form the basis of their weed control strategies. Optimization of weed control timing and equipment is vital to ensure crops are kept weed-free during the critical period of competition. This period needs to be identified so that effective weed removal can be targeted.

MATERIALS AND METHODS

The objective for the carrot trials work was to consider a range of weed management programmes comprising three main treatment effects. In 1999 the treatments were pre-emergence flaming (+ or -), time of weeding (3, 5 or 7 weeks after 50% emergence) and method of weed control (hand, brush weeded or steerage hoed). In 2000 pre-emergence flaming was used across the whole trial site, time of weeding was refined (3 or 5 weeks after 50% emergence or weeded twice at 3 and 5 weeks after 50% emergence) and method of control was updated to include a brush weeding and steerage hoe treatment. Two more treatments were added where brush weeded and steerage hoed plots were also hand weeded in the crop rows. The trials were a fractional factorial design, with two replicates in 1999 and four in 2000. Weekly assessments of weed flora were made with quadrats identifying, counting and estimating percentage cover of each weed monitored in and between the crop rows. Destructive assessments of the crop were taken at intervals through the season. The weight of weeds in and between the crop rows at harvest was also established by destructive quadrat sampling. Number and weight of carrot roots per plot were recorded at harvest.

In 2001 an experiment was designed to investigate efficacy of the steerage hoe, brush weeder and a flame weeder on *Poa annua*, *Chenopodium album* and *Stellaria media* at two different growth stages. The trial was a split-plot design with four replicates. Weed seedlings were raised and transplanted to the field with a sample area of 46 plants per plot. Machinery was adjusted to cover the entire test area and used at two dates to coincide with two contrasting, but relatively large, growth stages for each weed species. Growth stages were species specific, but >6 true leaves for the early weed and up to 40 true leaves for the later treatment. Percentage survival was assessed 2 and 9 days after treatment. Analysis of variance (ANOVA) was performed using Genstat 5 in all 3 years.

RESULTS AND DISCUSSION

Optimization of weed management programmes in organic carrots - 1999

There were no significant differences between final carrot weight for flamed or non-flamed plots. Without flaming the yield was 36.2 t/ha, with flaming, 37.7 t/ha. A difference of 1.5 t/ha was worth around £350/ha to the grower (based on wholesale values from Lampkin *et al.*, 2002). The cost of flaming can be estimated at £69/ha (Lampkin *et al.*, 2002) and would appear to be an important measure to include in a weed control programme, but would need further investigation as the differences were not statistically confirmed in this trial.

The data suggests that leaving weeding to the seventh week after 50% emergence would not be advisable even with such a low weed pressure. Weeds had reached 6 true leaves and were more difficult to remove or kill by mechanical means. This was seen with *Viola arvensis* where, after weeding, there was still a significantly greater area covered than on the earlier weeded plots. The destructive assessment samples showed that the fresh weight of carrots on the 7-week weeded plots was always slightly lower than the earlier weeded plots. This was confirmed at harvest when there was a significantly lower total number and weight of carrot roots ($P < 0.05$).

Weed assessments showed that the hand weeded plots tended to have a lower weight of weed, an increased percentage crop cover and an increased amount of bare ground between the crop rows in comparison to the mechanically weeded plots. However, the weed levels did not appear to have any direct effects on yield related to the method of weeding. The within-row weed, which was not removed by the mechanical options, did not significantly affect the final marketable yields, which were similar for the three weeding methods.

Optimization of weed management programmes in organic carrots - 2000

There was no clear advantage of weeding twice and there was no significant impact of timing i.e. choosing between 3 or 5 weeks after 50% emergence on percentage weed cover or harvest yield in this trial. A general pattern emerged from the weed percentage ground cover assessments that plots weeded twice tended towards less weed, although this was not significantly lower suggesting the single weeding was as effective in this season.

There was no significant impact of the steerage hoe or the brush weeder on final carrot yield. The brush weeded plots tended to have higher yields, but this was not statistically proven. There was also no beneficial effect of keeping the plots weed-free throughout the season, above a single or twice-weeded removal strategy. A once-weeded system proved as effective as two operations, there was no statistical significance between the two pieces of machinery tested or between plots weeded or not weeded in the crop row. In this experiment it would appear that the cultural control measures taken across the whole trial site (a stale seedbed and pre-emergence flaming) were very effective in this season.

The weather conditions in both trial years favoured the crop over the weeds; initial moisture when drilled to allow the crop to germinate and establish followed by dry weather preventing a large flush of weed emergence. The typically late sowing of the trial (early June) to avoid carrot fly had a positive impact on weed control and the main flush of spring germinating weeds was avoided. This coupled with ideal weather conditions for control produced poor coverage of indigenous weed flora on which to test the factors of weed control being investigated. The results produced in this year are typical of a low weed pressure situation or dry season and can form the basis of recommendations for such scenarios. Where weed pressure is low it can be seen that a ‘belt and braces’ approach is not necessary and would not be cost effective. For example, weeding once would be sufficient to avoid crop losses due to weeds rather than doubling the costs of mechanical weeding by performing the operation twice. These trials have emphasised that if cultural measures are employed effectively, one mechanical operation is sufficient to control weeds and getting it right is possible.

Efficacy of weeding implements on weed type and growth stage - 2001

Immediately after treatment, all of the physical methods employed had significantly reduced the percentage survival of the weeds as compared with the unweeded control ($P < 0.05$). Steerage hoeing was the most effective method leaving 33% of weeds, flame weeding the second most effective, reducing weed survival to 38% and brush weeding the least effective with 74% of weeds surviving treatment. The flaming, and the steerage hoeing were significantly more effective than the brush weeder ($P < 0.001$). Nine days after treatment, the plots were re-assessed in the same treatment area. The main effect was recovery of plants on plots that had been flamed, now having 48% survival as compared with 38% immediately after treatment.

At the first assessment immediately after weeding the treatments made at the earlier weed growth stages were significantly more effective than those made at the later growth stages ($P<0.01$). With the subsequent weed recovery over the next week the efficacy of the treatments were not significantly different by the second assessment.

The steerage hoe was the only weeder to have any real effect on reducing *P. annua* numbers (to 29% survival with all others treatments above 93%). The flame weeder was particularly effective on *C. album* and *S. media* (both controlled to about 9% survival). The hoe had a fairly similar effect on all species causing a reduction in weed survival of around 70%, regardless of weed type, habit and the size of weed. The weather conditions at the two treatment dates favoured the steerage hoe. It was hot and dry with a hard and crusted soil surface that the steerage hoe was able to penetrate and move freely through. The brush weeder was not able to work the hard soil surface failing to really uproot weeds. Due to the well established state and size of weeds the leaf stripping effect of the brushes only caused a set-back, rather than a kill effect. This was also found by Radics *et al.,* (2002) investigating different combinations of weed management methods in organic carrots: in Hungary in 2001, a particularly dry summer, the brush weeder was ineffective against annual weeds and greater attention to timing was required as compared to a weed cultivator. Pedersen (1990) also found that under very dry conditions the conventional hoe had a relatively better effect on the weeds than the brush weeder.

ACKNOWLEDGEMENTS

This work was part of studentship funded by HDRA, HRI and Coventry University. Thanks go to DEFRA who funded the 'Conversion to organic field scale vegetable production' (Projects OF0126T and OF0191) at which the trial site was located. Andrew Mead at HRI-Wellesbourne is thanked for his statistical assistance.

REFERENCES

BAUMANN D.T. (2001) *Competitive suppression of weeds in a leek-celery intercropping system - an exploration of functional biodiversity.* PhD thesis, Wageningen University, The Netherlands.

DEFRA (2002) Basic Horticultural Statistics 2001. Available from http://statistics.defra.gov.uk/esg/publications/bhs/2001/veg.pdf. Accessed 4/6/03.

FIRTH C. GEEN N. and HITCHINGS R. (2003). The UK Organic Vegetable Market (2001/02). *Report to DEFRA Project no. OF0307*, January 2003.

LAMPKIN N. MEASURES M. and PADEL S. (2002) *20002/03 Organic Farm Management Handbook.* 5th Edition. University of Wales, Aberystwyth and Elm Farm Research Centre, Newbury, UK.

LITTERICK A. (1999) Weed Strategies. *Grower*, April 1 1999. 131 (13) pp 20.

PEACOCK L. (1991) Effect on weed growth of short-term cover over organically grown carrots. *Biological Agriculture and Horticulture,* 7, 271-279.

PEDERSEN B.T. (1990) Test of the multiple row brush hoe. *Proceedings of the 3rd International Conference on Non-Chemical Weed Control*, Linz, pp 109-125.

RADICS L. GÁL I. and PUSZTAI P. (2002). Different combinations of weed management in organic carrot. *Proceedings of the 5th EWRS Workshop on Physical Weed Control*, Pisa, Italy, pp 137-146.

SOIL ASSOCIATION (2001) *The Organic Food and Farming Report 2001*. Bristol: Soil Association.

TAMET V. BOIFFIN J., DÜR C. and SOUTY N. (1996) Emergence and early growth of an epigeal seedling (*Daucus carota* L.): influence of soil temperature, sowing depth, soil crusting and seed weight. *Soil and Tillage Research,* 40, 25-38.

Soil Fertility – Changes during the Conversion Process

FRANCIS RAYNS and PHIL SUMPTION
HDRA-IOR, Ryton Organic Gardens, Coventry, CV8 3LG, UK

ABSTRACT
This paper describes changes in key parameters measuring aspects of soil fertility that have occurred throughout the conversion period and beyond (1996 to 2003) on ten commercial farms. The farms all produced field vegetables but otherwise they reflected a range of farming systems on varied soil types throughout England. There were very few changes in organic matter levels. The importance of the fertility building crops for providing nitrogen has been increasingly recognized by the farmers but there have been notable examples of deficiencies. Available phosphorus (P) and potassium (K) remained relatively stable overall, suggesting that the inputs of soil amendments (usually FYM) were broadly satisfactory. Problems with soil structure, particularly compaction, were apparent on several farms. Sometimes this was due to pre-existing problems which had been masked by the use of conventional fertilizers.

INTRODUCTION
Soil fertility is of fundamental importance in organic agriculture. Farmers considering conversion are often concerned about the transition from a system in which nutrients are supplied in a soluble form as fertilizers to one in which there is a much greater emphasis on the use of fertility building crops with limited use of soil amendments. Regular sampling and analysis is often advocated as the best method of monitoring fertility but routine analysis may not give the full picture, particularly with respect to the physical and biological aspects of the soil.

The work described in this paper was part of a wider project to monitor all agronomic and economic aspects of conversion to organic field vegetable production (HDRA 2000). Data was collected from commercial farms making their own management decisions in the light of 'best practice' management advice. The period of study covered the two year conversion phase itself followed by at least three years of organic cropping.

METHODS
Samples were taken from ten commercial farms in England beginning at the start of their conversion from conventional agriculture. The farms were selected to reflect a range of soil types and systems (focusing on field vegetable production in both stockless and livestock/arable rotations). Details of fertility building crops, soil amendments and cash crop yields were also recorded.

Soil from selected fields was sampled in March each year (between 1997 and 2003). At least sixteen soil cores (0-30 cm) were taken with an auger and pooled before being analysed by NRM Ltd (Bracknell). The analysis methods were those of Elm Farm Research Centre (EFRC, 1985); these were used since many organic farmers have their soils tested in this way and much of the relevant advice is tailored to interpretations of these results. Phosphorus (P) was measured by four different methods. Olsen P (bicarbonate extraction) is the standard ADAS technique and is suitable for all soils. The three acid extractants also used (acetic acid, lactic acid and citric acid) are not suitable for alkaline soils. Potassium (K) and magnesium (Mg) were measured using lactic acid extractions.

Visual observations were also made of soil structure and of crop growth.

RESULTS

Table 1 gives general information about soils on the selected farms together with some of the key results from the most recent soil sampling.

Table 1. Data relating to soil fertility on nine recently converted farms; samples were taken in spring 2003. The status of organic matter, P and K is based on the threshold values appropriate for the soil type that are used by Elm Farm Research Centre. The figures in brackets refer to the percentage of fields in each category on each farm.

Farm	Location	Soil type	Organic matter	Available P	Available K
1	S. Lincs.	Silty loam	OK (13) Low (87)	OK (100) Low (none)	OK (none) Low (100)
2	S. Lincs	Silty loam	OK (50) Low (50)	OK (100) Low (none)	Good (100) Low (none)
4	Warks	Clay loam	OK (20) Low (80)	OK (20) Low (80)	OK (20) Low (80)
5	N. Lincs	Sandy loam - peaty	Good (100) Low (none)	OK (9) Low (91)	OK (64) Low (36)
6	Cornwall	Loam l	OK (90) Low (10)	OK (none) Low (100)	Good (40) Low (60)
7	Devon	Clay loam	OK (39) Low (61)	OK (8) Low (92)	OK (8) Low (92)
8	Notts.	Sandy	OK (none Low (100)	OK (38) Low (62)	OK (63) Low (37)
9	Lancs	Loam/peat.	Good (91) Low (9)	OK (33) Low (67)	Good (92) Low (8)
10	Beds.	Sandy	OK (0) Low (100)	OK (100) Low (none	Good (100) Low (none)

Organic matter levels varied greatly between farms reflecting the variations in soil type, e.g. one site was on peaty soil with organic matter levels as high as 30%, another site on a sandy soil had only approximately 1.5% organic matter. On the individual farms there has been little change in organic matter levels since conversion began.

On most sites pH was managed by applications of lime; our data shows a lot of annual fluctuation because individual fields were limed at different times related to the occurrence of particular stages of the rotation. The target pH is the same in conventional and organic agriculture although some of the quicker acting lime amendments are not permitted under an organic management.

There was no overall general pattern for P availability. The initial values were very different between sites and six of the ten farms had fields considered (according to standard Elm Farm interpretations) to be low in phosphorus. On some farms there was a slight decline in levels over the period of study, whilst in others there was an increase in P status.

Initial levels of K availability were also variable between the various farms. Through the period of study, levels of available K tended to decline at all sites. At the most recent sampling, seven of the ten farms had fields low in potassium (according to standard Elm Farm interpretations). Magnesium levels were generally stable – low levels were only measured at two of the farms.

Six farms had high levels of at least one trace element (manganese, iron, copper or zinc). This was usually a reflection of basic rock type. Trace element availability can also indicate certain soil problems e.g. high manganese levels can be linked to compaction. Soil structural problems were seen on several of the farms. Compaction was associated with harvesting of vegetables in wet conditions (as dictated by marketing considerations) and also occurred on some sites as a result of repeated mowing of fertility building leys.

Nitrogen (N) is most commonly the limiting nutrient in organic farming but there is no routine test available to predict how much N may be made available during the growing season. Organic farmers often use a N budgeting approach to see if the fertility–building crops within the rotations will sustain the cash cropping. In practice this approach then relies on the fertility building crops performing as well as expected. Adverse weather can make it difficult to establish cover crops; other farm operations which give a direct financial return may also be prioritized ahead of the management of fertility-building crops. Often these crops are new to conventional vegetable farmers and in this study some had problems with their management (for example not mowing leys often enough). We also noted relatively poor utilization of winter green manures; sometimes these were established too late in the season to fulfil their full potential. The two farms on the lightest soils had the greatest problems with nitrogen availability. The situation was improved by the composting of FYM in one case and the application of pelleted chicken manure in the other.

DISCUSSION

Organic matter levels are widely considered to be higher under organic than under conventional management. However, this study showed very little change in organic matter status after the start of conversion. This is not surprising in view of the relatively short period of study and also because, depending on rotation, organic matter inputs are not necessarily greater than in conventional agriculture – lower yields also mean less crop residues are returned to the soil. Gosling and Shepherd (2002) also found that that, in paired organic and conventional farms, soil organic matter levels were very similar.

Regular analysis of available P and K in the soil showed little change overall through the period of study. This would suggest that there is no general 'mining' of soil reserves by organic farming. This is not unexpected since almost all the fields monitored (except on Farm 1 where available P and available K both declined slightly) received amendments of acceptable inputs. The picture is a complicated one as many of these materials provide several nutrients (as well as being a source of bulky organic matter). The most common input was FYM, either produced on the same farm or bought into the holding. In addition one farm used rock phosphate, one farm Silvinite (a source of K) and two farms pelleted chicken manure. It is necessary to demonstrate a need for such materials in order to obtain a derogation from the certifying body before they may be used by organic farmers.

Organic standards place a lot of emphasis on good soil management but problems with the soil physical condition were often seen. These were almost certainly present before conversion, but the poor growth which they caused was masked by the use of conventional farming practices. These problems were more serious on the larger farms where big machinery was used for cultivation and harvesting. Since organic farming relies much more on a healthy soil to supply nutrients to the crops it is important that such issues are addressed.

CONCLUSIONS

On the ten farms monitored there were relatively few changes in soil fertility during the conversion period and in the years immediately afterwards. Although there were specific problems, most farmers were able to develop strategies to address them. However, it is too early to detect the effect of the new regime on many soil properties and it is important that regular assessments continue to be made so that management can be modified in the light of any changes. Six of the farms in this study will continue to be monitored within new DEFRA-funded projects.

ACKNOWLEDGEMENTS

We thank DEFRA for funding this research (Projects OF0126T and OF0191).

REFERENCES

GOSLING P. and SHEPHERD M.A. (2002) Theory and reality of soil fertility – organic matter. *Proceedings of the UK Organic Research 2002 Conference, Aberystwyth*, 2002, pp 137-138.

EFRC (1985) *Elm Farm Research Centre Soil Analysis Service*. EFRC Research Notes Number 4.

HDRA (2000) Conversion to organic field vegetable production. Progress report to DEFRA on the first four years of work on projects OF0126T and OF0191.

Conversion Strategies for Stockless Organic Systems

D.L. SPARKES, S.K. HUXHAM and P. WILSON
University of Nottingham, Sutton Bonington Campus
Loughborough, Leicestershire, LE12 5RD, UK

ABSTRACT

Legume-containing leys are used to improve soil fertility in the two-year organic conversion period. While in-conversion land may be grazed, it is effectively out of production in stockless farming systems, potentially resulting in a reduction in income and pressure on cash flow. The effects of seven conversion strategies on a subsequent organic winter wheat crop were investigated on a sandy loam and a clay loam soil. The strategies were: 1) two-years' red clover-ryegrass green manure; 2) two years' hairy vetch green manure; 3) red clover for seed production then a red clover-ryegrass green manure; 4) spring wheat undersown with red clover, then a red clover green manure; 5) spring wheat, then winter beans; 6) spring oats, then winter beans; and 7) spring wheat undersown with red clover, then a barley-pea intercrop. Conversion strategies had a significant impact on organic wheat yield, which ranged from 2.8 to 5.3 t/ha. This was attributed to differences in organic wheat establishment, caused by variation in soil structure due to the different conversion strategies. Organic wheat yield was not related to weed abundance or soil mineral nitrogen.

INTRODUCTION

Organic farming is largely limited to mixed farming systems which use fertility- building leys and animal manure to enhance soil fertility. However, many UK farms are exclusively arable and it would not be feasible for these units to incorporate animals into the system in order to convert to organic production (Lampkin, 1990). If these farms are to convert to organic status then stockless organic systems need to be developed. Moreover, as the number of livestock-based organic farms continues to grow there will be an increasing market for organic cereals for livestock feed.

Organic farming systems are more dependent on the structure and nutrient status of the soil than conventional systems. In a stockless organic rotation, the system relies on biological nitrogen fixation and incorporation of crop residues to replenish nutrients taken off when crops are harvested. Therefore, fertility-building and fertility exploitative phases of the rotation must be carefully balanced to achieve a sustainable system. Work at Elm Farm Research Centre (Bulson *et al.*, 1996) has shown that stockless organic systems, using red clover as a green manure to build fertility, can be economically and agronomically viable. They also highlighted the importance of rotation design to maintain agronomic performance. However, work to date has focused on the rotation once organic status has been achieved rather than the optimum choice of cropping during the conversion period.

MATERIALS AND METHODS

A large-scale field experiment was established at Bunny Park Farm, University of Nottingham. The experimental area (approximately 1.5ha) was located within a 20ha block of land that was entered into organic conversion on 1 August 1999. A red clover/ryegrass mix was sown on the entire area on 4 September 1999, to prevent nitrate leaching over the winter period and to form the starting point for the different conversion strategies. The experimental plots (approximately 30m x 12m) were laid out in a randomized block design with four replicates. The soil texture on the experimental site fell into two broad categories of sandy loam (blocks 1 and 2) and clay loam (blocks 3 and 4), and was in the Dunnington Heath Association. In March 2000, spring-sown crops identified for year 1 of a conversion strategy were established (Table 1).

Table 1. The conversion strategies were two-year cropping sequences, all followed by wheat in the third year. † A cut and mulched green manure; u/s = undersown with red clover.

Strategy	**First Year Conversion**		**Second Year Conversion**
RCRC	† red clover-ryegrass	-	† red clover-ryegrass
VEVR	† hairy vetch	-	† hairy vetch-rye (cv Motto)
CSRC	red clover (seed) -ryegrass	-	† red clover-ryegrass
UWRC	u/s spring wheat (cv Paragon)	-	† red clover
WHBE	spring wheat (cv Paragon)	-	winter beans (cv Clipper)
OABE	spring oats (cv Solva)	-	winter beans (cv Clipper)
UWBP	u/s spring wheat (cv Paragon)	-	spring pea (cv Agadir) - spring barley (cv Static)

N.B. WHBE, OABE and UWBP are unlikely to be acceptable to certification bodies, but were included for experimental purposes.

In the third year of the experiments, a crop of winter wheat (cv. Hereward) was established across the entire experimental area to assess the effect of the different strategies on crop yield and quality once organic status had been achieved. Soil structure was assessed in March 2002 by inserting a sharp spade vertically into the soil to its full depth, levering out the soil and making a visual examination. Structure is scored on a scale of 1-10 where 1 = very big clods, smooth dense crack faces, reducing conditions, roots only in cracks; 10 = entirely porous crumbs. In addition, soil physical and chemical properties were monitored throughout the conversion period and first organic cropping season.

RESULTS AND DISCUSSION

There was a significant effect of conversion strategy on organic wheat grain yield (Table 2; P=0.002). Organic wheat following the three 'clover strategies', (containing a red clover green manure in the second year: RCRC, CSRC and UWRC) had the highest yields. These 'clover strategies' all produced greater yields than the UK average for organic wheat of 4 t/ha (Lampkin *et al.*, 2002). The remaining strategies produced yields below this average. Organic wheat yield following RCRC was similar to, or slightly lower than, those attained in other stockless organic rotation experiments following two years' red clover-ryegrass: 5.2 t/ha, 6.0 t/ha (Cormack, 1996; Stopes *et al.*, 1996). Organic wheat yield following RCRC was also similar to the yield achieved by the variety Hereward, 5.6 t/ha, when grown organically after one years' red clover-ryegrass green manure (Thompson *et al.*, 1993). Yields of wheat after winter beans compared favourably with previous reports of 4.05 t/ha (Bulson *et al.*, 1996).

Greater yields were produced on sandy soil than on clay soil. Although this difference in yield between soil textures was not significant (P=0.064), there was a significant interaction between soil texture and conversion strategy (P=0.024), such that wheat following WHBE and OABE performed much worse on the clay soil than sandy soil, compared to the other strategies. Wheat following WHBE and OABE had the most variable yields of all strategies, while wheat following CSRC had the least variable yield. The differences in yield between conversion strategies were mainly caused by variation in plant population. Establishment was poor at the 'clay' end of the experiment, particularly following the WHBE and OABE conversion strategies. This was related to the soil structure on those plots which, together with soil texture explained 89% of the variation in plant population (Figure 1).

Table 2. Combine-harvested grain yields (t/ha) (85% DM) of organic wheat crop as affected by conversion strategy and soil texture.

Conversion strategy	**Sand**	**Clay**	**Mean**
RCRC	6.07	4.49	**5.28**
VEVR	4.78	2.18	**3.48**
CSRC	5.14	3.92	**4.53**
UWRC	4.99	3.55	**4.27**
WHBE	5.11	1.05	**3.08**
OABE	5.64	1.12	**3.38**
UWBP	3.90	1.74	**2.82**
Mean	**5.09**	**2.58**	
	P value	SED	df
Conversion strategy	0.002	0.475	12
Soil texture	0.064	0.665	2
Conversion strategy x soil texture	0.024	0.911	6.24

Conversion strategy had a significant effect on soil mineral nitrogen (SMN) available to the wheat crop but, once the differences in plant population were accounted for, SMN did not explain a significant proportion of the variation in yield.

There were large differences in weed density within the organic wheat crop, but there were no significant differences between conversion strategies. This was surprising given the significant differences observed during the conversion period. There was no significant relationship between weed density and yield; thus weed density either had no impact on wheat yield, or the overriding effects of plant population masked any such impact.

CONCLUSIONS

Organic wheat yield following the seven conversion strategies ranged from 2.8-5.3 t/ha and was closely related to differences in plant population. Regression analysis showed that the variation in plant population could be explained by differences in soil structure created by the conversion strategies. Surprisingly, neither soil mineral nitrogen or weed density were related to organic wheat yield suggesting that, in this experiment, plant population was the main factor limiting yield, not nitrogen or weed competition.

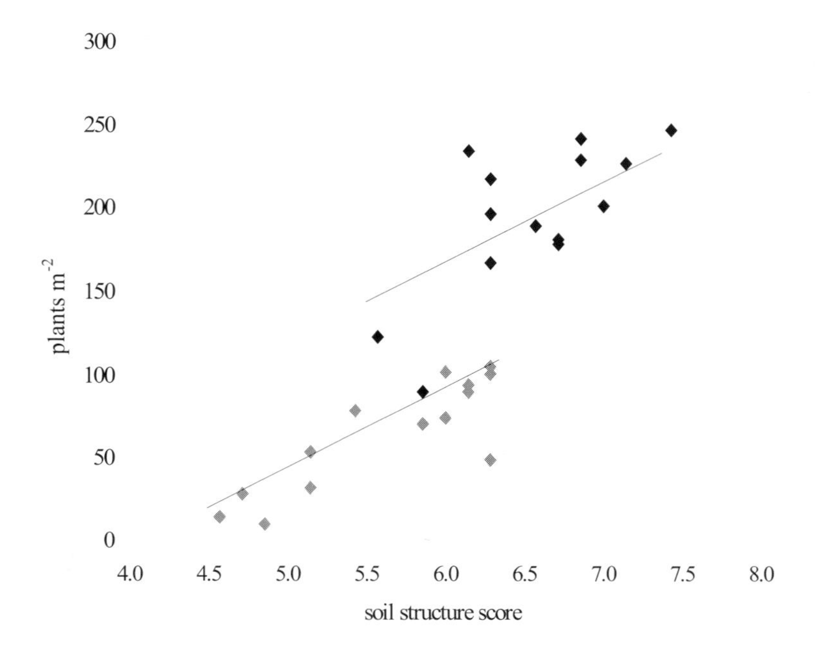

Figure 1. The relationship between soil structure, soil texture and establishment. Sand (◆), clay (◆). Curves fitted: Establishment = 52.42 x soil structure score - 82.0 x soil texture -148.6 ($R^2 = 0.89$; $P < 0.001$).

REFERENCES

BULSON H.A., WELSH J.P., STOPES C.E. and WOODWARD L. (1996) Agronomic viability and potential economic performance of three organic four year rotations without livestock, 1988-1995. *Aspects of Applied Biology*, 47, 277-286.
CORMACK W.F. (1996) Effect of legume species on the yield and quality of subsequent organic wheat crops. In: Younie D. (ed.) *Legumes in Sustainable Farming Systems, Occasional Symposium of the British Grassland Symposium* No. 30, pp. 126-127.
LAMPKIN N. (1990) *Organic Farming*. Ipswich: Farming Press.
LAMPKIN N., MEASURES M. and PADEL S. (2002) *Organic Farm Management Handbook*. 5th edn. University of Wales, Aberystwyth and Elm Farm Research Centre, Berkshire.
STOPES C., MILLINGTON S. and WOODWARD L. (1996) Dry matter and nitrogen accumulation by three leguminous green manure species and the yield of a following wheat crop in an organic production system. *Agriculture, Ecosystems and Environment*, 57, 189-196.
THOMPSON A.J., GOODING M.J. and DAVIES W.P. (1993) The effect of season and management on the grain yield and breadmaking quality of organically grown wheat cultivars. *Aspects of Applied Biology*, 36, 179-188.

Modelling Cereal Ideotypes for Optimizing Weed Suppression in Organic Farming

C.F.E. TOPP[1], D.H.K. DAVIES[2], S. HOAD[2], P. MASKELL[2]
[1] Land Economy, Research Division, SAC Edinburgh, West Mains Road, Edinburgh, EH9 3JG, UK
[2] Crops and Soils, Research Division, SAC Edinburgh, West Mains Road, Edinburgh, EH9 3JG, UK

ABSTRACT
Weeds impact on the growth and development of cereal crops, and hence on the yield achieved and crop quality. The growth and development of the weeds within the cereal crop is partially dependent on the light intercepted by both components, which is influenced by the plant structure and the management. Thus, in order to determine the impact of the key plant features on light interception, and hence weed suppression, a mechanistic model, based on Beer's law, has been developed. The model describes the radiation intercepted at key growth stages. Varieties of wheat have been defined by ideotypes, which describe extinction coefficient, leaf area index and height at these key growth stages. The model has been used to assess how the ideotypes interact with climate and key management factors on the light intercepted by the cereal crop, and hence that available for weed growth. The crop morphology impacts on the light intercepted by the crop, with planophiles tending to intercept more light than erectophiles. In addition, reducing the planting row width increases light interception, and there is an interaction between planting row width and morphology.

INTRODUCTION
In organic farming systems, minimizing the impact of weeds on cereal production is imperative. However, the impact of weeds on the growth and development of cereal crops is dependent on the competitiveness of the cereal crop. The key factors that influence competitiveness are the morphological characteristics of the crop (Bertholdsen and Jonsson, 1994; Lemerle *et al*, 1996; Eisele and Köpke, 1997; Gooding *et al.*, 1997), the quality of seed, sowing rate and row spacing, timing of emergence in relation to weeds and crop vigour and health in response to inputs and environmental factors (Davies and Welsh, 2002).

The European Union funded project, Weed Control in Organic Farming (WECOF), is evaluating the unknown relative importance of the different components of the wheat canopy morphology, crop height and speed of development at key stages on light penetration and this ability to reduce weed growth (Davies *et al.*, 2002; Davies *et al.*, 2004). In order to assess the impact of the morphological characteristics of the crop and crop management on light penetration and hence weed suppression, a mechanistic model of light interception by the crop has been developed. This paper outlines some preliminary results from the model.

THE MODEL
The model is designed to explore the impact of the management factors, namely: location of the site, drilling direction, and row width on the potential for weed suppression. The

model describes the light that is available to be intercepted at the soil surface. The input parameters are defined at key stages in the development of the crop, which have been identified within the project as third leaf (growth stage 13-21), stem elongation (growth stage 31), ear emergence (growth stage 49), and flowering (growth stage 65).

The inputs to the model include the latitude of the site and daily values of the climatic conditions as well as the management factors of drilling direction, row width, spacing and ideotype. The important characteristics of the ideotype at the key stages of crop development that are inputs into the model are extinction coefficient, height of the plant and the leaf area index.

The model is based on Beer's law, and describes the direct and diffuse light absorbed by the wheat canopy and that available for weed growth at the base of the canopy. It is recognized that this is a simplification as the weed canopy will also intercept light. In essence, the canopy produced by the crop is described by the inter-row spacing, width of the crop and crop height (Gizjen and Goudriaan, 1989). As the inter-row spacing is always wider than the intra-row spacing, intra-row spacing is ignored. The crop row is described, therefore, as a continuous rectangular 'hedgerow' within the model. As the direct light has an angle of incidence equal to the angle of the sun, the sowing direction of the crop will also impact on the light interception of the wheat and the weed, and this will be described within the model. For convenience the model may be divided into five sub-models, which are concerned with: (i) position of the sun; (ii) diffuse and direct photosynthetically active radiation flux; (iii) light interception by the canopy; (iv) leaf area traversed by the ray; and (v) reflection coefficient of the canopy.

METHODS

For the purpose of illustration of the model, it has been assumed that the crop rows are sown in a north-south orientation, and that the row spacings are 12, 17 or 24 cm, derived from a sub-set of the WECOF project variety trials. Typical data for erectophile and planophile ideotypes from the UK trials at key growth stages have been tested in the model (Table 1). In order to ascertain how much of the impact was due to differences in leaf area index between the erectophile and planophile ideotypes, two additional composite ideotypes (planophile2 and erectophile2) were tested. The planophile2 ideotype was defined by the hedge width and height, and extinction coefficient for planophiles and the leaf area index of erectophiles, while the erectophile2 ideotype was defined by the hedge width and height and extinction coefficient for erectophiles and the leaf area index of planophiles (Table 1).

RESULTS AND DISCUSSION

The results indicate, for the climatic conditions that were used for testing of the specified ideotype, that while full canopy closure had not been achieved, increasing row width reduced the light intercepted in proportion to the number of rows/m^2 for both erectophiles and planophiles (Table 2). Nevertheless, with the exception of growth stage 13-21, the planophiles ideotype (planophile) with the same leaf area index as the erectophile (erectophile2) intercepted more light. This impact was greater as row spacing increased. However, the erectophile had a greater leaf area index than the planophile and, with the exception of growth stage 65 at a row width of 24 cm, this resulted in the erectophile having a greater light interception than the planophile.

Table 1. The row width, crop height, leaf area index (LAI) and extinction coefficients (k) for the ideotypes at the key growth stages

Ideotype	Growth stage	Row width (m)	Crop height (m)	LAI (m^2 leaf)/ (m^2 ground)	k
Planophile	13-21	0.05	0.04	0.3	1.20
	31	0.09	0.20	0.9	1.14
	49	0.13	0.60	3.5	0.67
	65	0.24	0.85	4.5	0.66
Erectophile	13-21	0.03	0.06	0.3	1.00
	31	0.06	0.30	1.2	0.95
	49	0.10	0.80	4.0	0.61
	65	0.15	0.85	5.5	0.63
Planophile2	13-21	0.05	0.04	0.3	1.20
	31	0.09	0.20	1.2	1.14
	49	0.13	0.60	4.0	0.67
	65	0.24	0.85	5.5	0.66
Erectophile2	13-21	0.03	0.06	0.3	1.00
	31	0.06	0.30	0.9	0.95
	49	0.10	0.80	3.5	0.61
	65	0.15	0.85	4.5	0.63

Table 2. The percentage of light intercepted by the crop at the key growth stages

Ideotype	Growth stage	Light intercepted by the crop (%)		
		12 cm	17cm	24 cm
Planophile	13-21	0.06	0.04	0.03
	31	4.07	2.88	2.05
	49	47.62	36.08	25.71
	65	88.89	89.00	89.09
Erectophile	13-21	0.07	0.05	0.03
	31	4.62	3.28	2.33
	49	54.08	38.16	27.11
	65	100.00	91.92	65.20
Planophile2	13-21	0.06	0.04	0.03
	31	6.19	4.38	3.12
	49	55.67	42.19	30.06
	65	100.00	100.00	100.00
Erectophile2	13-21	0.07	0.05	0.03
	31	2.98	2.12	1.50
	49	45.98	32.45	23.05
	65	84.33	73.56	52.19

CONCLUSIONS

The morphological characteristics of the plant have a key role in determining the light intercepted by the crop, and hence the amount of light that is available for weed growth. However, the model confirms that at growth stage 13-21, morphology has little impact on the light intercepted by the crop. Nevertheless, until full canopy closure is established, row width has a major impact on light interception, and hence there is an interaction between crop morphology and row width. The results of the model suggest that planophiles intercept more light than erectophiles. Similarly, decreasing the inter-row spacing reduces the likelihood of weed infestation. In addition, leaf area index has a major impact on the light intercepted by the canopy and hence on the potential for weed growth. This model will be tested on further data from the trials conducted as part of the WECOF project.

ACKNOWLEDGMENTS

We would like to thank the European Union for the funding of the WECOF project (Contract EU-QLRT-1999-31418) and The Scottish Executive Environment and Rural Affairs Department for associated research funding, which match funds the WECOF project. We would also like to thank our colleagues at the University of Bonn (Professor U Köpke, Dr D Neuhoff), the Agricultural University of Warsaw (Professor S Gawronski), IMIA, Madrid (Dr C de Lucas Bueno, Sna.R Alarcon) and their colleagues. Also David Bickerton and Alistair Drysdale from SAC, and the site farmer, Mr Andrew Stoddart.

REFERENCES

BERTHOLDSEN N.O. and JONSSON R (1994) Weed competition in barley and oats. *Proceedings of the European Society of Agronomy,* Padova, pp 656-657.

DAVIES D.H.K and WELSH J.P. (2002) Weed control in organic cereals and pulses. In: Younie, D., Taylor B.R., Welsh J.P. and Wilkinson J.M. (Eds). *Organic cereals and pulses*, pp.77-114, Lincoln: Chalcombe Publications.

DAVIES D.H.K., HOAD S. and BOYD L. (2002) WECOF: A new project developing enhanced weed control through improved crop and plant architecture. *Proceedings of the UK Organic Research 2002 Conference,* Aberystwyth, pp. 299-302.

DAVIES D.H.K, HOAD S., MASKELL P. and TOPP C.F.E. (2004) Looking at cereal varieties to help reduce weed control inputs. *Proceedings of the Crop Science & Technology 2003 - The BCPC International Congress*, Glasgow (in press).

EISELE J.A and KÖPKE U. (1997) Choice of variety in organic farming: New criteria for winter wheat ideotypes. Vol. I: Light conditions in stands of winter wheat affected by morphological features of different varieties. *Pflanzenbauwissenschaften*, 1, 19-25.

GIJZEN H. and GOUDRIAAN J. (1989) A flexible and explanatory model of light distribution and photosynthesis in row crops. *Agricultural and Forest Meteorology*, 48, 1-20.

GOODING M.J, COSSER N.D, THOMSON A.J, DAVIES W.P. and FRED-WILLIAMS R.J. (1997) The effect of cultivar and Rht genes on the competitive ability, yield and bread making qualities of organically grown wheat. *Proceedings 3rd ENOF Workshop*, Ancona 1997: Resource use in organic farming, pp. 13-121.

LEMERLE E., VERBEEK B., COUSENS R.D. and COOMBES N.E. (1996) The potential for selecting wheat varieties strongly competitive against weeds. *Weed Research*, 36, 505-513.

Does Organic Farming Favour Arbuscular Mycorrhizal Fungi?

AYAKO OZAKI[1], FRANCIS W. RAYNS[1], PAUL GOSLING[1], GARY D. BENDING[2] and MARY K. TURNER[2]
[1]HDRA-IOR, Ryton Organic Gardens, Coventry, CV8 3LG, UK
[2]HRI Wellesbourne, Warwick, CV35 9EF, UK

ABSTRACT
Arbuscular mycorrhizal fungi (AMF) are important contributors to plant health and yield in low input systems. We investigated the relative size of the AM fungus community in paired organic and conventional fields from 12 farms in England on a range of soil types. It was found that AMF spore numbers and root colonization potential were significantly higher under organic relative to conventional management, and, for spore numbers, the difference between organic and conventional management increased significantly with time since conversion. There were no overall differences in total C, N, P, available P, Ca, Mg and K between conventional and organic management. AMF spore numbers and root colonization potential were not related to any of the soil chemical parameters.

INTRODUCTION
Arbuscular mycorrhiza (AM) are symbiotic structures formed between plant roots and fungi in the phylum Glomeromycota. Arbuscular mycorrhizal fungi (AMF) play an important role in plant health by improving nutrient (especially inorganic P) and water uptake by their host plant, and providing protection against soil-borne pathogens (Kurle and Pfleger, 1994). However, many management practices typical of conventional high input systems, particularly P fertilizer application and the use of biocides, are known to be deleterious to AM fungus communities (Kabir *et al.*, 1998; Thingstrup *et al.*, 1998). This study was conducted to determine the relative impact of organic and conventional management on the size of AM fungus communities.

MATERIALS AND METHODS
Soil was collected from paired organic and conventional fields on twelve English farms on a range of soil types (Table 1). All but one of the samples were from arable/field vegetable systems; Redesdale is an upland grassland system.

Within each field five 10x10 m areas were marked out and from within each area 20 soil cores (0-30 cm depth) were collected and pooled to provide a 5 kg sample. Each sample was analysed for total C, N and P, extractable (Olsen) P, and extractable Mg and K. AMF spores were extracted from a sub-sample using sucrose density gradient centrifugation and counted by microscopy. To determine root colonization potential by AM fungi, onion (*Allium cepa*) was grown in pots containing 100 g fresh weight of soil at 60 % water holding capacity for 14 weeks in the glasshouse; roots were stained with aniline blue (Grace and Stribley, 1991) and the percentage root colonized by AM fungi determined by microscopy. For statistical purposes each farm was considered as a block.

Table 1. Characteristics of the 12 sites examined for AMF population

Site	Soil type	Enterprise
Ryton, Warks.	Sandy loam	Vegetable
Wellesbourne, Warks.	Sandy loam	Vegetable/Arable
Terrington, Norfolk	Silt clay loam	Vegetable/Arable
Kirton, S. Lincs.	Silty loam	Vegetable
Sutterton, S. Lincs.	Silty loam	Vegetable
Epworth, N. Lincs.	Sandy loam/peat	Vegetable
Duggleby, N. Yorks.	Clay loam over chalk	Arable
Great Coxwell, Oxon.	Sandy silt loam over limestone	Arable
Cirencester, Gloucs.	Silty loam over limestone	Arable
Tarleton, Lancs.	Peat	Vegetable
Ormskirk, Lancs.	Peat	Vegetable
Redesdale, Northumberland	Fine sandy loam	Grassland

Table 2. AMF spores (per 100 g of soil) and percentage of onion root length colonized by AMF in soil from paired organic and conventional sites. Values shown are the means of 5 replicates (standard errors are shown in brackets).

Site	Years since conversion	Number of spores (per 100 g of soil)		Percentage of roots colonized	
		Con	Org	Con	Org
Ryton	15	45 (12)	116 (20)	25 (9)	43 (7)
Wellesbourne	4	162 (33)	143 (19)	34 (11)	39 (10)
Terrington	7	259 (36)	819 (125)	4 (2)	52 (8)
Kirton	4	832 (86)	1164 (228)	20 (7)	53 (11)
Sutterton	3	882 (209)	1746 (194)	35 (12)	58 (8)
Epworth	2	410 (100)	530 (82)	52 (17)	55 (5)
Duggleby	2	898 (195)	606 (85)	12 (3)	65 (10)
Great Coxwell	10	643 (31)	824 (138)	43 (2)	84 (4)
Cirencester	18	308 (53)	1955 (386)	20 (5)	72 (6)
Tarleton	4	396 (37)	504 (135)	16 (7)	34 (7)
Ormskirk	1	186 (28)	749 (112)	33 (4)	36 (7)
Redesdale	12	509 (107)	1250 (200)	50 (4)	62 (6)

RESULTS

There was no consistent effect of organic rather than conventional management on any of the soil chemical parameters (data not presented). However, both AMF spore numbers and root colonization potential were significantly higher ($P<0.01$) under organic than under conventional management (Table 2). The ratio of spores in organic: conventional management was significantly correlated with length of time following conversion ($r=0.6$, $P<0.05$).

DISCUSSION

There was no significant relationship between AMF spore numbers or root colonization potential and any of the soil chemical parameters. Some researchers (Jensen and Jacobsen, 1980; Kahiluoto *et al.*, 2001) have found a negative correlation between available P and AMF population. This was not the case here.

In a number of studies, organic management has been shown to stimulate AMF communities, with the effect attributed to reduced soil P under organic management (Ryan *et al.*, 1994; Mader *et al.*, 2000). However, differences in AMF inoculum between the organic and conventional farms included in our study cannot be ascribed to differences in soil chemistry. This suggests that other practices such as the use of fertility building crops, a greater variety of cash crops, non-chemical weed control, and non-use of fungicides may be important; all these factors are known to influence AMF populations (Kurle and Pfleger, 1994).

A wide variety of AMF spores were observed in this study. Over 150 species of AMF have been described, and both functional diversity and niche differentiation in AMF have been demonstrated (Newsham *et al.*, 1995). Work is ongoing to determine the impact of organic and conventional management on AMF diversity.

ACKNOWLEDGEMENTS

We thank Defra for funding this research, Rodney Edmondson for statistical advice and Gillian Goodlass for providing the soil samples from High Mowthorpe.

REFERENCES

GRACE C. and STRIBLEY D.P. (1991) A safer procedure for routine staining of vesicular-arbuscular mycorrhizal fungi. *Mycological Research*, 95, 1160-1162.

KABIR Z., O'HALLORAN I.P., FYLES J.W. and HAMEL C. (1998) Dynamics of the mycorrhizal symbiosis of corn (*Zea mays* L.): effects of host physiology, tillage practice and fertilization on spatial distribution of extra-radical mycorrhizal hyphae in the field. *Agriculture, Ecosystems and Environment,* 68, 151-163.

KAHILUOTO H., KETOJA E., VESTBERG M. and SAARELA I. (2001) Promotion of AM utilization through reduced P fertilization 2. Field studies *Plant and Soil,* 231, 65-79.

KURLE J.E. and PFLEGER F.L. (1994) The effects of cultural practices and pesticides on VAM fungi. In: *Mycorrhizae and Plant Health* Pfledger FL and Linderman RG. (eds) APS Press, Minnesota, pp 101-132.

MÄDER P., EDENHOFER S., BOLLER T., WIEMKEN A. and NIGGLI U. (2000) Arbuscular mycorrhizae in a long-term field trial comparing low-input (organic, biological) and high-input (conventional) farming systems in a crop rotation. *Biology and Fertility of Soils*, 31, 150-156.

NEWSHAM K.K., FITTER A.H. and WATKINSON A.R. (1995) Multifunctionality and biodiversity in arbuscular mycorrhizas. *Trends in Ecology and Evolution*, 10, 407-411.

RYAN M.H., CHILVERS G.A. and DUMARESQ D.C. (1994) Colonization of wheat by VA-mycorrhizal fungi was found to be higher on a farm managed in an organic manner than on a conventional neighbour. *Plant and Soil,* 160, 33-40.

THINGSTRUP I., RUBAEK G., SIBBESEN E. and JAKOBSEN I. (1998) Flax (*Linum usitatissimum* L.) depends on arbuscular mycorrhizal fungi for growth and P uptake at intermediate but not high soil P levels in the field. *Plant and Soil*, 203, 37-46.

Comparisons of Outputs and Profits from Different Organic Rotations in Northern Scotland

B. R. TAYLOR[1], D. YOUNIE[1], S. MATHESON[1], M. COUTTS[1] and C. MAYER[2]
[1]SAC, Craibstone Estate, Bucksburn, Aberdeen, AB21 9YA, UK
[2]BIOSS, Rowett Research Institute, Bucksburn, Aberdeen, AB21 9SB, UK

ABSTRACT

This paper presents the physical and financial performance of different crop rotations at two sites in NE Scotland. Total annual output of grain, swedes, potatoes and silage, and total livestock unit grazing days per rotation are compared and, by assigning values to outputs and variable costs to the crops, gross margins are compared. At a typically mixed farming site, rotations with 50% or 67% grass/clover ley gave similar outputs and gross margins. At a typically arable site outputs from rotations with 38% and 50% ley were similar, and gross margin marginally higher from the former. Where swedes for the domestic or table market could be grown, gross margins were greatly increased.

INTRODUCTION

The function of a rotation in an organic farming system is to maintain nitrogen fertility and to minimize weeds, pests and diseases. Fertility maintenance depends upon the balance of N-fixing legumes and fertility-depleting non-legumes in the rotation (Watson *et al.*, 1999).

In most organic systems N fertility is supplied by clover in grass/clover leys which are utilized by stock through grazing or conservation. It is generally considered that at least half the rotation should be devoted to such fertility-building crops (Soil Association, 2002). In areas of north east Scotland such as the Moray Firth coast, where arable cropping predominates, farmers look for rotations with a lower proportion of crops devoted to livestock feed, specifically with less grass/clover leys. The alternatives include vigorous N-fixers such as red clover, which can be cut and mulched but which give no financial return apart from a set-aside payment, and grain legumes, and although these fix N, most of the N is exported in the grain (Fisher, 1996).

Traditional rotations found in the north east of Scotland were typically 3 years of a grass/white clover ley followed by a cereal, a root crop and a cereal. Although these rotations are now largely abandoned on non-organic farms, they offer a good starting point for the design of organic rotations for the area. From the farmer's point of view, whole-farm value of output and costs of production are significant factors in determining the system to be adopted.

The trials reported here compare organic crop rotations at two typically mixed and arable farming areas of the north east of Scotland. The 6-year rotations at the Aberdeen site contain 50% and 67% fertility-building crops; the rotations at the Elgin site are 6 and 8 years in length and contain 50% and 38% fertility-building crops respectively.

MATERIALS AND METHODS

Rotation trials at Tulloch, Aberdeen (O.S. map ref. NJ843094; 160m above sea level; sandy loam soil; average rainfall 820mm), and Woodside, Elgin (O.S. map ref. NJ167625; 25m above sea level; loamy sand/sandy loam soil; average rainfall 730mm) were started in 1991. The objectives of the trials are described by Younie *et al.* (1996). The rotations are shown in Table 1.

Table 1. Rotations at Tulloch and Woodside.

Site	Rotation	Cropping Year							
		1	2	3	4	5	6	7	8
Tulloch	T1	GC →	GC →	GC →	C1 →	S →	C2 →		
	T2	GC →	GC →	GC →	GC →	C1 →	C2 →		
Woodside	W1	GC →	GC →	C1 →	P →	C2 →	GR →	S →	C3 →
	W2	GC →	GC →	GC →	C1 →	P →	C2 →		

(GC = grass/white clover, C1, C2, C3 = 1st, 2nd, 3rd cereal (spring oats) respectively after the main fertility-building phase, S = swedes, P = potatoes, GR = grass/red clover; 2nd and 3rd cereals undersown with appropriate grass/clover mixture)

There were two replicates of each rotation at each site and plot sizes were 27m x 30m at Tulloch and 28 x 30m at Woodside. All plots were fenced to allow grazing by sheep. Total annual manure applications to each rotation were calculated from an assumed stocking rate of 1.6 livestock units/ha on the grassland area multiplied by standard figures for manure production. For simplicity, organic manures used in the trials were taken from the parent organic farms rather than being produced specifically from straw taken from the trials.

Recordings included crop yields, including total dry matter yields of one or more silage cuts, and sheep grazing days converted to livestock unit grazing days (LUGD).

RESULTS

Mean grain yields from 1992 to 2003 are shown in Tables 2 and 3.

Table 2. Mean yields of oats Tulloch 1992-2003 (t/ha @ 85% DM)

	T1		T2		
	C1	C2	C1	C2	SED±
Mean	4.93	3.38	5.03	3.41	0.337 (within rotations)
					0.373 (across rotations)

Financial comparisons for the different rotations are given in Table 4. The lower part of the table shows how outputs and gross margins would be changed had the swedes been grown for the table market instead of as stock feed. Livestock Unit Grazing Days (LUGD) are valued at £3.00/LUGD, calculated from figures for spring-born store cattle, grass finished at 18 months (Lampkin *et al.*, 2002), animals equivalent to 0.41 LU being stocked at 1.25 LU/ha and gaining 0.84 kg LW/day, each kg LW valued at £1.17 (1.25/0.41*0.84*1.17 =£3.00/LUGD). Silage was valued at £80/tonne of dry matter. Published prices were used for oats, potatoes (70% of published price to allow for out-grades) and swedes (stock-feed and table market) (Anon, 2003).

Table 3. Mean yields of oats at Woodside 1992-2003 (t/ha @ 85% DM)

	W1			W2		
	C1	C2	C3	C1	C2	SED±
Mean	4.03	3.14	3.62	4.21	3.09	0.419 (within W1) 0.484 (within W2) 0.504 (between W1/W2)

Table 4. Total production, physical and financial output per rotation, and financial performance per hectare

		T1 (6 ha)	T2 (6 ha)	SED±	W1 (8 ha)	W2 (6 ha)	SED±
Stock feed swedes							
Oats (£125/t)	Yield t	8.31	8.44	0.24	10.80	7.30	
	Output £	1039	1055		1349	913	
Swedes (£25/t)	Yield t	50.90	-		47.59	-	
	Output £	1273	-		1190	-	
Potatoes (£140/t)	Yield t	-	-		27.1	26.6	
	Output £	-	-		3794	3724	
Silage (£80/t)	Yield t	15.02	15.23		15.22	9.17	
	Output £	1202	1218		1218	734	
LUGD (£3/LUGD)	LUGD	671.3	1096.7		560	587	
	Output £	2014	3290		1680	1761	
	Total £/ha*	921	927	4.3 *P*=0.40	1153	1189	62.0 *P*=0.66
	Variable costs £/ha	116	99		437	517	
	Gross Margin £/ha	805	828	4.3 *P*=0.12	716	672	62.0 *P*=0.61
Table swedes							
Swedes (£160/t)	Yield t	50.9	-		47.59	-	
	Output £	8144	-		7614	-	
	Total £/ha*	2067	927	17.3 *P*=0.01	1956	1189	117.8 *P*=0.10
	Variable costs £/ha	388	99		641	517	
	Gross Margin £/ha	1679	828	17.3 *P*=0.01	1315	672	117.8 *P*=0.12

* = total output divided by length of rotation (years).

Variable costs were estimated from Lampkin *et al.* (2002) (oats £147/ha, stock-feed swedes £147/ha, table swedes £1775/ha, potatoes £2552/ha, 2, 3, 4-year grass/white clover £211, £256, £301 respectively, 1-year grass/red clover £145/ha). Gross outputs, variable costs, and estimated gross margins for each rotation are shown in Table 4, with swedes as stock feed (£25/t) and as table swedes (£160/t allowing for 20% out-grades). Because of the likelihood of changes in the future, arable area payments for cereals were not included in the gross margins. SEDs and *P*-values were calculated by ANOVA based on the 24 (28) plot averages. Rotation was used as treatment factor and replicate/rotation as block factor (split-plot design).

DISCUSSION

The six-year rotations completed two complete cycles, including the second year of organic conversion, whilst the eight-year rotation completed only part of the second cycle. Problems associated with lack of balance in comparing rotations of different lengths and other practical considerations relating to these trials are discussed in Taylor *et al.* (2002).

The most important factor affecting yield potential of organic grain crops is the availability of soil N during key development periods (Philipps *et al.*, 2002). There is little indication from the Tulloch trials that four years instead of three years in grass/clover ley resulted in more N being available to the cereal crops in the rotation. Yields of the first oats after the break were similar in T1 and T2, as were those of the second oats. This may be because soil N is maximized after three years of ley. At Woodside oat yields were 0.18 t/ha less following two years of a grass/clover ley in W1 than after three years of a grass/clover ley in W2, and 0.41 t/ha less after one year than two years in W1, suggesting that soil N levels may have been marginal in this rotation. Average soil organic matter contents in the trials at Tulloch and Woodside were 10% and 6% respectively, indicating greater mineralizable soil N reserves on slightly heavier soils where mixed farming is traditional.

A number of assumptions have been made in the financial comparisons in Table 4 and changed prices might result in different conclusions. Nevertheless, within the constraints of the environment and available time, labour, capital etc., it is on financial grounds that farmers ultimately have to choose their rotations. The data used in this paper indicate similar outputs and gross margins from the two rotations at Tulloch, the value of the swedes in T1 being balanced by extra potential livestock production in T2. At Woodside gross margins from the two rotations were not significantly different when swedes were grown for stock feed, and the extra complexity of W1 (two root crops and establishing a ley twice) was only marginally more profitable (although the difference would have been greater had Government payments associated with growing cereal crops been taken into account). The substitution of a ley for a cash crop did not affect profitability where both alternatives gave a financial return. Olesen *et al.* (2002), in the first 4-year cycle of a rotation trial, found that the yield benefits from a green manure ley could not compensate for the yield reduction as a result of leaving one year out of production.

The original intention of the trials was that swedes would be grown for the table market, but the required quality was rarely achieved. Had table swedes been grown, gross margins from T1 would have been significantly greater than those from T2, and those from W1 greater than those from W2, although with less of a difference than at Tulloch. In practice, table swedes may not always be an option for growers if investment and management input are considerably more.

ACKNOWLEDGEMENTS

SAC receives funding from the Scottish Executive Environment and Rural Affairs Department. This work was carried out under SEERAD ROAME No. 606020.

REFERENCES

ANON (2003) Eye on the market. *Organic Farming,* Autumn 2003, pp 10-11. Bristol: The Soil Association.

FISHER N.M. (1996) The potential of grain and forage legumes in mixed farming systems. In: Younie, D. (Ed) *Legumes in Sustainable Farming Systems. Occasional Symposium of the British Grassland Society*, No. 30, pp 290-299.

LAMPKIN N., MEASURES M. and PADEL S. (2002) 2002/03 Organic Farm Management Handbook. Aberystwyth: University of Wales.

OLESEN J.E., RASMUSSEN I.A., ASKEGAARD M.A. and KRISTENSEN K.K. (2002) Whole-rotation dry matter and nitrogen grain yields from the first course of an organic farming crop rotation experiment. *Journal of Agricultural Science, Cambridge,* 139, 361-370.

PHILIPPS L., HUXHAM S.K., BRIGGS S.R. and SPARKES D.L. (2002) Rotations and nutrient management strategies. In Younie, D., Taylor, B.R., Welsh, J.P. and Wilkinson, J.M. (Eds) *Organic Cereals and Pulses*, pp 51-76. Lincoln, UK: Chalcombe Publications.

SOIL ASSOCIATION (2002) *Standards for organic food and farming.* Bristol: The Soil Association.

TAYLOR B.R., YOUNIE D., COUTTS M., MATHESON S. and MAYER C. (2002) Experiences with designing and managing organic rotation trials. *Proceedings of the UK Organic Research 2002 Conference, Research in Context*, Aberystwyth, 2002, pp 37-41.

WATSON C.A., YOUNIE D. and ARMSTRONG G. (1999) Designing crop rotations for organic farming: the importance of the ley/arable balance. In: Olesen, J.E., Eltun, R., Gooding, M.J., Jensen, E.S. and Kopke U. (Eds). *Designing and Testing Crop Rotations for Organic Farming*, DARCOF Report No. 1, pp 91-98.

YOUNIE D., WATSON C.A. and SQUIRE G.R. (1996) A comparison of crop rotations in organic farming: agronomic performance. *Aspects of Applied Biology* 47, *Rotations and Cropping Systems*, 379-382.

Author Index